Editors
Rina Rani Ray
Department of Biotechnology
Maulana Abul Kalam Azad University of Technology
Kolkata, West Bengal, India

Moupriya Nag
Department of Biotechnology
University of Engineering & Management
Kolkata, West Bengal, India

Dibyajit Lahiri
Department of Biotechnology
University of Engineering & Management
Kolkata, West Bengal, India

ISBN 978-981-16-0744-8 ISBN 978-981-16-0745-5 (eBook)
https://doi.org/10.1007/978-981-16-0745-5

This Springer imprint is published by the registered company Springer Nature Singapore Pte Ltd.
The registered company address is: 152 Beach Road, #21-01/04 Gateway East, Singapore 189721, Singapore

Rina Rani Ray • Moupriya Nag •
Dibyajit Lahiri
Editors

Biofilm-Mediated Diseases: Causes and Controls

Springer

Biofilm-Mediated Diseases: Causes and Controls

Preface

A disease is a particular abnormal condition that adversely affects the steady state of the body and is often associated with specific symptoms. Onset of disease may result from the entry of a pathogen, which is generally a microbial organism. But the condition becomes more aggravated when the microbes form a biofilm. Discovery of the fact that most of the microbial diseases are biofilm associated has shifted the attention of many researchers towards this particular field.

Biofilms are the group of sessile microbial colonies that remain attached to the abiotic or biotic surfaces with the help of self-secreted polymeric substances comprising of polysaccharides, proteins, and nucleic acids. These self-secreted materials are known as extracellular polymeric substances (EPS). Biofilm is the consortia of different microbial cells resulting in the development of polymicrobial colonies. In general, during growth and proliferation, most of the bacteria show two distinct life forms. The first form is free moving, single, and independent planktonic form, whereas the other is sessile and remains as an aggregated conglomerate in biofilm.

Actually, biofilm-associated growth is one of the effective survival strategies adopted by the microbes as they remain safe within the EPS matrix both from the antimicrobials and from aggressive immune response of the host. Inside the biofilm matrix, the microbe can withstand various environmental hazards like anoxia and non-availability of nutrition, by careful alteration of metabolic pathway and expression of gene. Such shifting results in the commencement of chronic diseases.

Acute infections caused by pathogenic bacteria are caused by the planktonic forms. These can be treated with antibiotics, after quick and accurate diagnosis. Most of the disease-related researches, in last few decades, have been focused on acute infections. However, in cases where the microbe succeeds in forming a biofilm within the human body, the infection becomes untreatable and will turn into a chronic disease. Biofilm-mediated infections represent a stern health problem, accounting for 65–80% of all infections. The significant symbols of chronic biofilm-based infections are extreme resistance to antibiotics and many other regular and commonly used antimicrobial agents, and an extreme capacity for evading the host defenses.

Biofilms can be located to almost every human body tissue and on exogenous devices such as intravenous catheters, prosthetics, dentures, and implants that pose a serious threat to the success of surgery and implantation technology.

The book "**Biofilm-Mediated Diseases: Causes and Controls**" aims to combine different communities to effectively advance the knowledge of microbial domains and specifically explore the crossing point between these disciplines. The chapters first describe sequentially the composition of biofilm, the quorum sensing, and signaling mechanism behind its formation, normal microbiota, and commensals in biofilm in various parts of our body. Various deadly diseases concerned with human natural microbiota are discussed. Next, the biofilm-associated acute and chronic diseases will be elaborately discussed with special reference to nosocomial diseases, the root cause of high percentage of morbidity, and mortality worldwide. Since the failure of host immune system indirectly accelerates the persistence of the biofilm-associated pathogenic cells and their gradual manifestations as chronic diseases, the changed scenario of immune response in the presence of biofilm is discussed with a hint that sometimes the biofilm-associated cells can change the action of immune response even in favor of biofilm formation. As the chance of postoperative safe-keeping of the patient is severely hampered by the profuse growth of biofilm on implants, prosthetics, and medical appliances, it needs to be illustratively discussed. The cardinal feature of these pathogens is their resistance to antimicrobials; hence, before designing a befitting drug, the possible mechanism of antimicrobial resistance offered by the particular pathogen, including its genetic basis is required to be found out. The last part of the book deals with the control and obviously included the inhibition of quorum sensing pathway and discussion on new strategies for the treatment of biofilm-mediated diseases that includes application of phytomedicine, nanomedicine, anti-QS enzyme treatment, phage therapy, and so on. Finally, a case report dealing with the isolation and identification of a biofilm forming bacteria from central venous catheter-associated bacteria from a patient is recorded, which confirms the severity of the biofilm in production of various diseases.

This book will be potentially useful for both clinicians and researchers who are dealing with infections associated with biofilms. This book will attempt to compile the different information available on recent advancements on various functional aspects of microbial biofilms, its pathogenesis, and present-day treatment strategies. Finally, the book will also elucidate a comprehensive yet a representative description of a large number of challenges associated with bacterial biofilm, viz., virulence, pathogenesis, antibiotic resistance, cell signaling, immune evasion, and is also a take home for researchers to develop effective strategies to combat these threatening diseases.

Kolkata, West Bengal, India

Rina Rani Ray
Moupriya Nag
Dibyajit Lahiri

Contents

Editors and Contributors

About the Editors

Rina Rani Ray is working as an Associate Professor and Head, Department of Biotechnology, Maulana Abul Kalam Azad University of Technology, West Bengal. Her research interest focuses on the diverse fields of microbial biotechnology. She has published around 80 research papers in various peer-reviewed national and international journals and is the editorial board member of various international journals. She is the member of various scientific associations and is Fellow of Zoological Society, India, and Academy for Environment and Life Sciences, India.

Moupriya Nag is presently working as an Assistant Professor at the Department of Biotechnology, University of Engineering & Management, Kolkata. Her research focuses on analysis of functional amyloids in bacteria and their relation in forming biofilms towards the development of novel antimicrobial and anti-biofilm compounds from natural sources including plants, microbes, and green synthesized nanoparticles. She has also published many research articles in the peer-reviewed international journals and authored or co-authored numerous book chapters. She is also a member of many scientific societies like Indian Biophysical Society and Indian Science Congress Association.

Dibyajit Lahiri is working as an Assistant Professor at the Department of Biotechnology, University of Engineering & Management, Kolkata. His research interest is precisely on biofilm isolated from human prosthesis and its inhibition by various novel phyto- and nanocompounds. He is a life Member of Indian Science Congress Association and Institution of Engineers (India).

Contributors

Anindita Ballav Department of Microbiology, College of Medicine and Sagore Dutta Hospital, Kolkata, West Bengal, India

Prateek Banerjee Department of Biotechnology, University of Engineering & Management, Kolkata, West Bengal, India

Ritwik Banerjee Department of Biotechnology, University of Engineering & Management, Kolkata, West Bengal, India

Tanay Bhagwat Department of Chemical Engineering, Northeastern University, Boston, MA, USA

Abarna Chatterjee Department of Biotechnology, University of Engineering & Management, Kolkata, West Bengal, India

Deboleena Das Department of Biotechnology, University of Engineering & Management, Kolkata, West Bengal, India

Ankita Dey Department of Biotechnology, Maulana Abul Kalam Azad University of Technology, West Bengal, Kolkata, West Bengal, India

Bandita Dutta Department of Biotechnology, Maulana Abul Kalam Azad University of Technology, West Bengal, Kolkata, West Bengal, India

Sayantani Garai Department of Biotechnology, University of Engineering & Management, Kolkata, West Bengal, India

Anushka Ghosh Department of Biotechnology, University of Engineering & Management, Kolkata, West Bengal, India

Sougata Ghosh Department of Microbiology, School of Science, RK University, Rajkot, Gujarat, India
Department of Chemical Engineering, Northeastern University, Boston, MA, USA

Sreejita Ghosh Department of Biotechnology, Maulana Abul Kalam Azad University of Technology, West Bengal, Kolkata, West Bengal, India

Dibyajit Lahiri Department of Biotechnology, University of Engineering & Management, Kolkata, West Bengal, India

Dipro Mukherjee Department of Biotechnology, University of Engineering & Management, Kolkata, West Bengal, India

Moupriya Nag Department of Biotechnology, University of Engineering & Management, Kolkata, West Bengal, India

Smaranika Pattnaik Department of Biotechnology and Bioinformatics, Sambalpur University, Burla, Odisha, India

Bhisma Narayan Ratha Department of Phytopharmaceuticals, School of Agriculture and Bioengineering, Centurion University of Technology and Management, R. Sitapur, Odisha, India

Rina Rani Ray Department of Biotechnology, Maulana Abul Kalam Azad University of Technology, West Bengal, Kolkata, West Bengal, India

Aihik Saha Department of Biochemistry, Nil Ratan Sircar Medical College and Hospital, Kolkata, West Bengal, India

Raymond J. Turner Department of Biological Sciences, University of Calgary, Calgary, AB, Canada

Thomas J. Webster Department of Chemical Engineering, Northeastern University, Boston, MA, USA

Introduction to Bacterial Biofilm and Acute Infections

1

Bandita Dutta, Dibyajit Lahiri, Moupriya Nag, Dipro Mukherjee, and Rina Rani Ray

Abstract

Bacterial cells form a consortium, which can adhere to the surface and can develop biofilm. The biofilm can be distinguished from their suspended counterparts or the other normal microbial cells by the presence of an extracellular polymeric substance (EPS) matrix. The biofilm formation process is a complex process and generally is regulated by a combination of different variables present in nature. It is dependent on the growth medium, the substratum, and the cell surface. There are several bacterial biofilms, which are mainly pathogenic in nature and can cause the development of nosocomial diseases. The National Institutes of Health revealed that most of the chronic and microbial infections are associated with the formation of the biofilm. The bacterial biofilms develop resistance against the host immune system along with antibiotics. This has resulted in the development of health-related concerns as they potentially cause various device- and non-device-associated infections. This chapter provides a detailed view of biofilms and various biofilm-associated acute infections.

B. Dutta · R. R. Ray (✉)
Department of Biotechnology, Maulana Abul Kalam Azad University of Technology, West Bengal, Kolkata, West Bengal, India

D. Lahiri · M. Nag · D. Mukherjee
Department of Biotechnology, University of Engineering & Management, Kolkata, West Bengal, India

R. R. Ray et al. (eds.), *Biofilm-Mediated Diseases: Causes and Controls*,
https://doi.org/10.1007/978-981-16-0745-5_1

1.1 Introduction

Humankind for centuries has suffered from acute bacterial infections and life-threatening diseases caused by various types of pathogens that include *Vibrio cholerae*, *Streptococcus pneumoniae* and *Yersinia pestis*. From the time of the discovery of antibiotics and vaccines, it has resulted in the massive reduction in the burden related to such infections which are mainly caused by individualized groups of pathogenic bacterial cells (Costerton et al. 1999; Donlan (2002)). From the era of antibiotic discovery, physicians confronted two major challenges: spread of antibiotic-resistant bacterial cells and the rise of chronic infections (Davies 2007). During this era of research, some scientists established the concept of biofilm upon natural environment where they established the difference between the planktonic and sessile groups of the bacterial cells (Geesey et al. 1977, 1978; Henrici 1933). The present chapter will deal with the architecture, composition of biofilm in general and the serious infections caused by it.

1.2 Description of Biofilm

A biofilm is a multilayered community of the sessile cells that form a syntrophic association that remains embedded in hydrated extracellular polymeric substances (EPS). This EPS helps the cells to colonize upon the living, inert or upon the boundary surface. The composition of the matrix contains various nutrients like carbohydrates, proteins, lipids, nucleic acids, and other minerals that provide nutrients to the dwelling cells. This influences the organisms that are living in the matrix of the biofilm to become virulent, as this encapsulation gives rise to the antimicrobial resistance, and is also associated with the phenotypic and genotypic changes within the organisms. This biofilm matrix is the preferred way for the bacteria to live in as it provides the cells with optimal conditions for the exchange of genetic materials involving the process of horizontal gene transfer, and hence biofilm becomes the natural state of its existence (Hall-Stoodley et al. 2004). The microbial community remains distributed through matrix or glycocalyx upon any other hard non-shedding surface material. At the bottom layers of a biofilm, microbes are bound together through a polysaccharide matrix containing organic and inorganic materials. The above layer is amorphous in nature and it is extended towards the surroundings.

The sessile microcolonies dwelling within the biofilm develop intimate connection by cell-to-cell communications known as quorum sensing (QS). Quorum sensing is a density-dependent communication system existing between the sessile cells which help in establishing the biofilm. The QS involves various chemical inducers that vary from Gram-positive to Gram-negative bacterial cells. Autoinducer (AI) molecules present within the EPS layer diffuse freely across the cell membrane and regulate the quorum sensing. At the initial stage of the biofilm formation, the AI concentration is very low, but with the increase of the cell population, the AI value reaches to the threshold level in order to activate or repress target genes.

EPS, forming the biofilm, comprises glycocalyx and is chemically made up of carbohydrates, proteins and nucleic acid. EPS helps those adherent cells by embedding them within a slimy layer and provides nutrients to the developing cells within the biofilm, which also helps in pathogenesis of biofilm associated with infection and resistance due to the formation of impermeable layers to the penetration of antibiotics and drugs. Biofilm can be formed by the microbes depending on various cellular and environmental factors including cellular recognition for specific or non-specific attachment sites, nutritional level or exposure of planktonic cells to subinhibitory concentration of antibiotics.

The biofilm formation is regulated by environmental factors, nutrient supplied and the components present inside the biofilm layer. The EPS matrix of the biofilm layer provides the architectural integrity to the bacterial colonies present within the biofilm to ensure the stability of the biofilm in negative conditions and enhances cell division (Sehar and Naz 2016). It also provides essential nutrients which enables genetic and intracellular transfer through quorum sensing of the biofilm-forming bacterial species (Ongenae 2017).

The major component of any EPS matrix is water which comprises about 95–97% of the total space. Apart from that, there are 2–5% of microbial cells of different species, extracellular proteins and also the proteins which resulted from the lysis of the bacterial cells.

1.3 History of Discovery of Biofilm

Bacterial assemblage in the form of biofilm on teeth enamel was first observed by Antonie van Leeuwenhoek with his simple microscope (Donlan (2002)) in the seventeenth century. But for the detailed study of biofilm, we had to wait quite a long time, till the discovery of electron microscope. The photomicrograph of slimy layer (Jones et al. 1969a, b) revealed the cell morphology. Later by using a special stain, ruthenium red coupled with osmium tetroxide fixative, scientists could show the presence of polysaccharide in the biofilm matrix. It was found that those bacterial cells, associated with the consortium of microorganisms, can adhere to the surface and are able to develop biofilm. According to Costerton et al. (1978), microbes can stick to both biotic and abiotic surfaces to form biofilm. Later it was established that the biofilm formation is a complex process and generally is regulated by a combination of different variables present in nature, which is dependent on the growth medium, the substratum, and the cell surface (Jones et al. 1969a, b). A well-formed biofilm is generally composed of microbial cells and EPS and possesses a surrounding which is used for the exchange of genes or genetic material between the cells (Characklis et al. 1990). The biofilm-forming microbial cells transport chemical substances among themselves in order to communicate between the cells for exchange of fluids and nutrients by the mechanism of quorum sensing. This has an effect on different processes of biofilm such as separation of the microbial cells from the consortium to detach the biofilm. Biofilm is of the utmost importance to public health because it possesses certain functionalities in the treatment of different infectious diseases and also plays an important role in a variety of device-related infections. The biofilm is protected from antibiotic, antiviral, antimalarial, antifungal and anthelmintic drugs (Corpe 1980). Because of this, many of the medicines turn out to be not effective, and the infections dominate the body, increasing the risk of spreading of the infections (Rosenberg et al. 1982). Since then, study of bacterial biofilm got a significant role to play in the arena of healthcare, industrial process and environment.

1.4 Composition of Bacterial Biofilm

Biofilm is an aggregation of sessile microbial cells living within extracellular polymeric substances (EPS) which help in the irreversible attachment with living or non-living surfaces whose removal becomes difficult until the application of physical stresses (Hurlow et al. 2015; Costerton et al. 1994). The development of EPS occurs as soon as the bacterial cells attach to an inanimate or solid surface that provides the shield and strength of interactions between the sessile communities existing within the biofilm (Brandas et al. 2005; Costerton et al. 1995a, b; Miron et al. 2001). The thickness of the EPS ranges from 0.2 to 10 μm, and the total size of the biofilm does not exceed more than 10 nm (Sleytr 1997). It has been observed that the total volume of the biofilm comprises 5.35% of sessile communities and the remaining volume is EPS. The major components (Table 1.1) being available within the EPS are proteins and polysaccharides that make the basic structural framework of the matrix (Sun et al. 2005). It also helps in the entrapment of various nutrients by the mechanism of scavenging activity from the surrounding environment (Henrici 1933). It also comprises extracellular DNA (e DNA) that forms another important component within the EPS.

1.5 Planktonic to Biofilm-Growing Bacteria

Genetic studies show that the formation of biofilm is a multistep process. The process of biofilm formation requires a specific signalling mechanism, known as quorum sensing, occurring in between the bacterial cells. This process also involves the transcription of various genes with respect to the planktonic form of microbial cells of the same organism (Donlan 2002). The existing channels within the biofilm help in separating the microcolonies. The viscoelastic feature of the EPS matrix provides mechanical stability to the indwelling biofilm (Shaw et al. 2004). The process of the biofilm development involves events such as initial attachment or contact of the sessile communities with a surface, development of microcolonies, maturation and formation of the architecture of the biofilm and dispersion of the sessile communities resulting in the spread of the biofilm-associated infections (Sutherland 2001a, b).

Stage 1. Initial contact and reversible attachment to the surface: The attachment of the sessile cells upon the biotic and abiotic surface occurs with the help of flagella

Table. 1.1 Integral components of bacterial biofilm

SL no.	Components of biofilm	Overall composition of microbial biofilms	References
1.	Extracellular polymeric substances (EPS)	• Cationic groups present in amino sugars and proteins (e.g. NH_3 þ) • Anionic groups of uronic acids, proteins and nucleic acids (e.g. COO; HPO_4) • A polar group from proteins (present in aromatic amino acids), (phospho) lipids and humic substances	Grkovic et al. (2002)
2.	Microbial cell outer membrane	• Lipopolysaccharides of gram-negative bacterial cells • Cell wall consisting of *N*-acetylglucosamine and *N*-acetylmuramic acid, offers cationic and anionic sites and the lipoteichoic acids in Gram-positive cells	Grkovic et al. (2002)
3.	Cytoplasmic membrane, offering a lipophilic region	• Cytoplasm, as a water phase separated from the surrounding water minerals • Precipitates (sulphides, carbonates, phosphates, hydroxides) • Free and bound metals (Ca2+, Fe3 +, Mg2+)	Hellstrtom (1938)
4.	Biogenic particulate materials (degradation products) environmentally relevant substances	• Organic pollutants (e.g. biocides, detergents, xenobiotics) • Inorganic pollutants (e.g. heavy metals)	Nickel et al. (1987)

and pili that provides them with physical forces like that of the electrostatic, van der Waals forces. Other factors which greatly influence the process of attachment involves the type of surface on which the attachment would take place and the cohesive forces existing between the sessile communities and the surface (Garrett et al. 2008). The two factors which also influence the attachment of the bacterial cells are the adhesion, which leads the attachment of cells to a solid biotic and abiotic surface, and cohesion leading to the interaction and attachment of cells that occur at the time of the biofilm formation (Garrett et al. 2008). The interface between solid and the liquid can also be the potent cause for the biofilm formation and microbial growth (Costerton et al. 1999). The initial attachment of the motile cells to the surface includes the formation of the conditioning layer that mainly comprises organic (proteins, electrolytes, surface-active compounds and cholesterol) as well as inorganic (salts and ionic materials) compounds. After this initial step, biofilm formation occurs rapidly. The primary colony interacts to the surface in two different ways, either due to different forces like Brownian motion, gravity or diffusion or the flow of the liquid or air or due to positioning mechanisms like flagella motility or surface appendages. The bacterial adherence to the surface may be reversible due to the interactive forces (hydrophobicity, electrostatic forces, charges interactions) applied in the single pole of the bacteria. Irreversible attachment is much more stable compared to the previous one as adherence proteins and extracellular proteins are expressed to cement the bacteria to the surface as the long axis of the bacterial cell is positioned parallel to the surface.

Stage 2. Cell accumulation and microcolony formation: The accumulation of cells involves the mechanism of cell-to-cell adhesion and provides them stability for the multiplication and division of the microbial cells, which are initiated by the cell signalling mechanism originating with the EPS. This leads to the development of microcolonies within the cells (Costerton et al. 1999; Mckenney et al. 1998). The microcolonies existing within the biofilm play an important role in exchanging substrate, distributing the metabolizing and excreting products. A multilayered bacterial microcolony is formed as mid-late colonizers adhere to primary colonizers. This occurs over a period of a few hours by the help of signalling molecules and quorum sensing pathways. After the attachment to the biotic and abiotic surface, cell divisions and multiplications of the microbial cells start. The microbial cells coordinate among themselves by several aspects, including exchange of the substrates, distribution of important metabolic products and excretions of metabolic end products. Biofilm provides a suitable environment to the microorganisms, which helps them to develop the syntrophic association between the metabolically different bacteria, depending on the utilization of the certain substrates.

Stage 3. Extracellular polymeric substance production: After cell accumulation and adherence to the surface, the bacterial cells develop extracellular and multilayered microcolonies which cover themselves with a layer of extracellular polymeric matrix (EPS) (Sutherland 2001a, b). This extracellular polymeric matrix consists of polysaccharides, proteins, lipids, nucleic acid, multivalent compounds and inorganic substances. EPS is one of the major components of biofilm formation and can produce 50–90% of total biofilm mass Donlan (2002). It helps the bacterial colonies to communicate with each other and attach on any biomaterial surfaces.

In Gram-negative bacteria, the outside of EPS is anionic in nature due to the presence of negatively charged compounds such as uronic acids and pyruvate, whereas, in the inner side of EPS, the compounds are positive in nature like calcium and magnesium ions. The major component present in EPS is extracellular DNA which provides the structure of biofilm. eDNA serves a number of different functions, depending on the type of bacteria. It helps in formation of the architecture

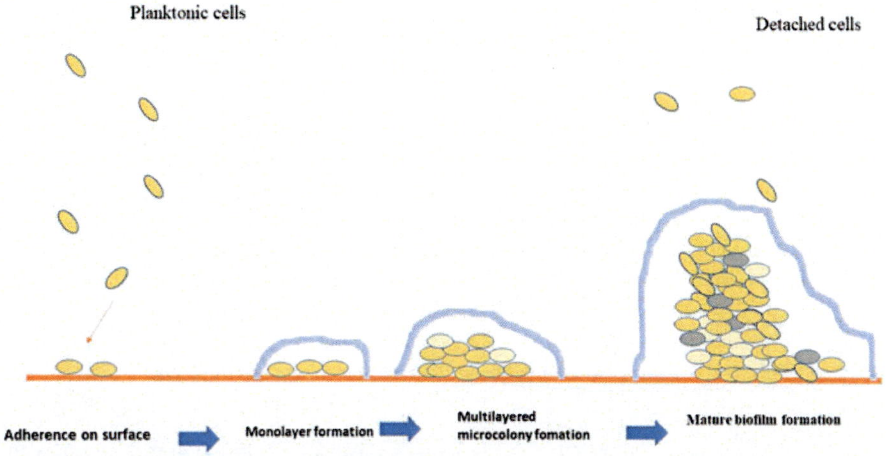

Fig. 1.1 Different stages of biofilm formation

of the biofilm; it is reported to act as food source for sessile bacteria and also helps in gene transfer, DNA repair and quorum sensing.

Stage 4. Biofilm maturation: It is the fourth stage where the biofilm gets matured. Biofilm is a complex architecture and has pores of different sizes through which bacteria can freely move within the EPS. As the biofilm mature, more void spaces are produced through which nutrients, oxygen and other inorganic salts can freely move into the biofilm and the waste by-products are removed through the void space (Costerton et al. 1994).

Stage 5. Detachment: It is the separation of the bacterial cells from the biofilm layer by the physical and chemical mechanisms. Physical mechanisms like shear force can cause erosion of biofilm. Chemical factors may stimulate detachment, for example, substrate changes, nutrient changes and changes in the EPS (Fig. 1.1).

1.6 Quorum Sensing and Its Role in Bacterial Biofilm

Quorum sensing relies on the small signal molecules (Table 1.2) that act as transcriptional activators in order to initiate gene expression within the biofilm layer. Quorum sensing mechanism is classified into three categories depending upon the signal molecules which help in cell-to-cell communication.

1. LuxI/LuxR-type quorum sensing, which is facilitated by signal molecules acyl-homoserine lactones (AHL) in Gram-negative bacteria.
2. Oligopeptide-two-component-type quorum sensing is only for the Gram-positive bacteria where bacterial cells use small peptides as signal molecules.
3. LuxS-encoded autoinducer 2 (AI-2) quorum sensing where signal molecules are found in both Gram-positive and Gram-negative bacteria in the gene regulatory mechanism.

Table 1.2 Regulatory mechanisms in bacterial quorum sensing

SL no.	Microorganisms	Regulatory system	Group-derived benefits	Major signal molecules	References
1.	*Bacillus subtilis*	ComP/ ComA, Rap proteins	Competence, sporulation, biofilm formation, antibiotic production	ComX CSF (PhrC) PhrA-E, PhrA-F, PhrA-K, PhrA-H	
2.	*Myxococcus xanthus*	SasSRN	Fruiting body formation or sporulation	A-signal, C-signal	
3.	*Pseudomonas aeruginosa*	LasI/LasR, RhlI/RhlR, OscR (orphan)	Structured biofilm formation, virulence factors	3O-C12-HSL, C4-HSL	Dufour et al. (2011)
4.	*Staphylococcus aureus*	AgrC/ AgrA	Biofilm formation, virulence factors	AIP-I, AIP-II, AIP-II, AIP-IV	Vendeville et al. (2005); Davies et al. (1998)
5.	*Streptococcus mutans*	ComD/ ComE, ComR	Bacteriocins, biofilm formation, competence	CSP (ComC), XIP (ComS)	Lemme et al. (2011)
6.	*Streptococcus pneumoniae*	ComD/ ComE	Competence, fratricide, biofilm formation, virulence	CSPs	
7.	*Vibrio harveyi*	LuxLM/ LuxN, LuxP/ LuxQ	Bioluminescence emission, symbiosis	HAI-1, CAI-1 AI-2	Nadell et al. (2009)

1.6.1 Quorum Sensing in Gram-Positive Bacteria

Quorum sensing mechanism in the Gram-positive bacteria is quite different from the Gram-negative bacteria as the former uses post-transcriptionally synthesized peptide signal molecules instead of using AHLs.

There is a two-component signal transduction system in which histidine kinase acts as sensor element and interacts with the peptide signals. The bacterial competence is developed among *Bacillus subtilis* and *Streptococcus pneumoniae* along with conjugation in *Enterococcus faecalis* and virulence in *Staphylococcus aureus* are regulated by the quorum sensing mechanism (Dunny et al. 2016; England et al. 1999).

The autoinducer peptides are encoded as precursors (pro-AIPs), which are secreted by the specialized transporter as the bacterial cell membrane is impermeable to peptides. These pro-AIPs get modified post-translationally and become 5 to 17 amino acids long linear or cyclic molecules (Magnuson et al. 1994). These extracellular AIPs are detected by the membrane-bound sensor kinase system. When conserved histidine present within the sensor kinase binds to the AIP, the sensor kinase is autophosphorylated. The phosphoryl group within the system is passed to a conserved aspartate from the histidine in order to control the expression of QS-target gene, by phosphorylating the response regulator. An operon system encodes these QS circuits containing the pro-AIP, histidine kinase receptor, transporter and response regulator. The expression of the whole operon system gets activated by this phosphorylation of the response regulator.

1.6.1.1 Quorum Sensing in *Staphylococcus aureus*

Staphylococcus aureus is a common Gram-positive bacterium, mainly found in the normal human skin, which may cause skin infections when the epithelial barrier is compromised. These infections may cause diseases like pneumonia, bacteraemia and sepsis. The immune system gets affected by the expression of the array of the molecules like toxins that cause diseases. Quorum sensing regulates the genes encoding the virulence factors. In the *S. aureus* the *agr* locus, having the P2 promoter which drives expression of a transcript (RNAII), also encodes the four components of the QS system (Novick et al. 1995). The pro-AIP, which is encoded by *agrD*, is processed to the final AIP, which is secreted by the transmembrane transporter protein AgrB. 45–47 residues long pro-AIP is converted to the 7–9 residues peptide. This conversion is processed by the cyclization of a five-membered peptide ring via a thiolactone bond between a central cysteine residue and the carboxyl terminus (Thoendel et al. 2011). After getting accumulated, AIP binds to membrane-bound histidine kinase AgrC, which transfers the phosphate group to an aspartate on the response regulator AgrA by autophosphorylation of kthe conserved histidine residues (Lina et al. 1998). The agr operon is auto-induced when phosphorylated AgrA binds to the upstream of the P2 promoter. AgrA can activate the P3 promoter, which controls expression of RNAIII. RNAIII has the dual functions in regulating gene function by encoding the virulence factor d-haemolysin by hld gene and also in activating α-toxin production and repressing rot as well as some other proteins like the fibronectin-binding protein A and B, protein A, coagulase and other surface proteins (Morfeldt et al. 1995). All these mechanisms actually regulate the cascade of quorum sensing. The surface virulence proteins (protein A) are downregulated, or the secreted virulence factors (alpha toxin) are upregulated. The virulence of *S. aureus* is directly or indirectly regulated by RNAIII. The two additional virulence genes which are activated by the phosphorylated AgrA encode phenol-soluble modulins which are also regulating the virulence of *S. aureus* (Queck et al. 2008). This quorum sensing cascade has a major role in biofilm development by *S. aureus*. There are some other regulatory factors (converge on the P2 and P3 promoters) which are regulating the QS of the *S. aureus*. In *S. aureus* there are agr regulators, helping the microorganism to respond to extracellular environmental signals along with the AIPs of the QS system. When there is an extracellular stress, an interaction occurs between the alternative sigma factors with core RNAP in order to direct transcription of surface proteins and pigments production, and this inhibits the expression of the secreted toxins and proteases. The QS system of the *S. aureus* has cross competition among AIP specificity types. The variability of the bacteria relies on the agrD and agrB gene (Dufour et al. 2002). Depending upon the AIPs, *S. aureus* are classified into four specific groups (I–IV). The AIP receptor interactions with the non-cognate AIPs are actually regulating the inhibition of the QS. The binding of the incorrect AIPs stabilizes the inhibitory conformation of the AgrC gene in order to halt cell signalling through the QS (Lyon et al. 2002; Geisinger et al. 2009).

1.6.2 Quorum Sensing in Gram-Negative Bacteria

The QS circuits of the numerous Gram-negative bacteria are regulated by the factors or proteins like LasI/LasR, RhlI/RhlR, SmaI/SmaR, CviI/CviR, etc. The synthesis of the main signal molecules in QS system, acyl-homoserine lactones (AHL), is regulated by the LuxI like protein. AHLs, which diffuse across the cell membrane, increase in concentration when cell density is high. A cognate LuxR protein recognizes the AHL molecules, binds to it and activates transcription of the target gene by binding to the specific promoter of that. The biochemical actions of each of the protein pairs are conserved. The AHL molecules are produced by the LuxI enzymes and regulated by the LuxR. LuxI used to couple with the specific acyl-acyl carrier protein (acyl-ACP) present in the fatty acid biosynthetic machinery through the acyl-side chain of it to the homocysteine moiety of *S*-adenosylmethionine (SAM). Methylthioadenosine is released when acyl-ACP lactonizes to form acyl-HSL (Parsek et al. 1999).

The specific quorum sensing molecules are controlling the wide range of the cellular process for species-specific recognition. A unique AHL is produced by each species, which is recognized and utilized by only the members of a particular species (Fig. 1.2).

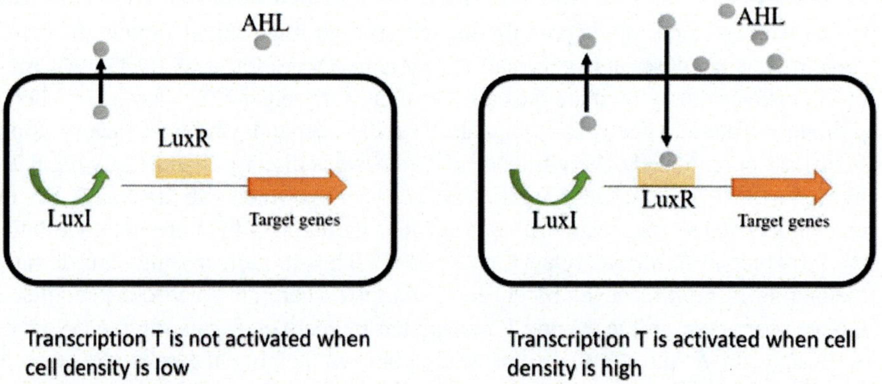

Fig. 1.2 Mechanism of quorum sensing

1.6.2.1 Quorum Sensing in *Pseudomonas aeruginosa*

Pseudomonas aeruginosa is a ubiquitous Gram-negative bacterium that can cause both the acute and chronic infections in humans, having immunosuppressed or immunocompromised systems. People with cystic fibrosis mainly become affected by the *P. aeruginosa* infections in their lungs (Lyczak et al. 2002). The QS system of *P. aeruginosa* produces a group of virulence factors that cause disease to the host. Mainly three types of the QS system are present in *Pseudomonas*, two LuxI/LuxR-type QS circuits regulating the expression of the virulence factors and one non-LuxI/LuxR-type system which is regulating *Pseudomonas* quinolone signal (PQS).

In the high cell density, LuxI and LuxR, a homolog to the LasS and LasR, play a significant role in synthesis and transcriptional regulation of the 3-oxo-C12-homoserine lactone (3OC12HSL). The LuxI also detects the AI and also encodes the virulence factors like elastase, protease and exotoxins A (Gambello et al. 1993).

The LasR-3OC12HSL regulates luxI, which is homologous to the rhlI and synthesizes butanoyl homoserine lactone (C4HSL). At HCD, AI binds to RhlR (homolog to LuxR). These RhlR-C4HSL activate target genes encoding pyocyanin, proteases, elastase, siderophores, etc. (Schuster et al. 2003).

The non-LuxI-LuxR QS system is controlling the virulence of the bacteria by the different genes. PqsA, PqsB, PqsC, PqsD and PqsH genes act together and produce 2-heptyl-3-hydroxy-4-quinolone which is detected by the PqsR regulator gene. LasR-3OC12HSL activates the expression of the genes PqsH and PqsR, whereas RhlR-C4HSL represses PqsABCD and PqsR. The PQS autoinducers activate and regulate RhlI and RhlR expression and along with the LasI/LasR influences virulence factor production (Gallagher et al. 2002; Diggle et al. 2007) (Table 1.2).

1.7 Biofilm-Associated Infection and Pathogenesis

Around 65% of all the bacterial infections (including device-associated and non-device-associated infections) are associated with the bacterial biofilms. A statistical analysis showed that device-associated infections are more infectious compared to the non-device-associated infections. Among those diseases, 2% are for joint prostheses; 2% for breast implants; 4% for mechanical heart valves; 10% for ventricular shunts; 4% for pacemakers and defibrillator; and about 40% for ventricular assist devices. The vascular endothelium and pulmonic valves of the heart are the results of streptococci and staphylococci bacterial infections. In this infection, the microbial cells attack the heart and blood through the gastrointestinal tract. Microorganisms damage the endothelium valve to attach to it and cause the infection at the site of injury (Darouiche 2004; Kokare et al. 2009).

There are many microorganisms which affect the soft tissues and cause the infections (Table 1.3). Periodontitis is a gum infection, with red or swollen tender as well as bleeding gum, having painful chewing and sensitive teeth (caused by *P. aerobicus* and *Fusobacterium nucleatum*) which damages the soft tissues as well as the supporting bones of the teeth and causes poor oral hygiene and tooth loss (Kokare et al. 2007). Microorganisms colonize upon the teeth surface, invade

Table 1.3 Infections caused by microbial biofilms

S. no.	Common biofilm-forming bacterial species	Nature of species	Infections caused by them	In vitro biofilm formation	References
1.	*E. coli*	Rod-shaped Gram-negative bacteria	Nosocomial and community infections such as urinary tract infections (UTIs) and prostatitis	Form biofilm in vitro	De Kievit et al. (2001)
2.	*P. aeruginosa*	Gram-negative notorious opportunistic pathogen	Responsible for chronic infections	Form biofilm in vitro	Grkovic et al. (2002)
3.	*S. epidermidis*	Opportunistic pathogen	Causes chronic infections	Doesn't form biofilm in vitro	Nickel et al. (1987)
4.	*S. aureus*	Gram-positive bacteria	Multidrug resistant bacteria causing a number of nosocomial infections	Form biofilm in vitro	Hellstrtom (1938)
5.	*Staphylococcus epidermidis*	Opportunistic pathogen	Leads to the pathogenesis	Form biofilm in vitro	Falanga (2000)
6.	*E. cloacae*	Gram-positive bacteria	Causing a range of nosocomial infections in human, i.e. lower respiratory tract infection, bacteraemia	Doesn't form biofilm in vitro	
7.	*K. pneumoniae*	Gram-negative bacterium	Causing nosocomial infections	Form biofilm in vitro	Neut et al. (2007)
8.	*Actinomyces israelii*	Gram-positive, rod-shaped bacteria	Causes actinomycosis	Doesn't form biofilm in vitro	Mendoza (2004)
9.	*Haemophilus influenzae*	Bacillus bacteria	Some of these illnesses, like ear infections, are mild while others, like bloodstream infections, are very serious	Doesn't form biofilm in vitro	Hellstrtom (1938)
10.	*Burkholderia cepacia*	Aerobic Gram-negative bacillus	Causes various types of infections, including catheter-associated infections and respiratory tract infections	Doesn't form biofilm in vitro	Lewis (2008)

mucosal cells and alter the flow of the calcium to the epithelial cells and release toxins. A plaque is formed upon the teeth surface, and this mineralizes calcium and phosphate ions and forms tartar. Bacteria enter into the bones through the bloodstream, and metaphysis of the bones happens which causes bone infections, known as osteomyelitis. This leads to the recruitment of the white blood cells (WBCs) at the site of infections. These WBCs secrete enzymes to lyse the pathogens through phagocytosis. These enzymes cause bone lysis resulting in the formation of the pus that spreads through the bone-blood vessels and stops the proper flow of the blood by damaging the tissues (Overman 2000; Ziran 2007).

Lungs of the cystic fibrosis (CF) patients are the arena of biofilm formation by bacteria which cause lung infection. The sessile cells of *P. aeruginosa* establish the biofilm in the lungs of CF patients, and it is facilitated by hypersecretion of viscous mucus. This mucus adheres to the surface, and persistent mucin secretion generates the formation of mucus plaques and plugs (thick in nature), leading to the bacterial infection along with symptoms like very salty-tasting skin; persistent coughing at times with phlegm; frequent lung infections including pneumonia or bronchitis; wheezing or shortness of breath; poor growth or weight gain in spite of a good appetite; frequent greasy, bulky stools; or difficulty with bowel movements. During the course of infection, nonmucoid converts itself to mucoid. This mucoid mutant phenotype produces exopolysaccharide which makes colonies mucoid and the bacterial resistance get increased against phagocytosis and the tolerance to the antibiotics.

The biofilm (benign or commensal in nature) mainly forms upon the open wound as they lack the protective covering of the skin, leading to the destruction of the host immune system. The adhesion of microorganism on the surface of wound is facilitated by fimbriae and pili. Large numbers of micro-organisms have evolved with an array of different mechanisms to adhere on the surface and form biofilm. For example, *Pseudomonas aeruginosa* and *Staphylococcus epidermidis* form the biofilm on the skin wound by polysaccharide intercellular adhesion (PIA) and develop the community by encasing individual cells in extracellular polysaccharides

Biofilm has both the positive and negative effects, and the ill effect of biofilm is fatal as they develop antimicrobial resistance. Biofilms may originate on indwelling medical devices from the skin of the patients or healthcare personalities, tap water to which the devices are exposed to and other environmental sources. The Gram-negative bacteria *Vibrio parahaemolyticus* causing seafood-derived food poisoning are resistant to the antibiotics like bacitracin, chloramphenicol, rifampin, ampicillin, vancomycin, nalidixic acid, penicillin and spectinomycin. The strain of *Vibrio parahaemolyticus* ATCC 17802 is resistant towards the detergent D1 (linear alkylbenzene based) (Elexson et al. 2014). Sometimes during the surgical implantation of prosthetic heart valves, tissue damage occurs which leads to the accumulation of platelets and fibrins making a greater possibility for the formation of biofilms. Colonized microorganisms can be found on all the venous catheters, for example, *S. epidermidis*, *Candida albicans*, *S aureus*, *Pseudomonas aeruginosa*, *Klebsiella pneumoniae*, *Enterococcus faecalis*, *S. epidermidis*, *Proteus mirabilis*, etc. are some of the commonly isolated bacteria found on catheters. High pH condition prevailing at biofilm-urine interface leads to precipitation of minerals such as hydroxyapatite and struvite, thereby completely blocking the inner lumen. Hence the increase of the pH of the urine represents the microscopic aggregation of the cells and the crystal formation in the urine which settle on the polymeric surface and initiate the crystalline biofilm formation Jacobsen et al. 2008).

1.8 Biofilm-Associated Acute Infections

Within the biofilm, the bacteria adapt to environmental conditions and nutrient limitation by exhibiting an alteration of the metabolism, gene expression and protein production, which leads to a lower metabolic rate and a reduced rate of cell division. By staying dormant and hidden from the immune system, bacteria may cause local tissue damage and later cause an acute infection (Vestby et al. 2020).

Acute infection refers to the infection type where the microbial cells live within the host for a small period of time that can be even less than a period of 6 months. Although they are short lived, they have a long-lasting impact upon the host, providing an onset to the beginning of the chronic infections. There can be a marked difference between the acute and chronic infections as the pathogens causing the acute infections have faster rate of growth, having symptoms which are more profound and are short lived. It marks the onset of chronic infections to a person and affects the person for the rest of the life. A question always revolves in the context of the acute and chronic infections that whether the organisms differ in their molecular mechanisms to cause acute and chronic diseases. It has been further observed that a single bacterial cell possesses the ability of causing both acute and chronic infections. Studies showed that *P. aeruginosa* is a potent organism to cause pneumonia by virtue of disintegrating the lungs defence and spreading into the blood causing morbidity. *P. aeruginosa* possess type III (TTSS) secretory system that secretes various types of extracellular toxins that play an important role in causing various types of acute infections like that of pneumonia (Ghosh 2004. Goodman et al. 2004) (Fig. 1.3). Studies showed that TTSS being present within the *P. aeruginosa* play an important role for the survival of the organism within the blood stream (Vance et al. 2005). The strain possessing the defective TTSS system especially within the exsA mutant type of *P. aeruginosa* causes acute corneal diseases (Lee et al. 2003).

1.8.1 Biofilm-Associated Acute Infection of the Skin

Patients with acute bacterial skin and skin structure infections (ABSSSI) are commonly referred to emergency departments (EDs) where physicians encounter a wide spectrum of disease severity (Golan 2019). The prevalence of community-acquired methicillin-resistant *Staphylococcus aureus* (CA-MRSA) has increased in the past decade, and CA-MRSA is now a predominant cause of purulent ABSSSI in the United States. Acute skin infections where mixed population of the Gram-positive

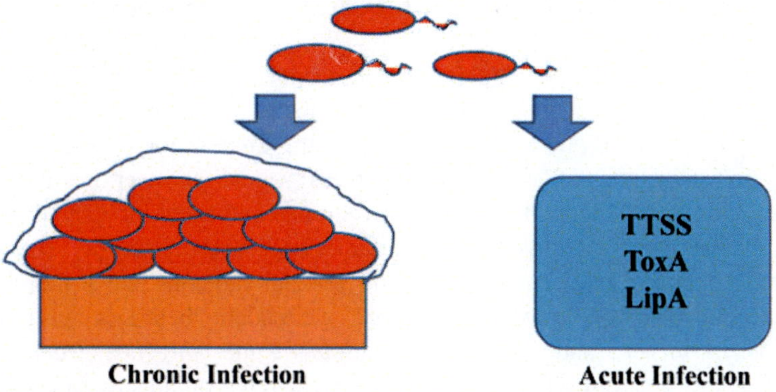

Fig. 1.3 Chronic and acute infection formation by *P. aeruginosa*

and Gram-negative pathogens are present can also be considered having the association with discordant antimicrobial therapy. The increasing prevalence of CA-MRSA is considered to be the cause of ABSSSI (Moran et al. 2013).

1.8.2 Biofilm-Associated Respiratory Tract Infections

Respiratory tract infection is a global health concern which also shows a very high percentage of morbidity. It especially occurs within elderly individuals and among young children (Dumas et al. 2018). Acute lower respiratory tract infections are a persistent and pervasive public health problem. Accumulation of neutrophils occurs in acute inflammation, leading to the exudation of the plasma outside of blood vessels in the pulmonary capillaries of uninfected lungs. The accumulation of extravascular plasma fluids, in the form of noncardiogenic pulmonary oedema, is a defining feature of acute lung injury. Perhaps due to this, lung infection is a common underlying cause of acute respiratory distress syndrome. *Streptococcus pneumoniae*, *Pneumocystis carinii*, *Legionella pneumophila*, influenza viruses, few Gram-negative bacteria and *Aspergillus fumigatus (*Mizgerd 2008) *are the major causative agents.*

The common types of respiratory tract infections include pneumonia and otitis media that are mostly caused by pathogenic organisms like *Haemophilus influenzae* and *Streptococcus pneumoniae.* Depending upon the duration of the disease rhinosinusitis (RS), it can be called or classified as acute when lasting less than 12 weeks or chronic when lasting more than 12 weeks. Up to 80–90% of acute RS are accounted by the viruses, and the most commonly involved viruses are rhinovirus, respiratory syncytial virus, influenza virus, coronavirus, parainfluenza virus, adenovirus and enterovirus (Mizgerd 2008).

1.8.3 Biofilm-Associated Acute Digestive Tract Infection

Acute cholangitis and acute cholecystitis are common conditions that may result in progressively severe infection and death when not treated appropriately. In the report on the threat of antibiotic resistance in 2013, the Centers for Disease Control and Prevention (CDC) have classified pathogens as ‘urgent’ and ‘serious threats’ based on their clinical and economic impact, current and estimated incidence, transmissibility, availability of effective antibiotics and available barriers for prevention, which are well known to gastroenterologists and visceral surgeons. *Clostridium difficile*, carbapenem-resistant *Enterobacteriaceae*, extended beta-lactamase-producing *Enterobacteriaceae* and drug-resistant *Salmonella*, *Shigella*, *Campylobacter*, *Pseudomonas* and *Candida* species are the pathogenic organisms causing biofilm-oriented diseases. Typhoid fever is an acute food-borne illness, caused by *Salmonella enterica* predominantly, causing the symptoms like high fever, weakness, headache, abdominal pain and constipation. After the acute phase of illness, 3–5% of the typhoid fever patients become chronic carriers. Antibiotic treatment is generally used to resolve the acute infection, but it may be often ineffective against the chronic colonization of the gall bladder by *S. typhi*.

1.8.4 Biofilm-Associated Urinary Tract Infections

Acute cystitis is a sudden inflammation of the urinary bladder. Most of the time, it is caused by bacterial infection. This infection is commonly referred to as a urinary tract infection (UTI). Irritating hygiene products, a complication of certain diseases, or a reaction to certain drugs can also cause acute cystitis. *P. mirabilis* produces two toxins, haemolysin (HpmA) and proteus toxic agglutinin (Pta), which are implicated in tissue damage and dissemination to the kidneys, initiating acute pyelonephritis (Flores-Mireles et al. 2015). Urinary tract infections (UTIs) have complex dynamics, often related to uropathogenic *Escherichia coli* (UPEC). This is the major causative agent, capable of colonization from the urethra to the kidneys in both extracellular and intracellular niches while also producing chronic persistent infections and frequent recurrent disease (Flores-Mireles et al. 2015).

1.9 Antimicrobial Resistance of the Biofilm

Antibiotic resistance of the bacteria present in the biofilm communities cause the chronic infections to the host cells. Resistance mechanism of the biofilm-forming bacteria is quite different from the planktonic group of bacteria. Though there are several aspects like low cell permeability, target site mutations, efflux pump, drug neutralizing protein and drug modifying enzymes, depending upon which antibiotic resistance developed, there are other conventional antibiotic resistance mechanisms which cannot be explained. Antibiotic resistance phenomenon follows both mechanisms during spreading the chronic infections. The repeated exposure of a particular antibiotic can develop the antibiotic resistance in the biofilm infections. It has been reported that repeated exposure of the ceftazidime in biofilm-forming bacteria *Pseudomonas aeruginosa* developed an intrinsic antibiotic resistance (Bagge et al. 2000). Slow or incomplete penetration of antibiotics into the biofilm or an altered chemical microenvironment or subpopulation of the microorganisms in a biofilm cause the antibiotic resistance to the microorganisms present inside the biofilm layer. Biofilm has the multicellular layers which is the key factor for the development of antibiotic resistance to the biofilm communities. The extracellular polymeric substances (EPS) hold the bacterial cells together within the biofilm layer and lead to the development of the heterogeneous multicellular consortia. Biofilm development is mainly regulated by the intercellular and intracellular signalling process. The upregulation and downregulation of the panel of genes are done by the quorum sensing mechanism (Davies et al. 1998).

1.10 Conclusion

Biofilm-associated bacterial infections frequently caused by *Staphylococcus epidermidis*, *Staphylococcus aureus*, *Escherichia coli* and *Pseudomonas aeruginosa* are found in the majority of human diseases. Effectiveness of many antimicrobial drugs has been lost due to the evolution of pathogenic resistance. Many of the microorganisms are no longer susceptible to most of the existing antibiotics and therapeutic agents. Therefore, an alternative way of reducing biofilm is very essential. The anti-adherence and anti-quorum sensing compounds can be used to eradicate the bacterial biofilm by help in enhancing susceptibility towards the bacteria. On the other side, the biofilm can be used in bioremediation purposes, or it can also be used for wastewater treatment. So a proper approach should be chosen to eradicate the biofilm as well as to use the biofilm in different environmental purposes.

References

Bagge N, Ciofu O, Skovgaard LT, Høiby N (2000) Rapid development in vitro and in vivo of resistance to ceftazidime in biofilm-growing Pseudomonas aeruginosa due to chromosomal beta-lactamase. APMIS 108:589–600

Brandas SS, Vik A, Friedman L, Koiter R (2005) Biofilm: the matrix revisited. Trends Microbiol 13:20e6

Characklis WG, McFeters GA, Marshall KC (1990) Physiological ecology in biofilm systems. In: Characklis WG, Marshall KC (eds) Biofilms. John Wiley & Sons, New York, pp 341–394

Corpe WA (1980) Microbial surface components involved in adsorption of microorganisms onto surfaces. In: Bitton G, Marshall KC (eds) Adsorption of microorganisms to surfaces. John Wiley & Sons, New York, pp 105–144

Costerton JW, Geesey GG, Cheng KJ (1978) How bacteria stick. Sci Am 238(1):86–95

Costerton JW, Lewandowski Z, Caldwell DE, Korber DR, Lappin-Scott HM (1995a) Microbial biofilms. Annu Rev Microbiol 49:711–745

Costerton JW, Lewandowski Z, Caldwelld DE, Korber DR, LappinScott HM (1995b) Microbial biofilms. Ann Rev Microbiol 49:711e45

Costerton JW, Lewandowski Z, DeBeer D, Caldwell D, Korber D, James G (1994) Biofilms, the customized microniche. J Bacteriol 176:2137e42

Costerton JW, Stewart PS, Greenberg EP (1999) Bacterial biofilms: a common cause of persistent infections. Science 284:1318–1322

Darouiche RO (2004) Treatment of infections associated with surgical implants. N Engl J Med 350:1422e9

Davies DG, Parsek MR, Pearson JP, Iglewski BH, Costerton JW, Greenberg EP (1998) The involvement of cell-to-cell signals in the development of a bacterial biofilm. Science 280:295–298

Davies J (2007) Microbes have the last word. A drastic re-evaluation of antimicrobial treatment is needed to overcome the threat of antibiotic resistant bacteria. EMBO Rep 8:616–621

De Kievit TR, Parkins MD, Gillis RJ, Srikumar R, Ceri H, Poole K, Iglewski BH, Storey DG (2001) Multidrug efflux pumps: expression patterns and contribution to antibiotic resistance in *Pseudomonas aeruginosa* biofilms. Antimicrob Agents Chemother 45(6):1761–1770

Diggle SP, Matthijs S, Wright VJ, Fletcher MP, Chhabra SR, Lamont IL, Kong X, Hider RC, Cornelis P, Cámara M, Williams P (2007) The Pseudomonas aeruginosa 4-quinolone signal molecules HHQ and PQS play multifunctional roles in quorum sensing and iron entrapment. Chem Biol 14(1):87–96

Donlan RM (2002) Biofilms: microbial life on surfaces. Emerg Infect Dis 8(9):881-890. https://doi.org/10.3201/eid0809.020063. PMID: 12194761; PMCID: PMC2732559.

Dufour D, Cordova M, Cvitkovitch DG, Levesque CM (2011) Regulation of the competence pathway as a novel role associated with a streptococcal bacteriocin. J Bacteriol 193:6552–6559

Dufour P, Jarraud S, Vandenesch F, Greenland T, Novick RP, Bes M, Etienne J, Lina G (2002) High genetic variability of the agr locus in Staphylococcus species. J Bacteriol 184:1180–1186

Dumas A, Bernard L, Poquet Y, Lugo-Villarino G, Neyrolles O (2018) The role of the lung microbiota and the gut-lung axis in respiratory infectious diseases. Cell Microbiol 20(12): e12966

Dunny GM, Berntsson RP (2016) Enterococcal sex pheromones: evolutionary pathways to complex, two-signal systems. J Bacteriol 198(11):1556–1562

Elexson N, Afsah-Hejri L, Rukayadi Y, Soopna P, Lee HY, Tuan Zainazor TC, Ainy MN, Nakaguchi Y, Mitsuaki N, Son R (2014) Effect of detergents as antibacterial agents on biofilm of antibiotics-resistant Vibrio parahaemolyticus isolates. Food Control 35(1):378–385

England R, Hobbs G, Bainton N, Roberts DM (1999) Microbial signalling and communication. University Press, Cambridge, pp 117–138

Falanga V (2000) Classifications for wound bed preparation and stimulation of chronic wounds. Wound Repair Regen 8:347–352

Flores-Mireles AL, Walker JN, Caparon M, Hultgren SJ (2015) Urinary tract infections: epidemiology, mechanisms of infection and treatment options. Nat Rev Microbiol 13(5):269–284

Gallagher LA, McKnight SL, Kuznetsova MS, Pesci EC, Manoil C (2002) Functions required for extracellular quinolone signaling by Pseudomonas aeruginosa. J Bacteriol 184:6472–6480

Gambello MJ, Kaye S, Iglewski BH (1993) LasR of Pseudomonas aeruginosa is a transcriptional activator of the alkaline protease gene (apr) and an enhancer of exotoxin A expression. Infect Immun 61:1180–1184

Garrett TR, Bhakoo M, Zhang Z (2008) Bacterial adhesion and biofilms on surfaces. Prog Nat Sci 18:1049e56

Geesey GG, Mutch R, Costertin JW, Green RB (1978) Sessile bacteria: an important component of the microbial population in small mountain streams. Limnol Oceanogr 23:1214–1223

Geesey GG, Richardson WT, Yeomans HG, Irvin RT, Costerton JW (1977) Microscopic examination of natural sessile bacterial populations from an alpine stream. Can J Microbiol 23:1733–1736

Geisinger E, Muir TW, Novick RP (2009) agr receptor mutants reveal distinct modes of inhibition by staphylococcal autoinducing peptides. Proc Natl Acad Sci 106:1216–1221

Ghosh P (2004) Process of protein transport by the type III secretion system. Microbiol Mol Biol Rev 68:771–795

Golan Y (2019) Current treatment options for acute skin and skin-structure infections. Clin Infect Dis 68(Suppl 3):S206–S212

Goodman AL, Kulasekara B, Rietsch A, Boyd D, Smith RS, Lory S (2004) A signaling network reciprocally regulates genes associated with acute infection and chronic persistence in Pseudomonas aeruginosa. Dev Cell 7:745–754

Grkovic S, Brown MH, Skurray RA (2002) Regulation of bacterial drug export systems. Microbiol Mol Biol Rev 66(4):671–701

Hall-Stoodley L, Costerton JW, Stoodley P (2004) Bacterial biofilms: from the natural environment to infectious diseases. Nat Rev Microbiol 2(2):95–108

Hellstrtom J (1938) The significance of staphylococci in the development and treatment of renal and urethral stones. Br J Urol 10:348–372

Henrici AT (1933) Studies of freshwater bacteria. I. A direct microscopic technique. J Bacteriol 25:277–287

Hurlow J, Couch K, Laforet K, Bolton L, Metcalf D, Bowler P (2015) Clinical biofilms: a challenging frontier in wound care. Adv Wound Care 4:295e301

Jacobsen SM, Stickler DJ, Mobley HL, Shirtliff ME (2008) Complicated catheter-associated urinary tract infections due to Escherichia coli and Proteus mirabilis. Clin Microbiol Rev 21 (1):26–59

Jones HC, Roth IL, Sanders WM (1969a) 3rd electron microscopic study of a slime layer. J Bacteriol 99(1):316–325

Jones HC, Roth IL, Saunders WMIII (1969b) Electron microscopic study of a slime layer. J Bacteriol 99:316–325

Kokare CR, Chakraborty S, Khopade AN, Mahadik KR (2009) Biofilm: importance and applications. Indian J Biotechnol 8:159e68

Kokare CR, Kadam SS, Mahadik KR, Chopade BA (2007) Studies on bioemulsifier production from marine Streptomyces sp. S1. Indian. J Biotechnol 6:78e84

Lee EJ, Cowell BA, Evans DJ, Fleiszig SM (2003) Contribution of ExsA-regulated factors to corneal infection by cytotoxic and invasive Pseudomonas aeruginosa in a murine scarification model. Investig Ophthalmol Vis Sci 44:3892–3898

Lemme A, Grobe L, Reck M, Tomasch J, Wagner-Dobler I (2011) Subpopulation-specific transcriptome analysis of competence-stimulating peptide-induced Streptococcus mutans. J Bacteriol 193:1863–1877

Lewis K (2008) Multidrug tolerance of biofilms and persister cells. Bacterial biofilms. Antimicrob Agents Chemother 51:1934–1938

Lina G, Jarraud S, Ji G, Greenland T, Pedraza A, Etienne J, Novick RP, Vandenesch F (1998) Transmembrane topology and histidine protein kinase activity of AgrC, the agr signal receptor in Staphylococcus aureus. Mol Microbiol 28:655–662

Lyczak JB, Cannon CL, Pier GB (2002) Lung infections associated with cystic fibrosis. Clin Microbiol Rev 15:194–222

Lyon GJ, Wright JS, Muir TW, Novick RP (2002) Key determinants of receptor activation in the agr autoinducing peptides of Staphylococcus aureus. Biochemistry 41:10095–10104

Magnuson R, Solomon J, Grossman AD (1994) Biochemical and genetic characterization of a competence pheromone from B. subtilis. Cell 77:207–216

Mckenney D, Hubner J, Muller E, Wang Y, Goldmann DA, Pier GB (1998) The ica locus of Staphylococcus epidermidis encodes production of the capsular polysaccharide/adhesin. Infect Immun 66:4711e20

Mendoza MT (2004) El papel del biofilm en el proceso infeccioso y la resistencia. Nova 2:71–80

Miron J, Ben-Ghedaliad D, Morrison M (2001) Invited review: adhesion mechanism of rumen cellulolytic bacteria. J Dairy Sci 84:1294e309

Mizgerd JP (2008) Acute lower respiratory tract infection. N Engl J Med 358(7):716–727

Moran GJ, Abrahamian FM, Lovecchio F, Talan DA (2013) Acute bacterial skin infections: developments since the 2005 Infectious Diseases Society of America (IDSA) guidelines. J Emerg Med 44(6):e397–e412
Morfeldt E, Taylor D, von Gabain A, Arvidson S (1995) Activation of a-toxin translation in Staphylococcus aureus by the trans-encoded antisense RNA RNAIII. EMBO J 14:4569–4577
Nadell CD, Xavier JB, Foster KR (2009) The sociobiology of biofilms. FEMS Microbiol Rev 33:206–224
Neut D, van der Mei HC, Bulstra SK, Busscher HJ (2007) The role of small-colony variants in failure to diagnose and treat biofilm infections in orthopedics. Acta Orthopaedic 78:299–308
Nickel JC, Olson M, McLean RJ, Grant SK, Costerton JW (1987) An ecological study of infected urinary stone genesis in an animal model. Br J Urol 59(1):21–30
Novick RP, Projan SJ, Kornblum J, Ross HF, Ji G, Kreiswirth B, Vandenesch F, Moghazeh S (1995) The agr P2 operon: an autocatalytic sensory transduction system in Staphylococcus aureus. Mol Gen Genet 248:446–458
Ongenae, A. (2017). Is biofilm formation a critical step for the valorisation of plastic waste? http://hdl.handle.net/2268.2/3018
Overman PR (2000) Biofilm: a new view of plaque. J Contemp Dent Pract 1:18e29
Parsek MR, Val DL, Hanzelka BL, Cronan JE Jr, Greenberg EP (1999) Acyl homoserine lactone quorum-sensing signal generation. Proc Natl Acad Sci U S A 96:4360–4365
Queck SY, Jameson-Lee M, Villaruz AE, Bach TH, Khan BA, Sturdevant DE, Ricklefs SM, Li M, Otto M (2008) RNAIII-independent target gene control by the agr quorum-sensing system: insight into the evolution of virulence regulation in Staphylococcus aureus. Mol Cell 32:150–158
Rosenberg M, Bayer EA, Delarea J, Rosenberg E (1982) Role of thin fimbriae in adherence and growth of Acinetobacter calcoaceticus RAG-1 on hexadecane. Appl Environ Microbiol:105–144
Schuster M, Lostroh CP, Ogi T, Greenberg EP (2003) Identification, timing, and signal specificity of Pseudomonas aeruginosa quorum-controlled genes: a transcriptome analysis. J Bacteriol 185:2066–2079
Sehar S, Naz I (2016) Role of the biofilms in wastewater treatment. Microbial biofilms-Import Appl. https://doi.org/10.5772/63499
Shaw T, Winston M, Rupp CJ, Klapper I, Stoodley P (2004) Commonality of elastic relaxation times in biofilms. Phys Rev Lett 93:098102
Sleytr UB (1997) Basic and applied S-layer research: an overview. FEMS Microbiol Rev 20:5e12
Sun D, Accavitti MA, Bryers JD (2005) Inhibition of biofilm formation by monoclonal antibodies against Staphylococcus epidermidis RP62A accumulation-associated protein. Clin Diagn Lab Immunol 12:93e100
Sutherland IW (2001a) Biofilm exopolysaccharides: a strong and sticky framework. Microbiology 147:3–9
Sutherland IW (2001b) The biofilm matrix e an immobilized but dynamic microbial environment. Trends Microb 9:222e7
Thoendel M, Kavanaugh JS, Flack CE, Horswill AR (2011) Peptide signaling in the staphylococci. Chem Rev 111:117–151
Vance RE, Rietsch A, Mekalanos JJ (2005) Role of the type III secreted exoenzymes S, T, and Y in systemic spread of Pseudomonas aeruginosa PAO1 in vivo. Infect Immun 73:1706–1713
Vendeville A, Winzer K, Heurlier K, Tang CM, Hardie KR (2005) Making 'sense' of metabolism: autoinducer-2, LuxS and pathogenic bacteria. Nat Rev Microbiol 3:383–396
Vestby LK, Grønseth T, Simm R, Nesse LL (2020) Bacterial biofilm and its role in the pathogenesis of disease. Antibiotics (Basel, Switzerland) 9(2):59. https://doi.org/10.3390/antibiotics9020059
Ziran BH (2007) Osteomyelitis. J Trauma 62:559e60

Quorum Sensing

2

Moupriya Nag, Dibyajit Lahiri, Anushka Ghosh, Deboleena Das, and Rina Rani Ray

Abstract

The bacterial cells consist of a cell-to-cell communication system which is density dependent and is referred to as quorum sensing (QS) which helps the sessile communities to proliferate inside the biofilm. QS is a congregation of production and responding to the diffusible signals which varies from one bacterial cell to another. The virulence of the bacterial cells is dependent upon the mechanism of QS. This mechanism of QS allows the bacterial cells to work as a group, and the interactions appear as multicellular organism that helps in colonization and formation of biofilm. The mechanism of QS is regulated by gene regulons, signal structures, and various behavioral responses. The QS provides us the concept of the competitive and cooperative interactions between the same and other group of species existing within the biofilm. This chapter focuses on the mechanism of biofilm formation by QS and regulation of QS by various genetic pathways and with substantial examples of organisms.

M. Nag · D. Lahiri · A. Ghosh · D. Das
Department of Biotechnology, University of Engineering & Management, Kolkata, West Bengal, India

R. R. Ray (✉)
Department of Biotechnology, Maualana Abul Kalam Azad University of Technology, West Bengal, Kolkata, West Bengal, India

R. R. Ray et al. (eds.), *Biofilm-Mediated Diseases: Causes and Controls*,
https://doi.org/10.1007/978-981-16-0745-5_2

2.1 Introduction

For a long time, bacteria are considered to be individual prokaryotic cells that primarily needed nutrients to survive and multiply. With the advent of microbial research, bacteria are found to perform efficient intercellular communication leading to a coordinated activity similar to multicellular organisms. This coordinated behavior helps the bacteria to survive in an environment which is more suitable and also the nutrient supply is better. This collective behavior is also responsible for adapting new modes of growth, e.g., biofilm formation or sporulation which protects the bacterial cells from deleterious environments. These may include coping to the harsh antimicrobials and extreme environmental stress such as varying temperatures, pH, nutrient supply, etc. (Kievit De Teresa and Iglewski 2000).

The cell-cell communication occurs due to small, signaling molecules called autoinducers (AI) produced by the bacteria itself that sense the neighboring population density and behave accordingly. The phenomenon of cell-cell interaction by detecting and responding to the population density with the release of AI signaling molecules by a cascade of gene expressions is termed as quorum sensing (QS). Quorum sensing is associated with regulation of specific genes that control important biological functions in bacteria such as antibiotic production, sporulation, competence, motility, bioluminescence, virulence, and biofilm formation. In a nutshell, QS is all about production, detection, and response to AIs during bacterial cell-to-cell communication (Rutherford Steven and Bonnie 2012).

So far quorum sensing is observed in bacterial species that are directly related to plants or animals, thus affirming that QS is a technique adopted by the bacteria in either favorable (symbiotic bacteria) or adversarial (pathogenic bacteria) conditions. Thus, release of AIs into the environment is facilitated by the conditions in which the bacteria are dwelling for its survival, be it in favorable or adversarial conditions. AIs released by a single cell into the environment are too low to be detected. However, concentrations of AIs produced by multiple cells will lead to a threshold level which senses a critical cellular mass and thereby activates gene targets.

QS system involves a three-step mechanism for its proper functioning. First, bacterial cells produce signaling molecules, viz., AIs during stressful conditions. At low cell density (LCD), lower concentration of AIs is produced much below the threshold stimulatory concentration of an autoinducer needed for its successful detection. A higher concentration of AI during higher cell density (HCD) leads to the subsequent detection and response. Second, AIs are detected when above threshold only by the receptors existing in cytoplasm or membranes. Third, detection of AIs induces activation of gene expression profiles necessary for cooperative behavior by the bacterial cells as well as activates more AI production by a feedback mechanism (development of autoinduction loop by forward feed mechanism), promoting synchrony in population (Rutherford Steven and Bonnie 2012).

Even though the chemical nature of the signaling molecules, target genes, and the signal transduction mechanisms associated with various bacterial quorum sensing circuits differ, ability to cross talk with neighboring cells to coordinate the gene expression in the bacteria is a common scenario in every case, and hence the behavior, of bacterial society. Thus, the evolution of QS systems in bacteria may be defined as one of the preliminary steps in multicellular development.

2.2 Mechanism of Quorum Sensing in Gram-Negative Bacteria

The communication system prevailing in Gram-negative bacterial cells occurs by the signaling molecules known as autoinducers (AI) such as acetyl homoserine lactone (AHL) or other molecules whose production is dependent on S-adenosylmethionine (SAM). The cells produce AIs which diffuse easily in the inner and outer cell membrane. During high cell density (HCD), concentration of AIs increases which can bind to various transcription regulators for gene expression related to QS. In addition to AHL, Gram-negative bacteria contain other signaling molecules also. For example, *Ralstonia solanacearum* releases 3-hydroxy-palmitic acid methyl ester which is considered to be a novel signaling molecule along with AHLs which regulates virulence; *Xanthomonas campestris* produces a diffusible extracellular factor (DSF); *Pseudomonas aeruginosa* produces PQS (*Pseudomonas* quinolone signal), in the form of 2-heptyl-3-hydroxy-4-quinolone; *Pseudomonas aureofaciens* produces butyrolactones; and diketopiperazines (DKPs) are produced in cell-free supernatants of *Citrobacter freundii*, *Pseudomonas fluorescens*, *Enterobacter agglomerans*, *P. aeruginosa*, and *Pseudomonas alcaligenes.* A list of quorum sensing molecules secreted by various Gram-negative bacteria is tabulated in Table 2.1.

2.2.1 Cellular Communication in *P. aeruginosa* via Quorum Sensing

P. aeruginosa is considered to be the best studied model organism in terms of understanding virulence factor and quorum sensing mechanism. *P. aeruginosa* virulence is due to coordinated expression of a number of genes starting from *lasA*, *lasB*, *toxA*, and *aprA* including *lasI.* LasR are the group of transcriptional activators that act in synergism with *N*-(3-oxododecanoyl)-L-homoserine lactone (3-oxo-C12-HSL), AHL, which regulates the expression of those five genes, and are synthesized by a type of autoinducer synthase LasI.

P. aeruginosa also possesses another type of signaling system which is governed by *rhl* QS system. It comprises autoinducer synthase RhlI synthesizing AHL and transcriptional activator RhlR, *N*-butyryl-L-homoserine lactone (C4-HSL). The expression of *rhl*AB is regulated by RhlR-C4-HSL complex and in turn is needed for production of pyocyanin and cyanide lasB which is a type of secondary metabolite, aprA, rhamnolipid production, and the stationary phase sigma factor RpoS production. With the discovery of two QS circuits in *P. aeruginosa*, it was found that the two systems do not interact with each other as the genes which are primarily activated by one system are not activated by the other. Though a minimum cross talk is observed between the two QS circuits, it was also observed that the regulation of expression of both *rhlR* and *rhlI* by *las* system is positive, and this indicates *las*

Table 2.1 Summary of quorum sensing in some Gram-negative bacteria (Kievit De Teresa and Iglewski 2000)

Organism	Signal molecule	Regulatory proteins	Phenotypic traits	References
Pseudomonas aeruginosa	C_4-HSL 3-Oxo-C_{12}-HSL	Rh1I/Rh1R LasI/LasR	Formation of biofilm, production of secondary metabolites, various types of extracellular enzymes, RhR, Xcp	Gambello et al. (1993); Passador et al. (1993); Pierson et al. (1994)
Burkholderia cepacia	C_4-HSL	CepI/CepR	Siderophores, proteases	Lewenza et al. (1999)
Pseudomonas aureofaciens	C_6-HSL	PhzI/PhzR	Phenazine antibiotics	Pierson et al. (1994)
Agrobacterium tumefaciens	3-Oxo-C_8-HSL	TraI/TraR	Conjugation of Ti plasmid	Piper et al. (1993); Zhang et al. (1993); Hwang et al. (1994)
Rhizobium leguminosarum	C_6-HSL	RhiI/RhiR	Nodulation	Gray et al. (1996); Cubo et al. (1992); Rodelas et al. (1999)
Chromobacterium violaceum	C_6-HSL	CviI/CviR	Antibiotics, cyanide, exoenzymes	McLean et al.(1997)
Aeromonas salmonicida	C_4-HSL	AsaI/AsaR	Extracellular protease	Swift et al. (1997)
Yersinia enterocolitica	C6-HSL	YenI/YenR		Throup et al. (1995)
Yersinia pseudotuberculosis	C8-HSL	YenI/YenR		Atkinson et al. (1999)
Serratia liquefaciens	C_4-HSL	SwrI/SwrR	Extracellular protease	Eberl et al. (1996); Givskov et al. (1997)
Escherichia coli		SdiA	Leads to lesion formation, attachment, and cell division	Sharma et al. (1986); Withers and Nordstrom (1998); Sperandio et al. (1999)
Erwinia carotovora subsp. *carotovora*	3-Oxo-C6-HSL	ExpI/ExpR	Exoenzymes	Bainton et al. (1992); Chhabra et al. (1993); Jones et al. (1993); McGowan et al. (1995); Pirhonen et al. (1993)
Erwinia stewartii	3-Oxo-C6-HSL	EsaI/EsaR	Exopolysaccharide	Beck von Bodman and Farrand (1995)
Aeromonas hydrophila	C4-HSL	AhyI/AhyR	Production of exoprotease	Swift et al. (1997)
Rhodobacter sphaeroides	7-cis-C14-HSL	CerI/CerR	Produced from the congregation of the bacterial cells	Puskas et al. (1997)

system to be the dominant regulator of QS mechanism (Gambello and Iglewski 1991; Gambello et al. 1993).

Also, a third autoinducer molecule was detected and is known as 2-heptyl-3-hydroxy-4-quinolone (PQS) which is involved in lasB expression with the assistance of RhlR system. This pathway is used for the secretion of elastase and proteases which are quorum sensing-controlled enzymes (Latifi et al. 1996; Pesci et al. 1997).

2.2.2 QS in *Burkholderia cepacia*

The presence of *Burkholderia cepacia* in cystic fibrosis (CF) lungs mainly indicates co-localization with *P. aeruginosa*. It is thought that *P. aeruginosa* signaling molecules (AHLs) can be used for interspecies communication and this in turn enhances the pathogenicity of *B. cepacia*. There is a significant homology between cepR and cepI homologs of *B. cepacia* with luxRI homologs of *P. aeruginosa*. Thus, it is proposed that CepRI quorum sensing system can have both positive and negative regulatory role in *B. cepacia* by increasing the production of protease and by decreasing the synthesis of siderophore. In the CF lung, *B. cepacia* colonization that utilizes the exogenous AHLs secreted by *P. aeruginosa* is thought to be the reason for increase in C8-HSL-mediated quorum sensing in *B. cepacia*, thereby regulating its own pathogenicity. This indicates the communication between two different bacterial species helping each other's survival (Govan et al. 1996; Lewenza et al. 1999; Taylor et al. 1993).

2.2.3 QS in *Erwinia carotovora*

It is a phytopathogen causing soft rot of plant tissue by producing a wide range of plant tissue-degrading enzymes such as pectate lyases, cellulase, protease, and polygalacturonase. Low cell density causes release of low concentration of these enzymes which are incapable of causing plant tissue maceration. Moreover, these enzymes will also activate the phytodefense mechanism at these low concentrations. Thus, quorum sensing is a necessity in *E. carotovora* for exoenzyme production which can successfully destroy tissue and evade the plant defenses. It has been found that LuxRI homologs ExpR and ExpI release AHL which regulates the expression of the tissue-softening enzymes in a cell density-dependent manner, 3-oxo-C6-HSL, which in turn helps in exoenzyme regulation (Barras et al. 1994; Jones et al. 1993; Pirhonen et al. 1993).

2.3 Mechanism of QS in Gram-Positive Bacteria

Unlike Gram-negative bacteria, AHL molecules are not produced by Gram-positive bacteria for quorum sensing mechanism. They are found to use peptide molecules which are post-translationally modified that interact with the two-component signal transduction system of histidine kinase. Development of bacterial competence and its regulation is done by QS circuit in Gram-positive bacteria such as *Streptococcus pneumonia* and *Bacillus subtilis*. The conjugation in *Enterococcus faecalis* and *Staphylococcus aureus* virulence is also regulated by QS circuit.

2.3.1 QS in *S. aureus*

During the early infection of *S. aureus*, initial attachment of surface proteins such as collagen- and fibronectin-binding proteins occurs along with protein A which is responsible for defense mechanism for its survival. As the infection progresses, high cell density at the infection site causes a decrease in the concentration of surface proteins and increase in the secretion of other proteins. The regulation of the switch in the protein expression system is done by two pleiotropic regulatory loci called *sar* (staphylococcal accessory gene regulator) (Cheung et al. 1992) and *agr* (accessory gene regulator) (Morfeldt et al. 1988; Recsei et al. 1986).

The *agr* locus comprises two operons, viz., RNAIII and RNAII (Fig. 2.1). The operon of RNAII possesses the genes *agrBDCA* encoding a response regulator (AgrA), signal transducer (AgrC), and QS signal generator (AgrB and AgrD). The RNAIII operon codes for δ-hemolysin which is responsible for cell membrane interaction. Quorum sensing mechanism in *S. aureus* involves autophosphorylation of AgrC as the first step followed by phosphorylation of AgrA, and this in turn leads to the induction of transcription of RNAIII, thereby upregulating the expression of numerous proteins and *agrBDCA*. The second regulatory locus, SarA, helps in

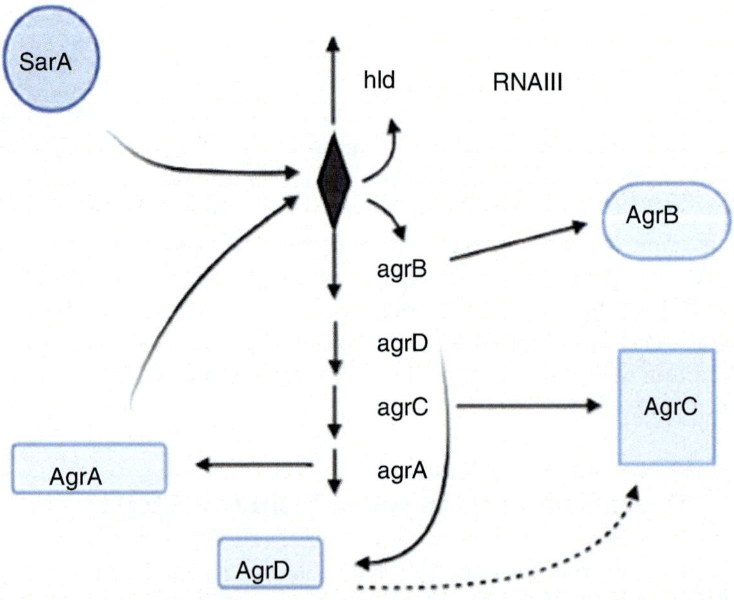

Fig. 2.1 Quorum sensing in *S. aureus* (adapted from Kievit De Teresa and Iglewski 2000)

expression of RNAIII and RNAII operons of *agr* locus (Ji et al. 1995, 1997; Lina et al. 1998).

2.4 QS and Virulence

Controlling the virulence factor expressions by pathogenic bacteria leading to host infections is one of the many traits of quorum sensing (Rutherford Steven and Bonnie 2012). It plays a vital role in expression of genetic virulence factors and has biochemical and structural traits. For example, in the disease citrus canker, caused by the bacterium *Xanthomonas citri*, virulence factor is controlled by a cascade of genes called rpf (regulation of pathogenicity factors), encoding elements of quorum sensing signaling molecules such as AIs from the DSF (diffusible signal factor). After the successful invasion and multiplication within the host tissue, the bacteria continue to produce autoinducer molecules to the extracellular environment owing to QS mechanism until it reaches a threshold value representing bacterial population density and detected by their receptors and triggers a chain of biological functions.

The QS system in *Xanthomonas* sp. is encoded by the gene cluster of rpf. The sensor RpfC combines with DSF synthase RpfF to form a complex at low cell density, whereas DSF attaches to RpfC and undergoes a change in conformation which initiates the autophosphorylation and phosphorelay to RpfG and then releases RpfC at high cell density. Activation of RpfG decreases the concentration of secondary messenger cyclic di-GMP and releases Clp which helps in the synthesis of EPS and extracellular enzymes, leading to biofilm formation, its subsequent dispersal, and cell motility and expressing virulence factors (Barber et al. 1997; Chan and Goodwin 1999). Another example of virulence mechanism involves the opportunistic pathogens causing infections in humans and having a compromised immune system by acquiring virulence genes via horizontal gene transfer (HGT) mechanism. This includes microorganisms such as *Enterococcus faecalis*, *Staphylococcus epidermidis*, *Streptococci*, *Burkholderia cepacia*, and *Bacteroides fragilis*. Capsule formation, escaping ability from the host's immune system, presence of higher concentrations at the site of infection in the host, and the resistance to antibiotics are some of the virulence mechanism taken up by these classes of bacteria (Viswanathan et al. 2015).

2.4.1 Virulence in *Staphylococcus aureus*

Many Gram-positive bacteria such as *S. aureus* use small peptides (autoinducing peptide, AIP) as signaling molecules for expressing virulent genes during QS (Fig. 2.2). AIPs are encoded by the accessory gene regulator (*agrD*). AIP is chopped by a membrane-bound protein AgrB only to be bound to AgrC, a sensor kinase which is membrane bound. This results into autophosphorylation of agrC and

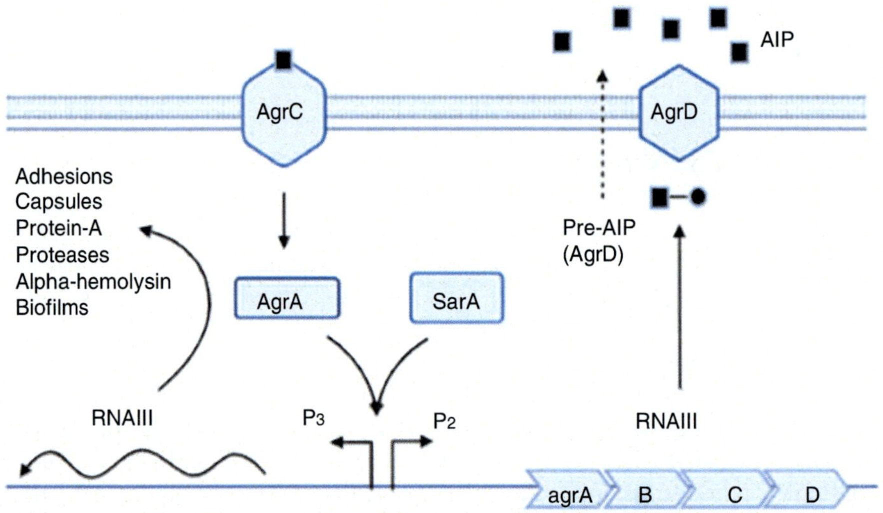

Fig. 2.2 *S. aureus* agr quorum sensing system (adapted from Antunes et al. 2010)

subsequent activation of AgrA. The phosphorylated AgrA leads to the induction of transcription at P2 and P3 promoters (Antunes et al. 2010).

Regulation and expression of virulence genes are intimately related to *agr* operon and promoters P2 and P3. Transcription from P3 produces RNAIII, which in turn goes on to produce toxins, proteases, and capsules. On the other hand, transcription from P2 produces RNAII which regulates the *agr* operon by regulating the production of agrA, agrB, agrC, and agrD (Fig. 2.2) (Roux et al. 2009; Novick and Geisinger 2008; Antunes et al. 2010).

In *S. aureus* the biofilm formation is believed to take place by two processes. The first process utilizes polysaccharide intercellular adhesin (PIA) and the extracellular polysaccharide, while the second process is a PIA-independent step dominated by adhesive proteins, *sarA* and *agr* regulators. It has been observed that *S. aureus* has the ability of adhesion (Vuong et al. 2000) during non-functional *agr*, thus providing the first stage of biofilm formation which is surface adherence. It has also been noticed that *agr* plays a dominant role in the detachment of cell clusters from the biofilm during the maturation stage (Kong et al. 2006).

2.4.2 Virulence in *Pseudomonas aeruginosa*

In case of Gram-negative bacteria such as *P. aeruginosa*, QS system involves productions of signaling molecules like acetyl homoserine lactones (AHLs), also known as autoinducers (AIs). One of the QS systems comprises *lasI* gene that codes for the AI *N*-3-oxododecanoyl-homoserine lactone (3OC12-HSL). When the *P. aeruginosa* cell density rises above the threshold value, released 3OC12-HSL binds to lasR which is the transcriptional activator; this dimerizes to bind the promoters for the respective gene expressions. Another QS system comprises of *rhl* that codes for the AI *N*-butyryl-homoserine lactone (C4-HSL) which binds to transcriptional regulator, RhlR controlling the activity of various promoters *las* and *rhl* systems and regulating the expression and production of multiple virulence factors, such as pyocyanin, exotoxin A, alkaline protease, rhamnolipids, elastase, lectins and superoxidase dismutase (Schuster et al. 2003; Smith and Iglewski 2003). It has been observed that QS helps in biofilm maturation and controls the functioning of efflux pumps of antibiotics, thus playing a major role in pathogenesis of *P. aeruginosa.*

Not only the autoinducer molecules are evident to be well associated with virulence, they are observed to interfere with the host immune responses as well. For example, 3OC12-HSL induces interleukin (IL)-8 secretion (DiMango et al. 1995) and COX-2 production (Smith et al. 2002) from bronchial epithelial cells of humans, and this leads to apoptosis in neutrophils and macrophages (Kievit De Teresa and Iglewski 2000; Smith et al. 2002; Skindersoe et al. 2009; Smith and Iglewski 2003; Tateda et al. 2003; Telford et al. 1998). It prevents lymphocyte proliferation, differentiation and prevents the synthesis of IL-12 and tumor necrosis factor alpha (TNF-α) and leads to production of gamma-interferon by activating T cells (Skindersoe et al. 2009; Smith et al. 2002). Another autoinducer molecule in *P. aeruginosa* is identified as 2-heptyl-3-hydroxyl-4-quinolone (*Pseudomonas* quinolone signal; PQS) produced by *pqsABCD* genes in the *pqsABCDE* operon. PQS is an important molecule in cell-to-cell communication and signal transduction (Fig. 2.3).

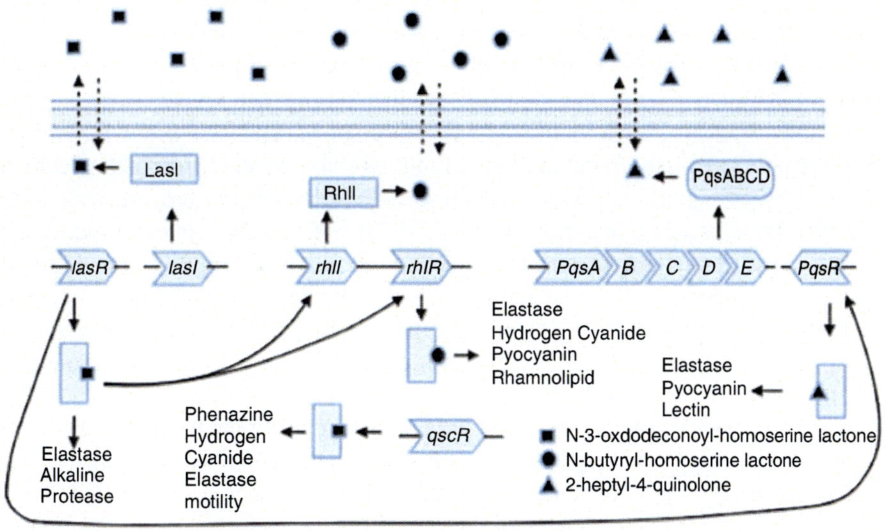

Fig. 2.3 Control of gene expression in *P. aeruginosa* by quorum sensing (adapted from Antunes et al. 2010)

2.4.3 Virulence in *E. coli*

E. coli produces quorum sensing signaling molecule AI-2 by the *luxS* gene which in turn activates the virulence factors' genetic circuit (Surette and Bassler 1998). LuxS in *E. coli* is responsible for controlling the initial attachment during biofilm formation (Surette et al. 1999). It is also known to be a regulator of 400 genes in enterohemorrhagic *E. coli* (EHEC) that determines the production of Shiga toxin, surface adhesion, and flagellar motility (Sperandio et al. 2001; DeLisa et al. 2001). Another signaling molecule AI-3 has also been detected in *E. coli* that is responsible for the regulation of gene QseBC determining motility gene expression. Expression of QseC is activated by AI-3 QseC and acts as the receptor for epinephrine and norepinephrine which are host catecholamine hormones (Clarke et al. 2006).

EHEC is responsible for causing acute and chronic pediatric diarrhea in humans by attaching and effacing (A/E) lesion formation in the epithelium of the intestine. A two-component regulatory system controlling AE lesions is composed of QseE which is the histidine kinase and QseF which is the response regulator. These help in the control at transcription level of the EspFu effector, which is translocated by EHEC into host cells, causing its pathogenicity. Moreover, regulators QseD and QseA are needed for controlling locus of enterocyte effacement (LEE) expression. This implies that expression of virulence gene with consequent quorum sensing in *E. coli* are linked by a complex regulatory system (Sperandio et al. 2002; Walters and Sperandio 2006; Reading et al. 2009) (Fig. 2.4).

2.5 Pharmacological Perspective of Quorum Sensing

Autoinducer molecules are well known for their role in QS circuit as signaling molecules. However, AIs also perform non-signaling functions in some cases. AI-2 can function as a metabolic by-product during detoxification process, for example, S-adenosylmethionine (SAM); precursor of AI-2 acts as a methyl donor in metabolic processes found in bacterial cells (Rezzonico and Duffy 2008; Schauder et al. 2001); *P. aeruginosa* PQS signal possesses iron-chelating properties and helps in iron homeostasis (Bredenbruch et al. 2006); and small peptides produced by Gram-positive bacteria (lantibiotics) possess antimicrobial properties in addition to signaling (Asaduzzaman and Sonomoto 2009), for example, production of nisin by *Lactococcus lactis* (Kuipers et al. 1995), streptin production by *Streptococcus pyogenes* (Wescombe and Tagg), subtilin production by *Bacillus*, and production of mersacidin (Kleerebezem 2004).

AI-3-mediated virulence in *E. coli* has led to the development of small molecules inhibiting this signaling system. A high-throughput screening from a library of compounds has identified a compound named *N*-phenyl-4-{[(phenylamino)-thioxomethyl]amino}-benzene-sulfonamide (LED209) which was found to be a bacterial virulence inhibitor, both in vivo and in vitro. LED209 stops QseC autophosphorylation, AE lesion formation by EHEC, and virulence factor

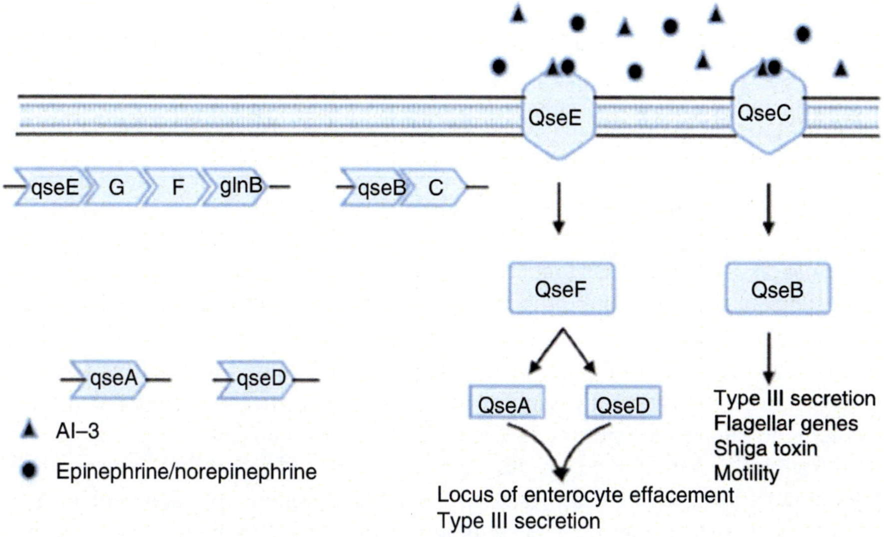

Fig. 2.4 Schematic representation of virulence gene expression and its control through quorum sensing and adrenergic signaling in *E. coli* (adapted from Antunes et al. 2010)

production. It can also prevent host colonization and virulence factor production by *F. tularensis* and *Salmonella typhimurium* (Rasko et al. 2008).

2.5.1 Targeting QS for Antimicrobial Therapy

Many organisms control expression of virulence factors by quorum sensing. Thus an attractive strategy of antimicrobial therapy involves targeting the quorum sensing signaling circuit. By blocking cell-to-cell signaling mechanism, pathogenic organisms could easily be turned avirulent. There are different possible ways of interfering with the QS mechanism which involves developing molecular analogs that bind to R proteins of *V. fischeri*, *Chromobacterium violaceum*, *A. tumefaciens*, and *Aeromonas salmonicida* (Milton et al. 1997; Schaefer et al. 1996; Swift et al. 1997; Zhu et al. 1998) as antagonists, thus preventing its further binding to autoinducers leading to shutting down the QS cascade. For example, 3-oxo-C12-HSL prevents RhlR activation in *P. aeruginosa*, and delay in the onset of bioluminescence in *V. fischeri* is observed presumably by competing C8-HSL with 3-oxo-C6-HSL for LuxR binding. The seaweed *Delisea pulchra* generates furanone compounds mimicking the structures of AHLs, thus interfering with the quorum sensing systems of *V. fischeri*, *Vibrio harveyi*, and *Serratia liquefaciens* (Givskov et al. 1996). Thus, preventing AHL synthesis, by the use of AHL precursor analogs, can be a plausible means of blocking the quorum sensing cascade. There are also reports of few enzymes that are involved in reducing AI production such as expression of the AiiA enzyme in *E. carotovora* reduces pectolytic enzymes, thus diminishing the soft rot disease of plants (Dong et al. 2000). Microorganisms having more than one QS system need to be targeted for all the systems that are responsible for virulence generation such as targeting both the *las* and the *rhl* quorum sensing systems in *P. aeruginosa* leads to decrease in virulence factors.

2.5.2 Strategies to Control Biofilm Formation

Formation of biofilm leads to increase in bacterial pathogenesis as well as antibiotic resistance. Long-term antibiotics are prescribed to prevent further growth of biofilm in patients where removal is impossible. It has been observed that premature biofilm is treated well with antibiotics as compared to the matured ones. However, inefficiency to detect premature biofilm makes it very difficult to start the diagnosis leading to clinical complications arising from mature biofilms.

Antibiotics are selected on the basis of sensitivity and ability to penetrate the biofilm matrix owing to the fact that bacterial biofilms are likely to be highly antibiotic resistant than their planktonic counterparts. It is preferably good to opt for the combinatorial therapy rather than the monotherapy as far as the selection of antibiotics is concerned. This is due to the difference in the mode of action, proper dispensation with regard to dosages, and duration of these antibiotics.

Some antibiotics are effective against growing bacterial cells, and others are against the dormant cells. Some antibiotics are coated with hydrophilic coatings such as PEG that build antifouling surfaces minimizing the microbial adhesion required for biofilm formation, whereas some are coated with nanoparticles to prevent formation of biofilm (Donelli et al. 2002, 2006). In addition, photodynamic therapy (PDT) is also widely used to dispense the bacterial cells in biofilm by carefully selecting and staining the bacterial cells by a photosensitizer dye (Percival et al. 2014a, b). Lastly, there exists a long list of molecules that interact with signaling pathways of the bacterial cells, in both types of bacterial cells. Polyphenol molecules, enzymes, or peptides may act as anti-biofilm molecules (Table 2.2).

2.5.2.1 Targeting the AHL-Mediated QS

AHLs or *N*-acyl homoserine lactones are the group of small signaling molecules used by Gram-negative bacterial cells to regulate the cell population density and swarming motility inherent during biofilm formation. Binding of these signaling molecules to Lux-R-type transcriptional regulator proteins helps in the target gene expression (Gambello and Iglewski 1991; Passador et al. 1993). Thus one plausible strategy to downregulate the biofilm-forming gene regulatory pathway is to search for the compounds that can compete with the AHL molecules during binding with the receptor proteins. For example, furanones produced by Australian microalga *Delisea pulchra* competes actively with AHL molecules. Furanones targets the *rhl* system of *P. aeruginosa* and helps in dispersal of bacterial biomass from the already formed biofilm. Furanone was found to displace AHL from LuxR protein which regulates gene expressions associated to quorum sensing and biofilm maturity (Givskov et al. 1996; Manefield et al. 1999). Some polyphenols such as tannic

Table 2.2 Summary of various anti-biofilm molecules with their target bacteria (Roy et al. 2018)

Source	Anti-biofilm molecules	Target bacteria	References
Camellia sinensis	Ellagic acid	*Streptococcus dysgalactiae*	Durig et al. (2010)
Chitin	Chitosan	*K. pneumonia*	Magesh et al. (2013)
Camellia sinensis (green tea)	Epigallocatechin gallate (EGCG)	*Acinetobacter baumannii*, *P. aeruginosa*, *S. aureus*, *Stenotrophomonas maltophilia*, *E. coli*	Vidigal et al. (2014)
Santolina oblongifolia, *Alchemilla speciosa*, *Tagetes lucida*	Esculetin	*S. aureus*	Durig et al. (2010)
Usnea longissima	Quercetin	*P. aeruginosa*, *K. pneumonia*	Gopu et al. (2015)
Curcuma longa	Curcumin	*S. mutans*, *K. pneumonia*	Hentzer et al. (2002)
Natural furanone derivative	Synthetic halogenated furanone (F-56)	*Helicobacter pylori*, *P. aeruginosa*, *Serratia liquefaciens*	Hentzer et al. (2002)
	Peptide (1018)	*P. aeruginosa*	Fuente-Nunez et al. (2014)
Berberis vulgaris, *B. aquifolium*, *B. aristata*	Berberine	*K. pneumonia*	Magesh et al. (2013)
P. aeruginosa	Pel polysaccharide, Psl polysaccharide	*S. aureus*	Rendueles et al. (2013); Qin et al. (2009); Liang (2015)
Paenibacillus polymyxa	Colistin (Polymyxin E)	*S. maltophilia*	Vidigal et al. (2014)
Lactococcus lactis	Nisin	*Staphylococcus epidermis*, *S. aureus*	Parisot et al. (2008); Saising et al. (2012)
Staphylococcus epidermidis Tu3298	Epidermin	*Lactococcus lactis*	Parisot et al. (2008); Saising et al. (2012)
Staphylococcus gallinarum Tu3928	Gallidermin	*S. aureus*	Parisot et al. (2008); Saising et al. (2012)
B. subtilis strain ATCC6633	Subtilin	*Lactococcus lactis*	Parisot et al. (2008); Saising et al. (2012)

(continued)

Table 2.2 (continued)

Source	Anti-biofilm molecules	Target bacteria	References
Gaegurin 5 is a type of synthetic analog	Lytic peptide (PTP-7)	*P. aeruginosa*, *S. aureus*	Park et al. (2011); Shadia and Aeron (2014)
Bacterial cells isolated from the specimen of pig's intestine	PR-39	*P. aeruginosa*, *E. coli*, *S. aureus*	Park et al. (2012)
Cationic host defense peptide in human	LL-37	*S. aureus*, *E. coli*, *P. aeruginosa*	Cho et al. (2009)
Isolate from the cytoplasm of the granules of bovine neutrophils	Indolicidin	*P. aeruginosa*, *S. aureus*, *E. coli*	Subbalakshmi and Sitaram (1998)
Peptides which are post-translationally modified produced by *E. coli* possessing plasmid-borne MccB17 operon	Microcin B17	*E. coli*	Vizan et al. (1991)
	Containing enzymes like glycoside hydrolase (dispersin B), deoxyribonuclease I	*Enterococcus* and *Staphylococcus*	Donelli et al. (2007); Shadia and Aeron (2014)
	Silver	*P. aeruginosa*, *S. proteamaculans*	Percival et al. (2014a, b)
	Octenidine hydrochloride	*P. aeruginosa*, *S. aureus*	Percival et al. (2014a, b)
	Tetrasodium EDTA, disodium-EDTA, and sodium citrate	*P. aeruginosa*, *S. aureus*	Shadia and Aeron (2014)
Rhus semialata, *Quercus infectoria*, *Rhus coriaria*, *Caesalpinia spinosa*	Tannic acid	*S. aureus*	Payne et al. (2013)
Secondary metabolite from lichen	Usnic acid	*C. albicans*, *S. aureus*	Francolini et al. (2004); Nithyanand et al. (2015)
	Cadexomer iodine	*P. aeruginosa*, *S. aureus*	Percival et al. (2014a, b)
	Polyhexamethylene biguanide	*P. aeruginosa*	Percival et al. (2014a, b)
	Chlorhexidine	*P. aeruginosa*, *S. aureus*	Percival et al. (2014a, b)

acid, ellagic acid, and EGCG are also found to follow the same mechanism as furanones. Flavonoids, e.g., quercetin, are also known to interfere with QS mechanism of *S. aureus* with reduced alginate production and a decrease in adherence to substratum. It also reduces the EPS production leading to swarming motility (Lee et al. 2013; Roy et al. 2018).

2.5.2.2 Preventing the Stringent Response in Bacteria

During nutritional stress conditions, bacteria produce signaling molecules known as alarmones, guanosine pentaphosphate and guanosine tetraphosphate, together called as (p)ppGpp. Change in (p)ppGpp pool affects the biofilm development in bacteria during starvation. There are various molecules that affect the functioning of (p)ppGpp by inhibiting their accumulation within the protoplasm. Amphipathic cationic peptide 1018 surpasses the bacterial cell membrane and directly binds to (p)ppGpp, thus disrupting the biofilm in three ways. First, when added before biofilm initiation step, it prevents formation of biofilm. Second, it eradicates the bacterial cells present within biofilm without having any effect on the planktonic cells. Third, it can collapse the established biofilm which can be even as old as 2 days (Fuente-Nunez et al. 2014). Peptide 1037 was found to reduce biofilms formed by various Gram- positive and Gram-negative bacteria (de la Fuente-Nunez et al. 2012). Peptide 1038 is known to induce twitching motility and prevent initial attachment and quorum sensing of *Pseudomonas* during biofilm creation, thus destroying the biofilm. Derivatives of the peptide 1018 such as HE4 and HE10 are known to be active against *B. cenocepacia* and *P. aeruginosa.* Moreover, in some cases, synergistic actions of these peptides along with antibiotics have led to interesting results. Some secondary metabolite polyphenols like eugenol are found to prevent the stringent response in bacteria such as *S. mutans* by downregulation of gene, *relA*, involved in the control of stringent response (Lemos et al. 2004).

2.5.2.3 Enzymatic Dispersion of EPS

EPS serves as a protective matrix providing nutrition and shelter to the bacterial cells within the biofilm. Thus, molecules that disperse the EPS layer will tend to expose the microorganisms to the antimicrobial agents. DNases and polysaccharide lyases enzymes are capable of disintegrating the EPS (Stewart 2015). DNase I possesses the ability of denaturing the extracellular DNA (eDNA) present within the biofilm structure (Kaplan 2009; Izano et al. 2008).

2.5.2.4 Disrupting the Peptidoglycan Layer

Peptidoglycan layer in the cell wall acts as a firewall preventing and helping the bacteria against antimicrobial agents. Thus cleaving this layer will inhibit the biofilm generation. Polyphenolic compounds like tannic acid and epigallocatechin gallate reduces biofilm formation by directly or indirectly affecting the peptidoglycan layer. While the former is known to increase the extracellular level of IsaA (*staphylococcal* antigen A) which brings about the catalysis of the β-1,4-glycosidic bond between *N*-acetyl glucosamine (GlcNAc) and *N*-acetyl muramic acid (MurNAc) resulting in the development of pore upon the peptidoglycan layer (Holtje et al. 1975), the latter is known to damage the bacterial cell wall through peptidoglycan binding (Yoda et al. 2004; Zhao et al. 2002), thus restricting the initial attachment phase of biofilm construction (Carpentier and Cerf 1993). Endolysins are certain peptidoglycan hydrolases which are encoded by bacteriophages (Shen et al. 2013) and are able to degrade the bacterial cell wall. After that, the progeny bacteriophages are released. They can cleave the cell wall after binding to it. A certain *streptococcal* bacteriophage endolysin, *PhyC* (Hoopes et al. 2009; Nelson et al. 2006), is able to cause damage, and it leads to disruption of in vitro biofilms.

2.5.2.5 Molecules Causing Biofilm Dispersal

Biofilm disassembly involves disruption of the EPS matrix by production of extracellular enzymes causing degradation and dissolving of the adhesive components being present with the matrix found within the biofilm. This leads to detachment of bacterial cells from the colony and its release into the environment. There are a variety of matrix-degrading genes involved in the production of surfactants, deoxyribonucleases (DNases), and proteases. The gene regulatory system comprises *agr* system that regulates the production of enzymes, which are capable of degrading the matrix of biofilm. Activation of the *agr* system causes release of several proteases (e.g., sarA, sigB, Esp) (Beenken et al. 2010; Lauderdale et al. 2009; Tsang et al. 2008) and small phenol-soluble modulins (PSMs) which are pore-forming toxins (Boles and Horswill 2008; Beenken et al. 2010; Vuong et al. 2000), thus preventing biofilm maturation and performing an inhibitory role. Compounds like AA-861 and parthenolide exhibited biofilm degradation in *Bacillus cereus*, *B. subtilis*, and *E. coli* that interferes with the polymerization of protein TasA responsible for formation of fibers resembling amyloid during biofilm growth. Other peptidomimetic compounds like FN075 and BibC6 of ring-fused 2-pyridones target the protein-protein interactions in biofilm formation, hindering the synthesis of curli in *E. coli*. In addition, various D-amino acids are also found to be responsible for biofilm dispersal. For example, D-tyrosine causes decrease in cell attachment in *B. subtilis* and *P. aeruginosa*; D-cysteine, D-histidine, and D-tryptophan prevent 35–86% biofilm formation in *A. baumannii*.

2.5.2.6 Disassembly of Lipopolysaccharides/Membrane Permeabilization

One effective way to stop biofilm formation is the use of antimicrobial peptides (AMP) as an alternative to conventional antibiotics. AMPs are low weight proteins that are evolutionary conserved possessing antimicrobial activity and can act effectively against bacteria, fungi and viruses. They possess hydrophobic and hydrophilic sides that help in inserting into the lipid bilayer or lipopolysaccharides, thereby solubilizing in aquatic environment (Izadpanah and Gallo 2005). This mechanism results in destabilization of lipid head groups by the formation of multiple pores, causing the disruption of cellular membrane integrity. PTP-7 is an example of lytic peptide that can enter deep in the biofilm and kill bacteria within the biofilm (Kharidia and Liang 2011). Polymyxin E or B and colistin can bind to LPS in Gram-negative bacteria, making the outer membrane permeabilized. Gramicidin S can distort the membrane integrity of the Gram-positive and Gram-negative bacteria. Alteration of membrane potential by pore formation also helps in biofilm disruption. The pore formation can take place by any of the following mechanisms: a toroidal pore (Mihajlovic and Lazaridis 2010; Gottler and Ramamoorthy 2009), barrel-stave (Oren and Shai 1998), or through carpet-like mechanism (Shai and Oren 2001), causing efflux of intracellular materials. Lantibiotics are another class of ribosomally synthesized peptide antibiotics that are modified post-translationally in Gram-negative bacteria and serve as anti-biofilm agents. Their mode of action involves damaging the bacterial membrane and preventing the production of enzymes. Common lantibiotics such as nisin and subtilisin induce leakage to the cytoplasmic membrane by forming pores that causes the cytoplasmic solutes to leak out of *B. subtilis* and *Staphylococcus simulans* (Bierbaum and Sahl 2009). In another study, biosurfactants such as sophorolipids show its efficacy against biofilm by enhancing the membrane permeability. The sophorolipids of *B. subtilis* help in disrupting the cytoplasmic membrane causing the leakage of various intracellular enzymes like malate dehydrogenase which in turn results in the efflux of their cytoplasmic contents (Rienzo et al. 2015).

2.5.2.7 Prevention of Cell Division and Adhesion Molecule Synthesis

Bacterial cell division within the biofilm is needed for its survival and spread in new regions. There are a wide range of molecules starting from metal ions, antibiotics, chelating agents, natural polymers, and antimicrobial peptides that are known to disturb the membrane potential of the plasma membrane, thereby preventing cell division. For example, accumulation of silver within the intracellular vacuoles leads to pore formation in plasma membrane (Tiwari et al. 2015; Percival et al. 2014a, b). Antimicrobial peptides (AMPs), e.g., apidaecin (Kragol et al. 2001), pyrrhocoricin (Kragol et al. 2002), and drosocin (Laszlo et al. 2000), can penetrate the membrane either by channel formation or by flip-flop mechanism and effect the functions of cytosolic proteins needed for cellular functions such as replication and translation. Chelating agents like EDTA can destabilize biofilms by sequestering metal ions such as iron, magnesium, zinc, and calcium (Finnegan and Percival 2015). Chitosan being a cationic natural polymer is known to disrupt negatively charged cell membranes of the microbes (Zhang et al. 2013). There are certain AMPs that directly interact with the nucleic acids without affecting the cell membranes' porosity. For example, peptide PR-39 can surpass the outermost membrane and affects the replication and translation during biofilm formation (Boman et al. 1993). The peptide, indolicidin, makes the cell membrane permeable without breaking the bacterial cells (Subbalakshmi and Sitaram 1998). The host defense peptide, LL37, in humans reduces bacterial adhesion, downregulates genes of QS, and promotes twitching motility mediated by type IV pili (Boles and Horswill 2008) AMPs like citropin (Cirioni et al. 2006) and melamine prevent bacterial adhesion on medical equipment, contact lenses, and catheters (Willcox et al. 2008). Small peptides known as bacteriocins, e.g., bovicin HC5 and nisin, alter the hydrophobicity of the outer surfaces of *S. aureus*, thus reducing the extent of adhesion to surfaces (Pimentel-Filho et al. 2014). A peptide named lactoferrin stops the initial attachment of *S. mutans*, *Porphyromonas gingivalis*, *Prevotella intermedia*, and *Streptococcus gordonii*, preventing the biofilm formation in oral cavity (Arslan et al. 2009). Peptides such as cadexomer iodine bind with the bacterial cell membrane by interacting with the membrane proteins and interfere with the DNA and protein synthesis causing lipid membrane disruption (Percival et al. 2014a, b). Antimicrobial peptides such as nisin and bovicin HC5 act by altering the hydrophobicity of food surfaces inhibiting the initial attachment of bacteria such as *S. aureus*, thereby preventing the contamination of food (Pimentel-Filho et al. 2014). This enables storage for prolonged time and preservation of packed food matters.

2.6 Conclusion

There has been lot of research work and studies in the field of biofilm formation by Gram-positive as well as Gram-negative bacteria. The greatest challenge in the medical world is to combat the antibiotic resistance of the bacterial biofilms. Therefore various effective and newer techniques have to be developed with main focus on different anti-biofilm molecules and modifying different signaling pathways related with quorum sensing. Polymicrobial model systems have been studied, and it is found out that quorum sensing is important for cooperation and also for competition among various bacterial species. These models also help us to understand social behavior and evolution of quorum sensing. Cyclic-di-GMP signaling pathway is absent in higher eukaryotes, and therefore this knowledge can be used to design anti-biofilm molecules. Another way to reduce the process of biofilm formation is by targeting the amyloids; this in turn leads to weakening of the adherence of the bacterial cells to the surface (Wu et al. 2015). The presence of virulence factor in pathogenic bacteria helps in spreading of infections within the host. An idea about the genetic and virulence factors may help us design drugs that can fight against the infections and also help in inhibiting the infection through QS mechanism. There may be different modes of action of every anti-biofilm molecule, but more than one mechanism can be followed by a single molecule, for example, the anti-biofilm molecule ECGC can perform their action either by disrupting the membrane and degrading the peptidoglycan layer or by hindering the AHL-mediated quorum sensing pathway. The antimicrobials which are derived from natural sources have more diversity regarding their structure and biochemical properties when compared with synthetic drugs. Therefore as a result, it has better binding capability to the target molecules and can be used for various in silico approaches in the field of pharmacy and also for creating alternative therapies. There are some disadvantages of using naturally derived anti-biofilm agents, that is, they are time-consuming, are expensive, can show various results when extricated from their sources, and are less sustainable. On the other hand, man-made drugs may be of nominal price and are relatively faster in action but can have adverse side effects.

References

Antunes MCL, Ferreira BRR, Buckner CMM, Finlay BB (2010) Quorum sensing in bacterial virulence—a review. Michael Smith Laboratories, The University of British Columbia, Vancouver, Canada Department of Microbiology and Immunology The University of British Columbia, Vancouver, Canada. Microbiology 156(pt 8):2271–2282

Arslan SY, Leung KP, Wu CD (2009) The effect of lactoferrin on oral bacterial attachment. Oral Microbiol Immunol 24:411–416

Asaduzzaman SM, Sonomoto K (2009) Lantibiotics: diverse activities and unique modes of action. J Biosci Bioeng 107:475–487

Atkinson S, Throup JP, Stewart GS, Williams P (1999) A hierarchical quorum-sensing system in Yersinia pseudotuberculosis is involved in the regulation of motility and clumping. Mol Microbiol 33:1267–1277

Bainton NJ, Stead P, Chhabra SR, Bycroft BW, Salmond GPC, Steward GSAB, Williams P (1992) N-(3-Oxohexanoyl)-L-homo-serine lactone regulates carbapenem antibiotic production in Erwinia carotovora. Biochem J 288:997–1004

Barber CE, Tang JL, Feng JX et al (1997) A novel regulatory system required for pathogenicity of Xanthomonas campestris is mediated by a small diffusible signal molecule. Mol Microbiol 24:555–566

Barras F, Gijsegem van F, Chatterjee AK (1994) Extracellular enzymes and pathogenesis of soft-rot Erwinia. Annu Rev Phytopathol 32:201–234

Beck von Bodman S, Farrand SK (1995) Capsular polysaccharide biosynthesis and pathogenicity in Erwinia stewartii require induction by a N-acyl-homoserine lactone autoinducer. J Bacteriol 177:5000–5008

Beenken KE, Mrak LN, Griffin LM, Zielinska AK, Shaw LN, Rice KC, Horswill AR, Bayles KW, Smeltzer MS (2010) Epistatic relationships between sarA and agr in staphylococcus aureus biofilm formation. PLoS One 5:e10790

Bierbaum G, Sahl HG (2009) Lantibiotics: mode of action, biosynthesis and bioengineering. Curr Pharm Biotechnol 10:2–18

Boles BR, Horswill AR (2008) Agr-mediated dispersal of Staphylococcus aureus biofilms. PLoS Pathog 4:e1000052

Boman HG, Agerberth B, Boman A (1993) Mechanisms of action on Escherichia coli of cecropin P1 and PR-39, two antibacterial peptides from pig intestine. Infect Immun 61:2978–2984

Bredenbruch F, Geffers R, Nimtz M, Buer J, Haussler S (2006) The Pseudomonas aeruginosa quinolone signal (PQS) has an iron-chelating activity. Environ Microbiol 8:1318–1329

Carpentier B, Cerf O (1993) Biofilms and their consequences, with particular reference to hygiene in the food industry. J Appl Bacteriol 75:499–511

Chan JWYF, Goodwin PH (1999) The molecular genetics of virulence of Xanthomonas campestris. Biotechnol Adv 17:489–508

Cheung AL, Koomey JM, Butler CA, Projan SJ, Fischetti VA (1992) Regulation of exoprotein expression in Staphylococcus aureus by a locus (sar) distinct from agr. Proc Natl Acad Sci U S A 89:6462–6466

Chhabra SR, Stead P, Bainton NJ, Salmond GPC, Stewart GSAB, Williams P, Bycroft BW (1993) Autoregulation of carbapenem biosynthesis in Erwinia carotovora ATCC 39048 by analogues of N-3-(oxohexanoyl)-L-homoserine lactone. J Antibiot 46:441–454

Cho JH, Sung BH, Kim SC (2009) Buforins: histone H2A derived antimicrobial peptides from toad stomach. Biochim Et Biophysica Acta 1788:1564–1569

Cirioni O, Giacometti A, Ghiselli R, Kamysz W, Orlando F, Mocchegiani F, Silvestri C, Licci A, Chiodi L, Lukasiak J et al (2006) Citropin 1.1-treated central venous catheters improve the efficacy of hydrophobic antibiotics in the treatment of experimental staphylococcal catheter-related infection. Peptides 27:1210–1216

Clarke MB, Hughes DT, Zhu C, Boedeker EC, Sperandio V (2006) The QseC sensor kinase: a bacterial adrenergic receptor. Proc Natl Acad Sci U S A 103:10420–10425

Cubo MT, Economou A, Murphy G, Johnston AWB, Downie JA (1992) Molecular characterization and regulation of the rhizosphere-expressed genes rhiABCR that can influence nodulation by Rhizobium leguminosarum biovar viciae. J Bacteriol 174:4026–4035

de la Fuente-Nunez C, Korolik V, Bains M, Nguyen U, Breidenstein EB, Horsman S, Lewenza S, Burrows L, Hancock RE (2012) Inhibition of bacterial biofilm formation and swarming motility by a small synthetic cationic peptide. Antimicrob Agents Chemother 56:2696–2704

DeLisa MP, Wu CF, Wang L, Valdes JJ, Bentley WE (2001) DNA microarray-based identification of genes controlled by autoinducer 2-stimulated quorum sensing in Escherichia coli. J Bacteriol 183:5239–5247

DiMango E, Zar HJ, Bryan R, Prince A (1995) Diverse Pseudomonas aeruginosa gene products stimulate respiratory epithelial cells to produce interleukin-8. J Clin Invest 96:2204–2210

Donelli G, Francolini I, Piozzi A, Di Rosa R, Marconi W (2002) New polymer-antibiotic systems to inhibit bacterial biofilm formation: a suitable approach to prevent central venous catheter-associated infections. J Chemother 14:501–507

Donelli G, Francolini I, Ruggeri V, Guaglianone E, D'Ilario L, Piozzi A (2006) Pore formers promoted release of an antifungal drug from functionalized polyurethanes to inhibit Candida colonization. J Appl Microbiol 100:615–622

Donelli G, Francolini I, Romoli D, Guaglianone E, Piozzi A, Ragunath C, Kaplan JB (2007) Synergistic activity of dispersin B and cefamandole nafate in inhibition of staphylococcal biofilm growth on polyurethanes. Antimicrob Agents Chemother 51:2733–2740

Dong Y-H, Xu J-L, Li X-Z, Zhang L-H (2000) AiiA, an enzyme that inactivates the acylhomoserine lactone quorum-sensing signal and attenuates the virulence of Erwinia carotovora. Proc Natl Acad Sci U S A 97:3526–3531

Durig A, Kouskoumvekaki I, Vejborg RM, Klemm P (2010) Chemoinformatics-assisted development of new anti-biofilm compounds. Appl Microbiol Biotechnol 87:309–317

Eberl L, Winson MK, Sternberg C, Stewart GSAB, Christiansen G, Chhabra SR, Bycroft BW, Williams P, Molin S, Givskov M (1996) Involvement of N-acyl-L-homoserine lactone autoinducers in controlling the multicellular behavior of Serratia liquefaciens. Mol Microbiol 20:127–136

Finnegan S, Percival SL (2015) EDTA: an antimicrobial and antibiofilm agent for use in wound care. Adv Wound Care 4:415–421

Francolini I, Norris P, Piozzi A, Donelli G, Stoodley P (2004) Usnic acid, a natural antimicrobial agent able to inhibit bacterial biofilm formation on polymer surfaces. Antimicrob Agents Chemother 48:4360–4365

Fuente-Nunez C, Reffuveille F, Haney EF, Straus SK, Hancock REW (2014) Broad-spectrum anti-biofilm peptidethat targets a cellular stress response. PLoS Pathog 10:e1004152

Gambello MJ, Iglewski BH (1991) Cloning and characterization of the Pseudomonas aeruginosa lasR gene: a transcriptional activator of elastase expression. J Bacteriol 173:3000–3009

Gambello MJ, Kaye S, Iglewski BH (1993) LasR of Pseudomonas aeruginosa is a transcriptional activator of the alkaline protease gene (apr) and an enhancer of exotoxin A expression. Infect Immun 61:1180–1184

Givskov M, de Nys R, Manefield M, Gram L, Maximilien R, Eberl L, Molin S, Steinberg PD, Kjelleberg S (1996) Eukaryotic interference with homoserine lactone-mediated prokaryotic signalling. J Bacteriol 178:6618–6622

Givskov M, Eberl L, Molin S (1997) Control of exoenzyme production, motility and cell differentiation in Serratia liquefaciens. FEMS Microbiol Lett 148:115–122

Gopu V, Meena CK, Shetty PH (2015) Quercetin influences quorum sensing in food borne bacteria: in-vitro and in-silico evidence. PLoS One 10:e0134684

Gottler LM, Ramamoorthy A (2009) Structure, membrane orientation, mechanism, and function of Pexiganan—A highly potent antimicrobial peptide designed from magainin. Biochim Et Biophysica Acta 1788:1680–1686

Govan JRW, Hughes JE, Vandamme P (1996) Burkholderia cepacia: medical, taxonomic and ecological issues. J Med Microbiol 45:395–407

Gray KM, Pearson JP, Downie JA, Boboye BEA, Greenberg EP (1996) Cell-to-cell signaling in the symbiotic nitrogen-fixing bacterium Rhizobium leguminosarum: autoinduction of stationary phase and rhizosphere-expressed genes. J Bacteriol 178:372–376

Hentzer M, Riedel K, Rasmussen TB, Heydorn A, Anderson JB, Parsck MR, Rice SA, Eberl L, Molin S, Høiby N et al (2002) Inhibition of quorum sensing in Pseudomonas aeruginosa biofilm bacteria by a halogenated furanone compound. Microbiology 148:87–102

Holtje JV, Mirelman D, Sharon N, Schwarz U (1975) Novel type of murein transglycosylase in Escherichia coli. J Bacteriol 124:1067–1076
Hoopes JT, Stark CJ, Kim HA, Sussman DJ, Donovan DM, Nelson DC (2009) Use of a bacteriophage lysin, PlyC, as an enzyme disinfectant against Streptococcus equi. Appl Environ Microbiol 75:1388–1394
Hwang I, Pei-Li L, Zhang L, Piper KR, Cook DM, Tate ME, Farrand SK (1994) TraI, a LuxI homolog, is responsible for production of conjugation factor, the Ti plasmid N-acylhomoserine lactone autoinducer. Proc Natl Acad Sci U S A 91:4639–4643
Izadpanah A, Gallo RL (2005) Antimicrobial peptides. J Am Acad Dermatol 52:381–390
Izano EA, Amarante MA, Kher WB, Kaplan JB (2008) Differential roles of poly-N-acetylglucosamine surface polysaccharide and extracellular DNA in Staphylococcus aureus and Staphylococcus epidermidis biofilms. Appl Environ Microbiol 74:470–476
Ji G, Beavis RC, Novick RP (1995) Cell density control of staphylococcal virulence mediated by an octapeptide pheromone. Proc Natl Acad Sci U S A 92:12055–12059
Ji G, Beavis RC, Novick RP (1997) Bacterial interference caused by autoinducing peptide variants. Science 276:2027–2030
Jones S, Yu B, Bainton NJ, Birdsall M, Bycroft BW, Chhabra SR, Cox AJR, Golby P, Reeves PJ, Stephens S, Winson MK, Salmond GPC, Stewart GSAB, Williams P (1993) The lux autoinducer regulates the production of exoenzyme virulence determinants in Erwinia carotovora and Pseudomonas aeruginosa. EMBO J 12:2477–2482
Kaplan JB (2009) Therapeutic potential of biofilm-dispersing enzymes. Int J Artif Organs 32:533–695
Kharidia R, Liang JF (2011) The activity of a small lytic peptide PTP-7 on Staphylococcus aureus biofilms. J Microbiol 49:663–668
Kievit De Teresa R, Iglewski BH (2000) Bacterial quorum sensing in pathogenic relationships. University of Rochester School of Medicine and Dentistry, Rochester, NY, p 14642
Kleerebezem M (2004) Quorum sensing control of lantibiotic production; nisin and subtilin autoregulate their own biosynthesis. Peptides 25:1405–1414
Kong KF, Vuong C, Otto M (2006) Staphylococcus quorum sensing in biofilm formation and infection. Int J Med Microbiol 296:133–139
Kragol G, Lovas S, Varadi G, Condie BA, Hoffmann R, Otvos L Jr (2001) The antibacterial peptide pyrrhocoricin inhibits the ATPase actions of DnaK and prevents chaperone-assisted protein folding. Biochemistry 40:3016–3026
Kragol G, Hoffmann R, Chattergoon MA, Lovas S, Cudic M, Bulet P, Condie BA, Rosengren KJ, Montaner LJ, Otvos L Jr et al (2002) Identification of crucial residues for the antibacterial activity of the proline-rich peptide, pyrrhocoricin. Eur J Biochem/FEBS 269:4226–4237
Kuipers OP, Beerthuyzen MM, de Ruyter PG, Luesink EJ, de Vos WM (1995) Autoregulation of nisin biosynthesis in Lactococcus lactis by signal transduction. J Biol Chem 270:27299–27304
Laszlo OJ, Insug O, Rogers ME, Consolvo PJ, Condie BA, Lovas S, Bulet P, Blaszczyk-Thurin M (2000) Interaction between heat shock proteins and antimicrobial peptides. Biochemistry 39:14150–14159
Latifi A, Foglino M, Tanaka K, Williams P, Lazdunski A (1996) A hierarchical quorum-sensing cascade in Pseudomonas aeruginosa links the transcriptional activators LasR and RhlR to expression of the stationary-phase sigma factor RpoS. Mol Microbiol 21:1137–1146
Lauderdale KJ, Boles BR, Cheung AL, Horswill AR (2009) Interconnections between Sigma B, agr, and proteolytic activity in Staphylococcus aureus biofilm maturation. Infect Immun 77:1623–1635
Lee JH, Park JH, Cho HS, Joo SW, Cho MH, Lee J (2013) Anti-biofilm activities of quercetin and tannic acid against Staphylococcus aureus. Biofouling 29:491–499
Lemos JAC, Brown TA, Burne RA (2004) Effects of RelA on key virulence properties of planktonic and biofilm populations of Streptococcus mutans. Infect Immun 72:1431–1440
Lewenza S, Conway B, Greenberg EP, Sokol PA (1999) Quorum sensing in Burkholderia cepacia: identification of the LuxRI homologs CepRI. J Bacteriol 181:748–756

Liang ZX (2015) The expanding roles of c-di-GMP in the biosynthesis of exopolysaccharides and secondary metabolites. Nat Prod Rep 32:663–683

Lina G, Jarraud S, Ji G, Greenland T, Pedraza A, Etienne J, Novick RP, Vandenesch F (1998) Transmembrane topology and histidine protein kinase activity of AgrC, the agr signal receptor in Staphylococcus aureus. Mol Microbiol 28:655–662

Magesh H, Kumar A, Alam A, Sekar U, Sumantran VN, Vaidyanathan R (2013) Identification of natural compounds which inhibit biofilm formation in clinical isolates of Klebsiella pneumoniae. Indian J Exp Biol 51:764–772

Manefield M, Nys R, Kumar N, Read R, Givskov M, Steinberg P, Kjelleberg S (1999) Evidence that halogenated furanones from Delisea pulchra inhibit acylated homoserine lactone (AHL)-mediated gene expression by displacing the AHL signal from its receptor protein. Microbiology 145:283–291

McGowan S, Sebaihia M, Jones S, Yu S, Bainton N, Chan PF, Bycroft BW, Stewart GSAB, Salmond GPC, Williams P (1995) Carbapenem antibiotic production in Erwinia carotovora is regulated by CarR, a homologue of the LuxR transcriptional activator. Microbiology 141:541–550

McLean KH, Winson MK, Fish L, Taylor A, Chhabra SR, Camara M, Daykin M, Lamb JH, Swift S, Bycroft BW, Stewart GSAB, Williams P (1997) Quorum sensing and Chromobacterium violaceum: exploitation of violacein production and inhibition for the detection of N-acylhomoserine lactones. Microbiology 143:3703–3711

Mihajlovic M, Lazaridis T (2010) Antimicrobial peptides in toroidal and cylindrical pores. Biochim Et Biophysica Acta 1798:1485–1493

Milton DL, Hardman A, Camara M, Chhabra SR, Bycroft BW, Stewart GSAB, Williams P (1997) Quorum sensing in Vibrio anguillarum: characterization of the vanI/vanR locus and identification of the autoinducer N-(3-oxododecanoyl)-L-homoserine lactone. J Bacteriol 179:3004–3012

Morfeldt E, Janzon L, Arvidson S, Lofdahl S (1988) Cloning of a chromosomal locus (exp) which regulates the expression of several exoprotein genes in Staphylococcus aureus. Mol Gen Genet 211:435–440

Nelson D, Schuch R, Chahales P, Zhu S, Fischetti VA (2006) PlyC: a multimeric bacteriophage lysin. Proc Natl Acad Sci U S A 103:10765–10770

Nithyanand P, Beema Shafreen RM, Muthamil S, Karutha Pandian S (2015) Usnic acid inhibits biofilm formation and virulent morphological traits of Candida albicans. Microbiol Res 179:20–28

Novick RP, Geisinger E (2008) Quorum sensing in staphylococci. Annu Rev Genet 42:541–564

Oren Z, Shai Y (1998) Mode of action of linear amphipathic alpha-helical antimicrobial peptides. Biopolymers 47:451–463

Parisot J, Carey S, Breukink E, Chan WC, Narbad A, Bonev B (2008) Molecular mechanism of target recognition by subtilin, a class I lanthionine antibiotic. Antimicrob Agents Chemother 52:612–618

Park SC, Park Y, Hahm KS (2011) The role of antimicrobial peptides in preventing multidrug-resistant bacterial infections and biofilm formation. Int J Mol Sci 12:5971–5992

Park JH, Lee JH, Cho MH, Herzberg M, Lee J (2012) Acceleration of protease effect on Staphylococcus aureus biofilm dispersal. FEMS Microbiol Letters 335:31–38

Passador L, Cook JM, Gambello MJ, Rust L, Iglewski BH (1993) Expression of pseudomonas aeruginosa virulence genes requires cell-to-cell communication. Science 260:1127–1130

Payne DE, Martin NR, Parzych KR, Rickard AH, Underwood A, Boles BR (2013) Tannic acid inhibits Staphylococcus aureus surface colonization in an IsaA-dependent manner. Infect Immun 81:496–504

Percival SL, Finnegan S, Donelli G, Vuotto C, Rimmer S, Lipsky BA (2014a) Antiseptics for treating infected wounds: efficacy on biofilms and effect of pH. Crit Rev Microbiol 42 (2):293–309

Percival SL, Suleman L, Francolini I, Donelli G (2014b) The effectiveness of photodynamic therapy on planktonic cells and biofilms and its role in wound healing. Future Microbiol 9:1083–1094
Pesci EC, Pearson JP, Seed PC, Iglewski BH (1997) Regulation of las and rhl quorum sensing in Pseudomonas aeruginosa. J Bacteriol 179:3127–3132
Pierson LS, Keppenne VDIII, Wood DW (1994) Phenazine antibiotic biosynthesis in Pseudomonas aureofaciens 30-84 is regulated by PhzR in response to cell density. J Bacteriol 176:3966–3974
Pimentel-Filho NJ, Martins MCF, Nogueira GB, Mantovani HC, Vanetti MCD (2014) Bovicin HC5 and nisin reduce Staphylococcus aureus adhesion to polystyrene and change the hydrophobicity profile and Gibbs free energy of adhesion. Int J Food Microbiol 190:1–8
Piper KR, Bodman von Beck S, Farrand SK (1993) Conjugation factor of Agrobacterium tumefaciens regulates Ti plasmid transfer by auto-induction. Nature 362:448–450
Pirhonen M, Flego D, Heikinheimo R, Palva ET (1993) A small diffusible signal molecule is responsible for the global control of virulence and exoenzyme production in the plant pathogen erwinia carotovora. EMBO J 12:2467–2476
Puskas A, Greenberg EP, Kaplan S, Schaefer AL (1997) A quorum-sensing system in the free-living photosynthetic bacterium Rhodobacter sphaeroides. J Bacteriol 179:7530–7537
Qin Z, Yang L, Qu D, Molin S, Tolker-Nielsen T (2009) Pseudomonas aeruginosa extracellular products inhibit staphylococcal growth, and disrupt established biofilms produced by Staphylococcus epidermidis. Microbiology 155:2148–2156
Rasko DA, Moreira CG, de Li R, Reading NC, Ritchie JM, Waldor MK, Williams N, Taussig R, Wei S et al (2008) Targeting QseC signaling and virulence for antibiotic development. Science 321:1078–1080
Reading NC, Rasko DA, Torres AG, Sperandio V (2009) The two-component system QseEF and the membrane protein QseG link adrenergic and stress sensing to bacterial pathogenesis. Proc Natl Acad Sci U S A 106:5889–5894
Recsei P, Kreiswirth B, O'Reilly M, Schlievert P, Gruss A, Novick RP (1986) Regulation of exoprotein gene expression in Staphylococcus aureus by agr. Mol Gen Genet 202:58–61
Rendueles O, Kaplan JB, Ghigo JM (2013) Antibiofilm polysaccharides. Environ Microbiol 15:334–346
Rezzonico F, Duffy B (2008) Lack of genomic evidence of AI-2 receptors suggests a non-quorum sensing role for luxS in most bacteria. BMC Microbiol 8:154
Rienzo MAD, Banat IM, Dolman B, Winterburn J, Martin PJ (2015) Sophorolipid biosurfactants: possible uses as antibacterial and antibiofilm agent. New Biotechnol 7:720–726
Rodelas B, Lithgow JK, Wisniewski-Dye F, Hardman A, Wilkinson A, Economou A, Williams P, Downie JA (1999) Analysis of quorum-sensing-dependent control of rhizosphere-expressed (rhi) genes in Rhizobium leguminosarum bv. viciae. J Bacteriol 181:3816–3823
Roux A, Payne SM, Gilmore MS (2009) Microbial telesensing: probing the environment for friends, foes, and food. Cell Host Microbe 6:115–124
Roy R, Tiwari M, Donelli G, Tiwari V (2018) Strategies for combating bacterial biofilms: a focus on anti-bio fi lm agents and their mechanisms of action. Virulence 9(1):522–554
Rutherford Steven T, Bonnie BL (2012) Bacterial quorum sensing: its role in virulence and possibilities for its control. Cold Spring Harb Perspect Med 2:a012427
Saising J, Dube L, Ziebandt AK, Voravuthikunchai SP, Nega M, Gotz F (2012) Activity of gallidermin on Staphylococcus aureus and Staphylococcus epidermidis biofilms. Antimicrob Agents Chemother 56:5804–5810
Schaefer AL, Hanzelka BL, Eberhard A, Greenberg EP (1996) Quorum sensing in Vibrio fischeri: probing autoinducer-LuxR interactions with autoinducer analogs. J Bacteriol 178:2897–2901
Schauder S, Shokat K, Surette MG, Bassler BL (2001) The LuxS family of bacterial autoinducers: biosynthesis of a novel quorum-sensing signal molecule. Mol Microbiol 41:463–476
Schuster M, Lostroh CP, Ogi T, Greenberg EP (2003) Identification, timing, and signal specificity of Pseudomonas aeruginosa quorum-controlled genes: a transcriptome analysis. J Bacteriol 185:2066–2079

Shadia MAA, Aeron A (2014) Bacterial biofilm: dispersal and inhibition strategies. SAJ Biotechnol 1:105

Shai Y, Oren Z (2001) From "carpet" mechanism to de-novo designed diastereomeric cell-selective antimicrobial peptides. Peptides 22:1629–1641

Sharma S, Stark TF, Beattie WG, Moses RE (1986) Multiple control elements for the uvrC gene unit of Escherichia coli. Nucleic Acids Res 14:2301–2318

Shen Y, Koller T, Kreikemeyer B, Nelson DC (2013) Rapid degradation of Streptococcus pyogenes biofilms by PlyC, a bacteriophage-encoded endolysin. J Antimicrob Chemother 68:1818–1824

Skindersoe ME, Zeuthen LH, Brix S, Fink LN, Lazenby J, Whittall C, Williams P, Diggle SP, Froekiaer H et al (2009) Pseudomonas aeruginosa quorum-sensing signal molecules interfere with dendritic cell-induced T-cell proliferation. FEMS Immunol Med Microbiol 55:335–345

Smith RS, Iglewski BH (2003) P. aeruginosa quorum-sensing systems and virulence. Curr Opin Microbiol 6:56–60

Smith RS, Harris SG, Phipps R, Iglewski B (2002) The Pseudomonas aeruginosa quorum-sensing molecule N-(3-oxododeca-noyl)homoserine lactone contributes to virulence and induces inflammation in vivo. J Bacteriol 184:1132–1139

Sperandio V, Mellies JL, Nguyen W, Shin S, Kaper JB (1999) Quorum sensing controls expression of the type III secretion gene transcription and protein secretion in enterohemorrhagic and enteropathogenic Escherichia coli. Proc Natl Acad Sci U S A 96:15196–15201

Sperandio V, Torres AG, Giron JA, Kaper JB (2001) Quorum sensing is a global regulatory mechanism in enterohemorrhagic Escherichia coli O157:H7. J Bacteriol 183:5187–5197

Sperandio V, Li CC, Kaper JB (2002) Quorum-sensing Escherichia coli regulator A: a regulator of the LysR family involved in the regulation of the locus of enterocyte effacement pathogenicity island in enterohemorrhagic E. coli. Infect Immun 70:3085–3093

Stewart PS (2015) Prospects for anti-biofilm pharmaceuticals. Pharmaceuticals 8:504–511

Subbalakshmi C, Sitaram N (1998) Mechanism of antimicrobial action of indolicidin. FEMS Microbiol Lett 160:91–96

Surette MG, Bassler BL (1998) Quorum sensing in Escherichia coli and Salmonella typhimurium. Proc Natl Acad Sci U S A 95:7046–7050

Surette MG, Miller MB, Bassler BL (1999) Quorum sensing in Escherichia coli, Salmonella typhimurium, and Vibrio harveyi: a new family of genes responsible for autoinducer production. Proc Natl Acad Sci U S A 96:1639–1644

Swift S, Karlyshev AV, Durant EL, Winson MK, Chhabra SR, Williams P, Macintyre S, Stewart GSAB (1997) Quorum sensing in Aeromonas hydrophila and Aeromonas salmonicida: identification of the LuxRI homologues AhyRI and AsaRI and their cognate signal molecules. J Bacteriol 179:5271–5281

Tateda K, Ishii Y, Horikawa M, Matsumoto T, Miyairi S, Pechere JC, Standiford TJ, Ishiguro M, Yamaguchi K (2003) The Pseudomonas aeruginosa autoinducer N-3-oxododecanoyl homoserine lactone accelerates apoptosis in macrophages and neutrophils. Infect Immun 71:5785–5793

Taylor RFH, Gaya H, Hodson ME (1993) Pseudomonas cepacia: pulmonary infection in patients with cystic fibrosis. Respir Med 87:187–192

Telford G, Wheeler D, Williams P, Tomkins PT, Appleby P, Sewell H, Stewart GS, Bycroft BW, Pritchard DI (1998) The Pseudomonas aeruginosa quorum-sensing signal molecule N-(3-oxododecanoyl)-L-homoserine lactone has immunomodulatory activity. Infect Immun 66:36–42

Throup JP, Camara M, Briggs GS, Winson MK, Chhabra SR, Bycroft BW, Williams P, Stewart GSAB (1995) Characterisation of the yenI/R locus from Yersinia enterocolitica mediating the synthesis of two N-acyl-homoserine lactone signal molecules. Mol Microbiol 17:345–356

Tiwari V, Roy R, Tiwari M (2015) Antimicrobial active herbal compounds against Acinetobacter baumannii and other pathogens. Front Microbiol 6:618

Tsang LH, Cassat JE, Shaw LN, Beenken KE, Smeltzer MS (2008) Factors contributing to the biofilm-deficient phenotype of Staphylococcus aureus sarA mutants. PLoS One 3:e3361

Vidigal PG, Musken M, Becker KA, Haussler S, Wingender J, Steinmann E, Kehrmann J, Gulbins E, Buer J, Rath PM et al (2014) Effects of green tea compound epigallocatechin-3 gallate against Stenotrophomonas maltophilia infection and biofilm. PLoS One 9:e92876

Viswanathan P, Suneeva SC, Rathinam P (2015) Quorum sensing in pathogenesis and virulence. In: Kalia VC (ed) Quorum sensing vs quorum quenching: a battle with no end in sight. Springer, Delhi

Vizan JL, Hernandez-Chico C, Castillo I, Moreno F (1991) The antibiotic microcin B17 is a DNA gyrase poison: characterisation of the mode of inhibition. EMBO J 10:467–476

Vuong C, Saenz HL, Gotz F, Otto M (2000) Impact of the agr quorum-sensing system on adherence to polystyrene in Staphylococcus aureus. J Infect Dis 182:1688–1693

Walters M, Sperandio V (2006) Quorum sensing in Escherichia coli and Salmonella. Int J Med Microbiol 296:125–131

Willcox MD, Hume EB, Aliwarga Y, Kumar N, Cole N (2008) A novel cationic-peptide coating for the prevention of microbial colonization on contact lenses. J Appl Microbiol 105:1817–1825; PMID:19016975

Withers HL, Nordstrom K (1998) Quorum-sensing acts at initiation of chromosomal replication in Escherichia coli. Proc Natl Acad Sci U S A 95:15694–15699

Wu H, Moser C, Wang HZ, Hoiby N, Song ZJ (2015) Strategies for combating bacterial biofilm infections. Int J Oral Sci 7:1–7

Yoda Y, Hu ZQ, Zhao WH, Shimamura T (2004) Different susceptibilities of Staphylococcus and Gram-negative rods to epigallocatechin gallate. J Infect Chemother 10:55–58

Zhang L, Murphy PJ, Kerr A, Tate ME (1993) Agrobacterium conjugation and gene regulation by N-acyl-L-homoserine lactones. Nature 362:446–448

Zhang A, Mu H, Zhang W, Cui G, Zhu J, Duan J (2013) Chitosan coupling makes microbial biofilms susceptible to antibiotics. Sci Rep 3:3364

Zhao WH, Hu ZQ, Hara Y, Shimamura T (2002) Inhibition of penicillinase by epigallocatechin gallate resulting in restoration of antibacterial activity of penicillin against penicillinase-producing Staphylococcus aureus. Antimicrob Agents Chemother 46:2266–2268

Zhu J, Beaber JW, More MI, Fuqua C, Eberhard A, Winans SC (1998) Analogs of the autoinducer 3-oxooctanoyl-homoserine lactone strongly inhibit activity of the TraR protein of Agrobacterium tumefaciens. J Bacteriol 180:5398–5405

Bacteria and Biofilms as Natural Inhabitants of Our Body

3

Rina Rani Ray, Dibyajit Lahiri, Abarna Chatterjee, and Prateek Banerjee

Abstract

Biofilms are the adherent group of sessile communities that remain attached to biotic or abiotic surfaces with the help of extracellular polymeric substances and pili. They are found within the diverse environmental conditions and protect the sessile communities from various types of environmental resistances. The human body is the storehouse of diverse microbial species with varied levels of spatial and temporal arrangements. This normal flora of human body performs numerous functions to maintain homoeostasis. Besides acting as the natural barrier for foreign bacteria, the commensals also act to coordinate the action of human immune reactions. Hence, the interaction between host immune system and commensal microbiota results in either the steady state or development of disease. Natural biofilm-forming microorganisms are often found in various parts of human body like the oral cavity, gut, skin, vagina, lung cardiac tissue, etc. Formation of dental biofilm and gastric biofilm formed due to dysbiosis may lead to a number of diseases.

R. R. Ray (✉)
Department of Biotechnology, Maulana Abul Kalam Azad University of Technology, Kolkata, West Bengal, India

D. Lahiri · A. Chatterjee · P. Banerjee
Department of Biotechnology, University of Engineering & Management, Kolkata, West Bengal, India

R. R. Ray et al. (eds.), *Biofilm-Mediated Diseases: Causes and Controls*,
https://doi.org/10.1007/978-981-16-0745-5_3

3.1 Introduction

Microbial cells are omnipresent and live within any types of environment that provides suitable conditions for their higher mode of living. Microorganisms are highly sensitive to the change in pressure, temperature, pH and salinity, but sometimes it has been observed that certain groups of microbes are able to thrive in these extremes of physical conditions (Horikoshi and Grant 1998). The ability of the microbial cells to live within various conditions is due to the phenotypic plasticity and metabolic versatility (Davey and O'Toole 2000). The microorganisms at different situations exhibit complex differentiation and collective behaviour (Shapiro and Dworkin 1997). Researchers have shown that the microbial cells possesses the ability to perform various types of intercellular interactions and communications which help them to exist in altered environmental conditions (Kaiser and Losick 1993). The knowledge about structure, functions and dynamics of the persistent human microbiome is provided by various metagenomic and high-throughput studies (Qin et al. 2010; Huttenhower et al. 2012). These molecular studies provided us with the concept that the human body possesses myriad of varieties of microbes that cluster within the body cavities. The microbial communities do not exist in the planktonic forms but they remain encompassed by a self-producing polymeric matrix that help in the adherence of the cells to the inner surfaces of the body (Costerton et al. 1995; Hall-Stoodley et al. 2004). It is necessary to understand the biofilm formed by multiple species as they exhibit different types of physiology in comparison to the planktonic cells that result in the development of resistances and virulence (Burmølle et al. 2014). Various recent studies have been performed on oral microbiome that has been characterized by its involvement in periodontitis, dental caries and oral cancer (Zarco et al. 2012). It has been observed that most densely populated microbial biofilms are located within the intestine and showed a paradigm importance and beneficial interaction with the host (Sommer and Bäckhed 2013). The intestinal microbes have direct impact upon the metabolic functions that result in the conversion of non-digestable foods into short fatty acid chains, degradation of vitamins and toxic components. The microbes within the intestine also help in regulating human immune system and development of physiology. Thus change in the intestinal microflora results in various types of diseases like type II diabetes and colorectal cancers (De Vos and de Vos 2012). This chapter will focus on various types of biofilm associated with the human system, genetic and its effect upon human body.

3.2 Oral Biofilms

The mouth being the gateway of the digestive tract harbours diverse microorganisms. Indeed, the oral cavity is unique in the level of microbial diversity and complexity, supporting up to 1000 different species of microorganisms (Grigalauskienė et al. 2015). They have been estimated to be the second most complicated part of the body, after the colon (Mosaddad et al. 2019). Human oral microbiome comprises a large number of microbiota that are specific to particular niches like the cheek, teeth, surface of the tongue, palate, gums, lingual tonsils and gingival pocket (Krzyściak et al. 2016).

3.2.1 Human Oral Microbiome

Biofilm formation in the oral cavity is most widespread on the teeth because the tooth provides a non-shredding, stagnant surface with possible food compaction. So far, various investigations suggested that the oral cavity comprises approximately 700 different bacterial species, and about 10–20 species constitute about 90–95% of the bacteria present in an individual. These microbial cells can exist upon the oral mucosa comprising the cheeks, palate, lips and dorsal side of the tongue. Bacterial cells also exist on the tooth, subgingival areas, surface of the root, pits and fissures and upon the surfaces of the smooth muscles.

A broad spectrum of varied bacterial species are found within the buccal cavity that is formed by complex interactions existing between the microbial populations that determine the normal pathological and physiological conditions of both systemic and local levels (Kriebel et al. 2018). Other major inhabitants of gingival sulcus area and gingival cavities are *Bacilli*, *Moraxella*, *Neisseria* and *Spirochaetes* like *Treponema* (*T. denticola*, *T. orale*, *T vincentii*). Peridontal pockets are infested by *Mycoplasma orale*, *M. pneumoniae* and *M. hominis*. A large number of fungal flora are found in the oral cavity, gingival areas, peridontal abscess and infected root canal. Among them *Candida albicans*, *Penicillium*, *Hemispora* and *Aspergillus* are noteworthy. Protozoa like *Entamoeba* can be found from patients with periodontitis, and some virus like mumps virus, EBV, influenza and measles virus can be found during the advanced stage of the disease.

The papilla on the upper surface of tongue provides an important shelter to oral microorganisms. *Micrococcus mucilaginosus* and *Streptococcus salivarius* are the predominant members which are generally not found on the teeth. Saliva also abodes a number of bacteria like *Streptococcus oralis* and *S. salivarius*.

Tooth surface, pits, fissures and root canal all are very lucrative sites for the bacteria. The uninfected dental surface is especially infested by *Streptococcus sanguinis* and *S. mutans*. The gum or gingival area consists of the mucosal tissue lying over the mandible and maxilla regions of mouth remains protected from mastication, movements of tongue or flushing of saliva. It is occupied predominantly by *Actinomyces* and streptococci. But there is a variation in the nature of inhabitants according to the location, as supra-gingival plaque is dominated by cocci, while subgingival plaque is infested by filamentous bacteria and spirochetes. *Bacteroides melaninogenicus* is a type of pathogenic organism that possesses the ability to exploit this habitat and bring about destruction to the gingival epithelium.

Comprehensive microbiome analysis of tonsillar crypts indicates that the predominant bacteria are *Fusobacterium* spp., *Prevotella* spp., *Treponema* spp., *Sphingomonas* spp., *Porphyromonas* spp. and *Haemophilus* spp. (Watanabe et al. 2017).

Haemophilus influenzae, *Streptococcus pneumoniae* and *Moraxella catarrhalis* are the groups of potential pathogens that are found exclusively in the adenoids of patients with pharyngotonsillitis; this is an indication that adenoids and platine tonsils are the storehouse of various microbial cells that are potentially pathogenic in nature (Helena Fagö-Olsen et al. 2019)

3.2.2 Ecological Determinants for the Nature of Inhabitants

Various factors like nutrient availability, pH, oxygen, presence of other organism, mechanical activities, amount of gingival crevicular fluid and presence of antagonistic factors determine the nature of the microbiome present in a particular niche of oral ecosystem (Krzyściak et al. 2016; Marsh and Zaura 2017). Microorganisms require hemin (*Bacteroides melaninogenicus*, *B. gingivalis* and *Capnocytophaga*), menaquinones (Bentley and Meganathan 1982), oestrogen and progesterone (certain oral bacteroids), while *Treponema denticola* need spermin as nutrients (Wyss 1992).

Serum contains large amount of nutrients that are absent in saliva and are found within the gums or gingival crevice where they come in contact with enriched crevicular fluid (Asikainen et al. 2010). Saliva comprises lysozymes, lactoferrin, lactoperoxidase and specific antibodies possessing an inhibiting effect upon the growth of bacteria. Production of H_2O_2, reduction of pH by acid secretion, synthesis of oxidizing enzymes and bacteriocin can inhibit the growth of certain species while promote the growth of others. Thus the niche of a particular organism will be determined by the interplay of these ecological determinants.

3.2.3 Establishment of Oral Biofilm

Biofilms are formed at several surfaces in the mouth such as the tongue and the teeth. The composition of the microbiota varies at different locations due to the changing ecological factors such as pH, energy supply and presence of oxygen. Moreover, the anatomical configuration, quality and quantity of saliva, food habit and personal hygiene determine the composition of oral microbiota. The oral microbiome of the babies born through normal parturition process is similar to that of the vaginal flora of the mother, whereas babies born via caesarean section have oral microbiome which resembles with the mother's skin. Hence the first group of oral microbiome is dominated by bacterial species, for example, *Prevotella*, *Lactobacillus*, *Corynebacterium*, *Streptococcus*, *Staphylococcus*, *Bacteroides*, *Enterococcus* and *Sneathia* spp., whereas bacterial species *Corynebacterium*, *Staphylococcus*, *Propionibacterium* spp., *Firmicutes*, *Proteobacteria* and *Bacteroidetes* are the predominant member of the second group (Dominguez-Bello et al. 2010).

Within few hours after birth, the buccal cavity is extensively exposed to microorganisms coming from the external world through respiration, feeding and contact, and hence within few hours after birth, permanent colonization of the oral cavity begins. *Staphylococcus epidermidis* and *Streptococcus* are detected to be the pioneer colonizers after birth (Edlund et al. 2013).

The oral cavity provides habitat to various bacterial communities and results in the stepwise development of the dental biofilm (Kriebel et al. 2018). Moreover, oral commensal bacteria, which are classified as "pathobionts", present in extra-oral lesions with distinct bacterial species found to be involved in the onset of biofilm formation. The natural dentition and dental prostheses, including dentures and implants, are substrates for biofilms (Berger et al. 2018).

The formation of the oral biofilm on the teeth starts almost immediately after dental cleaning. If the biofilm formation is left undisturbed, the tooth will be covered by bacteria after 2 days, and it will be visible to the eyes after 4 days. Actually, the existence of soft (gingiva) and non-shedding hard (teeth) tissues provides microorganisms with potential surfaces for adherence and subsequent interaction with various host cells. It is accomplished through "acquired pellicle formation" which is a thin protein-containing film derived from salivary glycoproteins of the host to a clean tooth surface (Huang et al. 2011). These interactive forces may range from 5 nm to 100 nm, involving a variety of forces like ionic interactions, Lewis acid-base interactions, electrostatic and hydrophobic interactions, covalent and hydrogen bonds and even dipole-dipole interactions, van der Waals forces and Coulomb interactions (Hannig and Hannig 2009).

3.2.4 Formation of Oral Biofilm Is Done Through the Following Steps

3.2.4.1 Adsorption of Host and Bacterial Molecules to the Tooth Surface

The most prevalent bacterial species found within oral environment immediately after a period of 8 h from birth is *Streptococcus salivarius* (Rotimi and Duerden 1981) followed by *S. oralis*, *S. mitis* and *S. sanguinis* which also congregate within the oral environment. Some groups of facultative organisms like *Actinomyces* are present which act as the receptor structure facilitating the colonization of the first bacterial cells entering within the oral environment (Kolenbrander et al. 2002). The receptor binds with the oral bacterial species which acts as a nutrient, thus helping in forming the biofilm within the oral environment. These organisms adhere to the surface of the host by degrading proline-rich protein nutrient (Kolenbrander et al. 2002).

Studies have shown that there are approximately 500 potential bacterial communities which have the ability of regulating the gene in response with various molecules produced by the host, interactions between various types of opportunistic microbes and the environment and various types of physical interactions taking place with other bacterial cells. Various types of biomolecules including carbohydrates, proteins and nucleic acids help in forming the architecture of the matrix (Mosaddad et al. 2019). Various factors including the sulcular epithelium, adhesins, fimbriae, sucrose and glucan influence the formation of the biofilm.

3.2.4.2 Passive Transport of Oral Bacteria to the Tooth Surface

The primary step of biofilm formation includes colonization of sessile communities by the mutual attractions (Fig. 3.1) and attachment to a surface (Chandki et al. 2011). The congregation of the bacterial cells results in the development of various morphological structures like rosettes and corncobs. The accumulation of cells results in the change of the internal environment from aerobic to facultative anaerobic. The sessile microcolonies proliferate by the development of the extracellular polymeric substances (EPS) which result in better attachment and maturation of the biofilm. The biofilm possesses the ability of incorporating newer members or microcolonies, thus resulting in the development of polymicrobial structure.

3.2.4.3 Co-adhesion of Later Colonizers to Already Attached Early Colonizers

Gradually, more bacteria adhere and the composition of the biofilm becomes more diverse. *Fusobacterium nucleatum*, a bacterium with elongated shape, is considered to have an important role (Brennan and Garrett 2019) since it has the ability to co-aggregate with all bacteria in the biofilm as a linker organism, helping in forming polymicrobial colonies with other bacterial cells including *Aggregatibacter actinomycetemcomitans* and *Porphyromonas gingivalis* (Kolenbrander et al. 2010). Studies have shown that organisms like *F. nucleatum* utilize the surface molecules like RadD, adhesion and fusobacterial apoptosis protein Fap2 for the development of the biofilm (Kaplan et al. 2010).

The biofilm formed initially consists mainly of Gram-positive facultative bacterial cells, followed by incorporation of Gram-negative rod and cocci. This mechanism enhances the number of anaerobic organisms within the biofilm. When the biofilm matures, the microenvironment within the biofilm changes, and therefore the composition also changes. Several of the bacterial species involved in

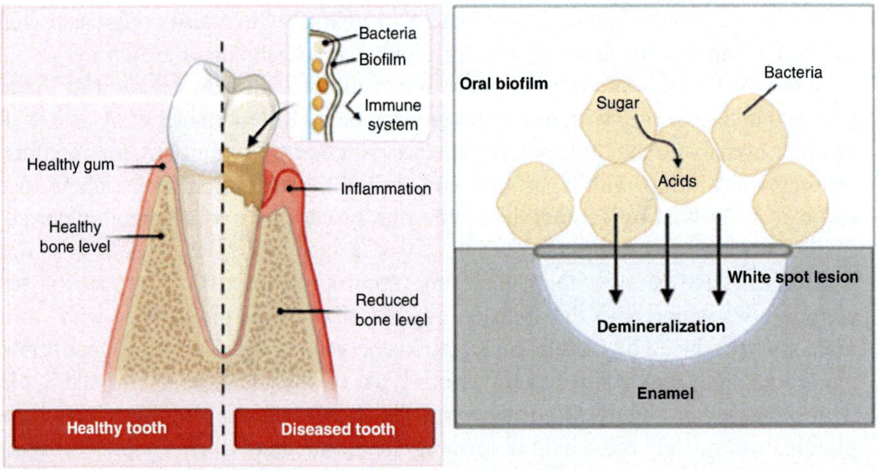

Fig. 3.1 Development of biofilm upon tooth surface

biofilm-mediated diseases are present in small numbers even under healthy conditions, but changes in the environment can favour growth of the more virulent species. This type of biofilm can be analysed by the use of combinatorial labelling and spectral imaging FISH (CLASI-FISH). Studies stated that the complex consortia of organisms existing within the biofilm are termed as hedgehog comprising various taxa of organisms such as *Streptococcus*, *Porphyromonas*, *Haemophilus/ Aggregatibacter*, *Corynebacterium*, *Leptotrichia*, *Neisseriaceae*, *Fusobacterium*, *Capnocytophaga* and *Actinomyces* (Kriebel et al. 2018). The plaque existing within the oral cavity is represented by radially organized structure possessing the major framework by *Corynebacterium*. The organisms existing in such biofilm show selective interactions like cooperative and antagonistic stratifications with other bacterial colonies existing within the dental biofilm depending upon various environmental factors and the availability of nutrients (Wessel et al. 2014).

The bacterial cells existing within the oral biofilm show complicated behaviours by communicating, competing and helping each other (Sintim and Gürsoy 2016). The bacterial cells existing within the oral cavity can maintain the communication with each other by chemical signal molecules like small signalling peptides which are more predominant in Gram-positive bacterial cells, whereas autoinducers-2 (AI-2) show their prevalence in both Gram-positive and Gram-negative bacterial cells (Suntharalingam and Cvitkovitch 2005).

3.2.4.4 Multiplication of the Attached Microorganisms

The mechanism of enhancement in biofilm formation upon a surface is mediated by the proliferation of the bacterial cells. The sessile cells encompass themselves with the help of a polymeric matrix (EPS) that helps in adherence of the cells as well as provides nourishment to the developing microcolonies. Apart from providing a three-dimensional scaffold, it also acts as a protective shield by the sessile microcolonies as it prevents the penetration of drugs, antimicrobials and antibiotics (Mah and O'Toole 2001). The growth of the bacterial cells continues upon the surface, resulting in the development of mushroom-shaped piles of highly congregated cells that remain enclosed by the matrix (Hall-Stoodley et al. 2004). The colonies existing within the mature biofilm are highly differentiated into large number of pockets which vary in their metabolism and reproductive ability. They result in the development of highly complex structure possessing its own microenvironment having a distinct cell density, pH, availability of nutrients and oxygen. The metabolically inactive and relatively old cells remain deep seated within the biofilm and develop a pile-like structure and remain protected from the action of antibiotics and other antimicrobial agents (Fux et al. 2004). The proliferation of the biofilm is also dependent on temperature, pH, oxidation-reduction potential of the gingival crevicular fluid and availability of nutrients. There are also various factors such as polyamines, fatty acids, lactate, vitamin K analogue and various other enzymes which help in promoting the growth of the cells. The local and environmental factors influence the structure and composition of the biofilm (Branda et al. 2005).

3.2.4.5 Active Detachment and Dispersal

The sessile cells detach from the biofilm into the saliva which helps in the mechanism of congregation and colonization, resulting in the development of supragingival plaque. The mass of the plaque is dependent on the rate of bacterial cell proliferation and the attachment of cells with one another and to the surface of the oral cavity (Kaplan et al. 2010). This process of dispersion plays an important role in the life cycle of the biofilm and results in the spread of chronic infections. The mechanism of dispersion involves various signal transduction pathways, environmental signals and effectors (Karatan and Watnick 2009). The detachment of the bacterial cells occur in three distinct phases comprising the detachment mechanism of the biofilm-forming colonies in a regular fashion also termed as sloughing; metastasis of the biofilm-forming cells from one position to a new location; and adherence of cells to a new location (Kaplan et al. 2010). The bacterial cells from the dental plaque get transported to a newer location via saliva either by direct contact or via vector. The dispersal of biofilm can be classified into two broad categories: active and passive. The mechanism of active dispersal is initiated by the sessile bacterial cells themselves, whereas the passive mechanism involves detachment in the presence of external forces like abrasion, shear forces, human intervention and the mechanism of predator grazing (Choi and Morgenroth 2003; Ymele-Leki and Ross 2007). The mechanism of sloughing and erosion of the sessile cells can be categorized under the passive process, whereas the dispersal is grouped under the active process.

3.2.5 Genetic Background of Oral Biofilm Formation

Studies have shown that *Streptococcus mutans* are the major causal organisms, associated with dental caries. The ability of the organism to form biofilm upon the surface of the tooth results in the virulence properties of the organism. It has been observed that glucosyltransferases and sucrose play an important role in the formation of the biofilm by *S. mutans*. Researches have shown that approximately seven varied types of mutans are associated with dental plaque infections (Yoshida and Kuramitsu 2002). The extracellular DNA (eDNA) plays an important role in the various phases of biofilm formation that include adhesion and congregation followed by mechanical stabilization of the structure formed by biofilm. It also plays an important role in the development of competence among the cells by the mechanism of horizontal transfer of genes (Fig. 3.2) among the polymicrobial colonies existing within the biofilm (Krzyściak et al. 2016).

3.2.6 Diseases Caused by Oral Biofilm

The normal flora of dental ecosystem maintains a complex equilibrium with each other by interspecific relations, which are required for maintenance of a steady healthy state. But if the homoeostasis is disturbed by the presence of a new species

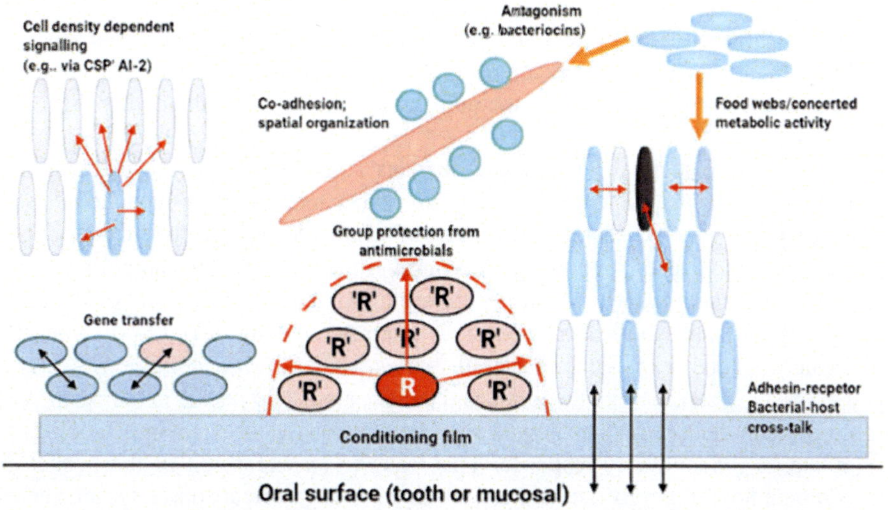

Fig. 3.2 Development of the oral biofilm

of bacteria or by the change of relative proportions of species, a pathogenic state appears which is known as dysbiosis. Due to hormonal changes brought about by ageing, pregnancy or onset of puberty; clinical conditions like salivary gland malfunction with reduced salivation with changed composition; gum infection; or change in food habit including smoking, the stable and healthy relationship between host and oral microbiota may be disrupted. This will induce the development of biofilm of a changed pattern on tooth and in various microenvironments of buccal cavity. Finally it will lead to the generation of plaque that will result in the formation of caries and destructive periodontitis. The plaque can only be removed by dental cleaning or tooth brushing as diverse microbes remain involved [nonspecific plaque hypothesis (NSPH), whereas treatment of caries by kanamycin reveals the involvement of only few species of microorganisms (specific plaque hypothesis).

The formation of cavities in the teeth is called caries. It is the demineralization of tooth substance which happens when pH drops below 5.5, caused by acids produced by microbial fermentation of sugars present in diet. Persisting acidic environment will favour the growth of acidophilic species like mutans *streptococci*, *lactobacilli* and *Actinomyces* species. They have a strong ability to adhere to the tooth surface, can produce extracellular polysaccharides and have the ability to metabolize sugars into mainly lactic acid. If the caries process is allowed to proceed, the demineralization of the tooth structure will create a pathway for the oral bacteria into the pulp where the vessels and nerves are located. At first, the tissue is not infected but the bacteria cause an inflammatory reaction with piercing pain/toothache. The initial lesion is dominated by the same bacteria that are involved in the caries process. If the pulp tissue becomes infected, necrosis will follow. When the pulp becomes necrotic, the available nutrients for the bacteria change from carbohydrates from the oral cavity to proteins from the necrotic tissue; further, the environment becomes anaerobic. This favours growth of proteolytic anaerobic bacteria. They can spread through the root canal system and can even reach the apical area of the tooth and spread into the jawbone. This can result in either an acute or chronic infection. The bacteria are forming biofilms in the pulp cavity which makes them very difficult to eliminate.

The microflora in the root canal changes according to the stage of infection (Lakshmi Narayanan and Vaishnavi 2010). Other major biofilm-mediated diseases are gingivitis which is characterized by swollen and bleeding gums and periodontitis which leads to loss of supporting tissue around the teeth and loosening of the teeth. Gingivitis is thought to be a nonspecific reaction linked to the amount of bacteria in the biofilm on the tooth. Periodontitis is more specifically linked to the concerted action of a number of certain bacterial species. Under healthy conditions, the gums are closely in contact with the tooth, and there is almost no pocket in between. When more and more bacteria are allowed to accumulate on the tooth, their metabolites can initiate an inflammatory response. This leads to an expansion of the pocket and increased flow of gingival exudate. This changes the nutrient source for the bacteria and favours proteolytic species. As a result of proteolysis, the pH of the subgingival environment increases slightly. A number of species have been linked to periodontal disease although they also can be found in healthy conditions in smaller proportions. Generally we see a greater diversity in the biofilm in periodontal disease with larger proportions of pathogenic species. The species most often associated with the development of periodontitis are *Fusobacterium nucleatum*, *Campylobacter*, *Prevotella species*, *Porphyromonas gingivalis*, *Tannerella forsythia* and *Treponema denticola. Aggregatibacter actinomycetemcomitans* is especially associated with juvenile periodontitis. The collective virulence of the bacteria is increased by intercellular communication. Many of the virulence factors of the periodontal pathogens are related to either direct tissue and cell destruction or upregulation of the inflammatory response causing an indirect tissue destruction. Besides the microbiological aspect of periodontal disease, there are several host-related factors such as smoking and genetic aspects influencing the progression of the disease. So the bacteria initiate the degrading of the periodontium by direct tissue destruction, but it is the host's inflammatory response that determines the severity of inflammation and makes the process chronic.

3.2.7 Systemic Consequences of Oral Dysbiosis

Dysbiosis may trigger the onset of bacteraemia which in turn facilitates systemic dissemination of oral bacteria. The microbial cells present within the oral cavity have shown their involvement in various systemic disorders that include rheumatoid arthritis, cardiovascular diseases, stroke, adverse pregnancy outcomes, colorectal cancers and inflammatory bowel diseases.

3.2.8 Advantages of Oral Biofilm

The oral microbiome shows various important roles like the production of cofactors and vitamins like vitamin K, riboflavin and biotin. The cofactors being produced by these organisms help in regulating the normal systemic functions associated with healthy living. Various types of digestive juices like lipase, amylase and protease produced by the microbial flora help in the digestion of food materials (Kilian et al. 2016).

3.3 Skin Microbiology

The skin is the largest organ in human body housing a community of millions of diverse types of microorganisms all over the body. Most of these are harmless or even beneficial to their host. Such colonization is influenced by the condition of the skin surface which is highly variable depending on geographical location; intrinsic host factors, including genetic makeup; and environmental factors. The skin contains more diverse bacteria than the gut and oral microbiota. The normal flora of the skin prevents transient pathogenic organisms from colonizing the skin surface, either by competing for nutrients, secreting chemicals against them or stimulating the skin's immune system. However, resident microbes can also cause skin diseases in immunologically challenged people.

3.3.1 Human Skin Microbiome

The human body is equipped with trillions of microbes. The microbiome existing upon the skin comprises various microbial cells which show the interrelation with one another (Belkaid and Segre 2014; Prescott et al. 2017). Studies have shown that the microbial communities persistent upon the skin have evolved with the evolution of human and have become an integral part of human body (Chiller et al. 2001). The skin is the myriad of a variety of microbial species comprising commensals having the ability of deriving its food from the skin and secretions like sebum and sweat. *Cutibacterium acnes* and *Staphylococcus aureus* are the most common types of bacterial cells that are found upon the skin specially in the sebum-rich areas, hair follicles and skin glands (Grice et al. 2009; Aubin et al. 2016; SanMiguel and Grice 2015; Feuerstein et al. 2020). Greater number of microbiota is observed at the moistened region of the skin such as nares and underarms than on the dryer regions of the skin surfaces like legs, arms, hands and top of the feet.

3.3.2 The Skin as an Ecosystem

The various physical and chemical features of the skin influence the composition of the existing microbial cells upon the surface. The microenvironment inhabiting the skin has a single set of dominant bacterial cells. The skin is generally acidic, dry and cool, and the variation of the skin texture determines the variability of the microbial density (Tagami 2008). It has been observed that the skin ecology comprises various groups of resident bacterial cells. The composition of the microbiota determines the health condition of the skin.

Extensive researches are ongoing to find out the role of the flora on human health and hygiene. As interaction between environmental factors and intrinsic factors determines the microbial composition of the skin, it is regarded as an ecosystem.

3.3.3 Composition of Skin Microbiota

The keratinocytes are present within the stratum corneum—the outermost layer of the skin which is composed of bilayers of lipids. The skin is a type of exfoliating organ which undergoes continuous shedding of cells from the surface. There are topographical variations of the skin due to the presence of various types of microbial cells upon the surfaces. The types of bacterial cells predominantly found in the sebaceous and dry regions of the skin are *Betaproteobacteria* and *Propionibacterium*. Some regions of the skin are with higher temperature and humidity, where moisture-loving microbes like some Gram-negative bacilli, coryneform and *S. aureus* are found to grow. As the density of sebaceous glands is a major influencing factor for growth of the skin microbiota, regions with high density of sebaceous glands, such as the face, chest and back, encourage the growth of lipophilic microorganisms like *Propionibacterium* spp. and *Malassezia* spp. The arm and leg skin, being relatively dry and exposed to large fluctuations in surface temperature, harbour quantitatively fewer organisms.

The commensal bacteria and fungi that live on and inside human body outnumber our own cells, and the viruses that crowd inside those cells may add another order of magnitude. The mouth microbiome is also suspected of influencing bacterial communities in the lungs.

In the moist regions, *Corynebacterium* and *Staphylococcus* are most commonly found. Viruses and fungi are also found on the skin, with *Malassezia* being the most common type of fungus found as part of the normal microbiota. The role and populations of viruses in the microbiota, known as **viromes**, are still not well understood, and there are limitations to the techniques used to identify them. However, *Circoviridae*, *Papillomaviridae* and *Polyomaviridae* appear to be the most common residents in the healthy skin virome.

The human skin microbiota is diverse and includes numerous pathogenic bacteria.

Staphylococcus aureus (*S. aureus*) are the most commonly isolated pathogens, accounting for 20–30% of SSI occurring in hospitals. This prevalence is related to the carriage of *S. aureus* in the healthy population (~20% persistent, ~60% intermittent) (Kluytmans et al. 1997). While topical antibiotics and antiseptics are often employed to reduce *S. aureus* colonization, these treatments may alter skin microbiota and reduce colonization by *S. aureus* competitors. The cutaneous innate and adaptive immune responses can modulate the skin microbiota, but the microbiota also functions in educating the immune system.

3.3.4 Variation by Skin Site

Molecular approaches examining bacterial diversity have underlined the concept that the skin microbiota is dependent on the body site. It is found that colonization of bacteria is dependent on the physiology of the skin site, with specific bacteria being associated with moist, dry and sebaceous microenvironments. In general, bacterial diversity seems to be lowest in sebaceous sites, suggesting that there is selection for specific subsets of organisms that can tolerate conditions in these areas. Sebaceous sites that contain low phylotype (Gannesen et al. 2018) richness include the forehead (6 phylotypes), the retroauricular crease (behind the ear) (15 phylotypes), the back (17 phylotypes) and the alar crease (side of the nostril) (18 phylotypes).

Microbial transplant experiments suggest that the microenvironment of sebaceous areas (such as the forehead) is a stronger force in determining microbial colonization than the microenvironment of dry areas (such as the forearm). These moist sites include the umbilicus (navel), the axillary vault, the inguinal crease (side of the groin), the gluteal crease (topmost part of the fold between the buttocks), the sole of the foot, the popliteal fossa (behind the knee) and the antecubital fossa (inner elbow). *Staphylococci* occupy an aerobic niche on the skin and probably use the urea present in sweat as a nitrogen source. Corynebacteria are extremely fastidious and slow-growing organisms in culture, and, as such, their role as skin microorganisms has been underappreciated until recently. Processing of apocrine sweat by corynebacteria and *staphylococci* (along with other axillary vault microorganisms) results in the characteristic malodour associated with sweat in humans. The most diverse skin sites are the dry areas, with mixed representation from the phyla *Actinobacteria*, *Proteobacteria*, *Firmicutes* and *Bacteroidetes* (Costello et al. 2009; Grice et al. 2009). A comprehensive analysis of skin microbiota across 20 sites (Table 3.1) including the forearm, buttock and various parts of the hand indicates microbiota of these sites with the abundance of Gram-negative organisms. The skin interestingly also harbours greater phylogenetic diversity than the gut or the oral cavity of the same individual (Costello et al. 2009).

3.3.5 Biofilms in Skin Microbiota

Various bacterial cells thriving upon the skin possess differential ability to form biofilm upon the surfaces of the skin. The bacterial microcolonies living within the biofilm show variations in their physiological and metabolic functions, making them virulent and resistant against various types of antibiotics and antimicrobial substances. The existent biofilm results in various biofilm-associated skin disorders (Brandwein et al. 2016). The various diseases related to the skin were initially considered to be non-infectious, but within time, recalcitration of infectious agents

Table 3.1 Microbial profile of various sites of the skin

Site of human body	Composition of the biofilm	References
Exterior of nose and ear Forehead and hair	*Propionibacterineae* (60–80%)	
Palms, forearm and fingers	*Propionibacterineae* (20–40%)	
Armpits (axillae) and soles of feet	*Staphylococcus* spp.	
Navel (umbilicus)	*Corynebacterium* spp.	
Backs of knees (popliteal fossae)	*Corynebacterium* spp.	
Antecubital fossa (inner elbow)	*Pseudomonas* sp., *Janthinobacterium* sp., *Proteobacteria, Actinobacteria, Firmicutes, Bacteroidetes, Cyanobacteria* and *Acidobacteria*	
	Betaproteobacteria and *Propionibacterium*	
Nare	*Micrococcineae* Other *Actinobacteria*	
Axillary vault, volar forearm, interdigital web space, hypothenar palm	*Proteobacteria* *Bacteroidetes*	
Inguinal crease, gluteal crease	*Corynebacterineae*	
Retroauricular crease	*Corynebacterineae 50* *Propionibacterineae 50*	
Back	*Propionibacterineae*	
Back of the knee, Plantar heel	*Staphylococcaceae*	
Core body and arm	*Malassezia* spp.	Byrd et al. (2018)
Foot	*Aspergillus* spp., *Cryptococcus* spp., *Rhodotorula* spp., *Epicoccum* spp. and others	Byrd et al. (2018)
Non-lipophilic yeasts of rectal origin	*Candida, Rhodotorula, Torulopsis, Cryptococcus*	Lihoreau and Agache (2017)
Superficial layer of the skin	*Trichophyton rubrum, T. interdigitale, Epidermophyton floccosum, Microsporum*	Lihoreau and Agache (2017)

may take place resulting in the development of the virulence and possesses the potent cause of various skin-associated infections (Brandwein et al. 2016). The biofilm existing upon the skin surfaces is also polymicrobial in nature, comprising various communities of bacterial cells. The biofilm found in nature comprises multispecies having a mutual coexistence within the architectural framework formed by the matrix being synthesized by the sessile communities (Røder et al. 2016).

The common type of bacterial species associated with the skin comprises *S. aureus*, *S. epidermidis* and *P. acnes* that are usually located at depth, yet *Malassezia* spp. biofilms remain relatively unstudied.

3.3.6 Diseases

The microbes persisting upon human skin are responsible for maintaining both health and diseases. Certain skin diseases are associated with chronic biofilm infections. A very common type of hyperproliferative skin disorder termed as seborrhoeic dermatitis is caused by fungal pathogen *Malassezia* spp. When these cells live solitarily, they are unable to cause infection, thus showing that some other factors make it more virulent which becomes predominant in causing such infections. The hyperproliferation and inflammation of the scalp occur by the release of the fatty acid due to the hyperactivity of the lipase-producing genes. *Cutibacterium acnes* is a commensal organism in the skin which causes acne, a very common problem associated with teenagers. This is characterized by the inflammation of the pilosebaceous glands. The secretions that are produced by these glands are attractants to various lipophilic bacterial cells as these secretions usually comprise lipases, proteases and hyaluronidase enzymes. Another common type of biofilm-associated skin problem is atopic dermatitis (AD) which is a chronic disorder that occurs by the colonization of the microbial cells. The microbial cells mostly associated with AD are *S. aureus*. It has been further observed that patients suffering from diabetes show serious and worse effects than the non-diabetic people.

Burn wound-associated infections are commonly caused by *S. pyogenes*, *Enterococcus* spp. or *Pseudomonas aeruginosa*; no particular organism can be detected in wounds caused by AD.

Staphylococcus epidermidis is a very common skin commensal, which can sometimes cause infection and disease if they invade some unnatural site. The treatment is becoming complicated due to antibiotic resistance, particularly to oxacillin or methicillin. It can even transfer the resistance gene to the closely related but more virulent organism, *S. aureus*.

3.4 Gut Microbiome

3.4.1 Introduction to Gut Microbiome

The gastrointestinal tract is the habitat of large number of microbial colonies that interacts with one another to form a complex network which results in the development of the biofilm. The sessile cells encompass themselves with the help of polymeric matrix comprising protein, polysaccharides, extracellular DNA (eDNA) and lipids. A large number of microbial cells comprising archaea and bacteria are located at different regions of the gastrointestinal tracts or within the gut that acts as commensals. Within the gastrointestinal (GI) tract, the microbiome exists in diverse conditions like the anaerobic condition persisting within the digestive tract where the existence of normal microbial flora is found from birth. One of the largest microbiota existing within the human body is gut microbiota comprising diverse groups of microbial organisms ranging from the beneficial to pathogenic ones.

3.4.2 Formation of Biofilm in Gut

The development of biofilm within the GI tract of the human body occurs at the time of birth when the first vaginal biofilm is delivered within the digestive canal. Studies also showed that the colonization of the biofilm within the infant occurs before birth which was understood by the similarity of microbiota persisting within the amniotic fluid and placenta with the meconium of the infant (Collado et al. 2016). The microbial biofilm existing within the body of the infant is also regulated by the type of diet, various bacterial species coming from the environment like siblings and pets, poor sanitary environments and exposure to antimicrobials (Martin et al. 2016). The development of biofilm within the gut occurs by the utilization of biofilm-promoting factors present within the intestinal epithelium. The commensal type of bacterial species utilizes proteins like surface adhesins and serine-rich protein repeats for the development of biofilm within the gut (Buret et al. 2019). Development of biofilm occurs by the binding interaction with mucin which is a type of glycoprotein being present within the intestinal mucosa, for instance, bacterial species like *Lactobacillus rhamnosus* possess mucus-binding pili that helps in the formation of the biofilm.

3.4.3 Composition of Gut Biofilm

Human gut is a storehouse of large microbiome, and among 1000 species studied, the predominantly inhabitant Gram-positive organism is *Firmicutes* and Gram-negative *Bacteroidetes*, whereas organisms like fungi, ascomycetes and methanogens are found in lower numbers (Eckburg et al. 2005). Some of the bacterial species like *Bifidobacteria* thriving in the gut play an important role in promoting inner gut health and comprise approximately 5% of the total microbiota. Various types of archaea like *Methanosphaera stadtmanae* and *Methanobrevibacter smithii* are also found within the gut. The polymicrobial existence within the gut in the form of biofilm helps in retaining enzymes, water and antimicrobial substances. These microcolonies existing within the biofilm proliferate by means of the density-dependent communication system known as quorum sensing and horizontal gene transfer. The microenvironment prevalent within the gut influences the bacterial composition. The oesophagus contains few bacterial species like *Streptococcus* due to low pH of the gastric lumen. Maximum diversity of microbial flora is observed within the large intestine.

The intestine possesses anaerobic condition, and thus the microbial species present within it are mostly anaerobic in nature. Although limited environmental conditions are available, these organisms thrive upon the human mucosa which helps in the formation of the biofilm. Biofilm formation within the gut is also dependent on peristaltic movements, viscosity and persistence of various types of indigenous microbes.

3.4.4 Architecture of Gut Biofilm

The development of biofilms upon the gut mucosa and on the particulate materials within the lumen of the gut causes various bacterial assemblages. The bacterial cells living in close proximity with the host tissues show efficient interactions with that of the immune system and the gut epithelium. The GI tract acts as a protective shield to the microbial flora as it prevents the exposure to various types of the physical barriers like the mucus and epithelial layer, antimicrobial peptides, immunological factors and enzymes (Hooper and Macpherson 2010). The microbial flora derives its energy by the process of sulphate reduction and fermentation of the host and dietary carbohydrates. Thus the microbial species living within the gut expresses its phenotypic trait to remain protected in the limited environmental condition. The ability of the polymicrobes to utilize the various nutrients like metabolizing sugars available within the small intestine can be studied with the help of metatranscriptomics (Zoetendal et al. 2012). The microbial flora existing within the colon is subjected to various microbiota-accessible carbohydrates which are found in the form of dietary fibres. The variations in the microbiota are also observed on the basis of animal-based or vegetable-based diets.

The cellular communication between the microbial species within the intestine of the host occurs via microbial compounds like metabolites or small signal molecules (Sommer and Bäckhed 2013). It has been observed that the microbial composition varies in the sample obtained from the faecal matters than those obtained from the lumen of the intestine as within the intestine, specific interactions exist between the intestinal cells and the microbes thriving within them (Zoetendal et al. 2002).

3.4.4.1 Biofilm in the Oesophagus

Studies showed that the microbial colonies existing within the oesophagus are simpler in comparison to other gut microbiomes, but they predominantly consist of anaerobic organisms like *Streptococcus* and *Lactobacillus* originating from the oral cavity. The microbiota predominantly found within the oesophagus is found to originate from the mouth or oral cavity Macfarlane and Macfarlane 2006).

3.4.4.2 Biofilm in the Stomach

A lesser number of organisms are observed within the stomach due to the very low pH of the gastric acid, defence mechanisms and shorter retention time of food materials within the stomach. Although the stomach was considered to possess sparse bacterial populations, recent studies have shown that it consists of large microbial diversity that has its origin from the food which is being consumed (Bik et al. 2006).

3.4.4.3 Biofilm in the Large Intestine

The functioning of the intestinal microbiome environment is greatly affected by the various biofilm or biofilm structures being present within human intestine. The intestine is a place where maximum microbial colonization occurs out of the entire gut. The predominant bacterial species like *Bifidobacteria* and *Bacteroides* are available within the intestine. The biofilm formed by polymicrobial colonies helps in efficient breaking down of the complex polysaccharides than the planktonic forms.

3.4.4.4 Biofilm in the Colon

Various studies showed that a large number of microbiomes exist at the colonic region, and mostly the evidences are the microscopic views of these organisms. It was observed that bacterial populations identified in the stool showed phenotypical resemblance with the planktonic communities (Macfarlane and Macfarlane 2006). It was also observed that certain specific groups of microbes like *Ruminococcus bromii* remain attached to solid food particles such as wheat bran and are resistant to starch. *R. bromii* comprises 10% of the total microbiome found at the coelomic regions.

3.4.5 Functions of Gut Biofilm

The microbiota existing within the gut helps in maintaining homeostasis and health by interacting with the mucosal surface. The microbiome existing within the intestine has varieties of impact on human health. The microbiome has direct impact on various types of metabolisms taking place within our body that includes the conversion of non-digestable forms of the food, i.e. complex sugar polymers, to simpler substances so that they can be easily utilized by the gut. It has been further observed that host's immune system is regulated by the signalling functions being performed by the microbes that are being present within the intestine. Change of intestinal microbiota may result in the development of various diseases like type II diabetes, colorectal cancer and inflammatory bowel diseases.

3.4.6 Diseases and Gut Biofilm

3.4.6.1 Crohn's Disease

Crohn's disease is a biofilm-associated chronic inflammatory bowel disease caused by colonization of the adherent microcolonies of *Escherichia coli* that undergoes proliferation and results in the spread of the disease (Chassaing and Darfeuille-Michaud 2013). The symptoms that are associated with such disorder comprise loss in weight, diarrhoea, fatigue and anaemia. Another very common type of biofilm-associated disorder is ulcerative infections commonly caused by *Helicobacter pylori*.

3.5 Bacterial Biofilms on Other Body Parts

3.5.1 Vaginal Biofilm

Vaginal epithelia are a common site for the formation of polymicrobial biofilm. There are many microbiomes existing within the vagina which are important in maintaining the vaginal health and are not associated with any symptoms of diseases. The bacterial species that show dominance in such environment but are fewer in number comprise *Gardnerella vaginalis*, *Prevotella* spp., *Atopobium vaginae*, *Sneathia* spp. and *Lactobacillus* spp. (van de Wijgert et al. 2014). The vagina comprises various types of low diverse groups having uniform congregation of *Lactobacillus* spp. The two most predominant groups of *Lactobacillus* sp. comprise *L. iners* and *L. crispatus*. *Lactobacillus* plays a vital role in the maintenance of the health of the genital organ in females by preventing any sort of genitourinary infections (Borges et al. 2014). The Gram-positive bacilli and various *Lactobacillus* spp. constitute the healthy vaginal microflora (Ravel et al. 2011). A very common type of biofilm-associated genital infection occurring within women during their reproductive age is known to be bacterial vaginosis that results in the development of serious health issues among the affected person. The organisms constituting such biofilm are *Atopobium vaginae* and *G. vaginalis* that show resistance to antibiotic treatments.

3.5.2 Middle Ear Biofilm

Patients suffering from chronic otitis media possesses biofilm within the middle ear. Various chronic diseases associated to the middle ear comprise cholesteatoma, otitis media with effusion and chronic suppurative otitis media. The researches showed that the mucosal otolaryngological infections are a biofilm-associated chronic infection (Rayner et al. 1998).

3.5.3 Lung Biofilm

The lower respiratory tract and lungs are the repository of diverse microbial species comprising fungi, bacteria, bacteriophages and viruses. The major types of bacterial species found in this region comprises *Prevotella*, *Veillonella*, *Sphingomonas*, *Streptococcus*, *Staphylococcus*, *Fusobacterium*, *Megasphaera*, *Acinetobacter* and *Pseudomonas* (Beck et al. 2012; Erb-Downward et al. 2011; Hilty et al. 2010). The organisms usually observed in these regions can be aerobic, aerotolerant and anaerobic in nature. Studies have shown that respiratory passage is an attractive site for colonization of large number of microbial colonies with the help of extracellular polymeric substances (EPS) comprising proteins, polysaccharides and nucleic acids (Gnanadhas et al. 2015).

3.5.4 Biofilm in the Pancreatic Duct

The pancreatic duct is composed of varieties of bacterial microcolonies that result in the development of biofilm. It comprises of varied thickness of biofilm at different regions of the duct (Swidsinski et al. 2005).

3.5.5 Biofilm on Renal Tissues

Pyelonephritis is a common biofilm-associated renal infection where the bacterial cells get adhered within the uroepithelium followed by invasion to the renal tissue. Another predominant type of biofilm-associated bacterial infection is urinary tract infection (UTI) (Delcaru et al. 2016).

3.5.6 Biofilm in the Eye

The growth of biofilm at the posterior region of the eye results in intraocular inflammation causing endophthalmitis. The introduction of organisms usually occurs by various types of exogenous trauma that includes intravitreal injections, injury within the eye and intraocular surgery. This type of eye infections can also be observed due to the metastasis of the pathogenic cells from other sites of infections.

3.6 Conclusion

Various studies showed the occurrence of biofilm within the gut regulates the normal metabolic activities. Microbial biofilms are found throughout the human body (Fig. 3.3) which plays an important function in regulating metabolism and internal health of a person. It has been observed that drift in this microbial flora results in the development of various types of disorders. Thus this chapter tried to analyse the availability of various microbiomes at various body parts of the human being, their importance and the associated diseases.

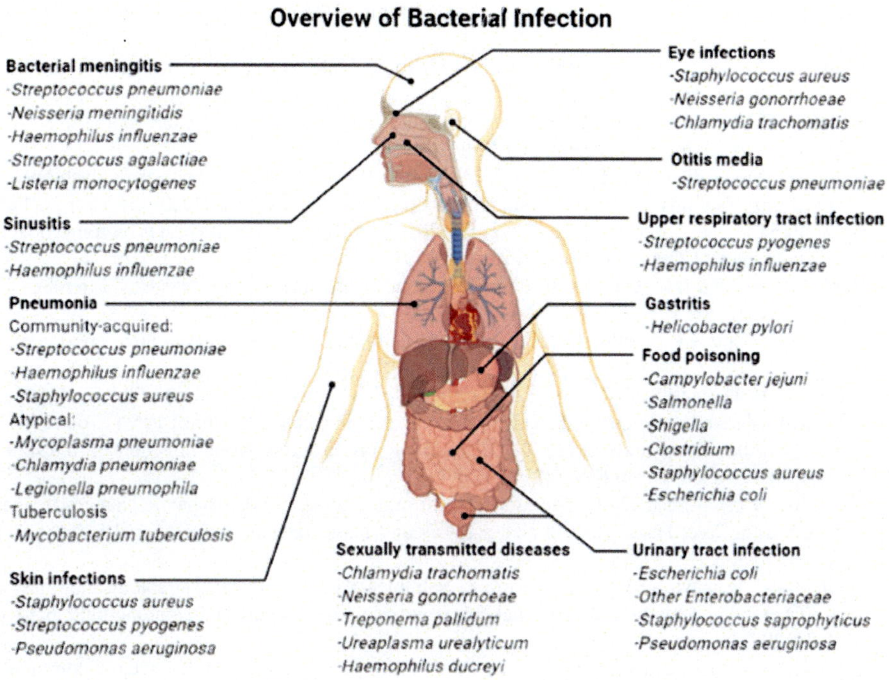

Fig. 3.3 Various types of bacteria-associated infections at various parts of the body

References

Asikainen S, Doğan B, Turgut Z, Paster BJ, Bodur A, Oscarsson J (2010) Specified species in gingival crevicular fluid predict bacterial diversity. PLoS One 5(10):e13589. https://doi.org/10.1371/journal.pone.0013589

Aubin GG, Kambarev S, Guillouzouic A, Lepelletier D, Bémer P, Corvec S (2016) Draft genome sequences of four propionibacterium acnes strains isolated from implant-related infections. Genome Announc 4(6):e01395–e01316

Beck JM, Young VB, Huffnagle GB (2012) The microbiome of the lung. Transl Res 160 (4):258–266

Belkaid Y, Segre JA (2014) Dialogue between skin microbiota and immunity. Science 346 (6212):954–959

Bentley R, Meganathan R (1982) Biosynthesis of vitamin K (Menaquinone) in bacteria. Microbiol Rev 46:241–280

Berger D, Rakhamimova A, Pollack A, Loewy Z (2018) Oral biofilms: development, control, and analysis. High-Throughput 7(3):24

Bik EM, Eckburg PB, Gill SR, Nelson KE, Purdom EA, Francois F, Perez-Perez G, Blaser MJ, Relman DA (2006) Molecular analysis of the bacterial microbiota in the human stomach. Proc Natl Acad Sci U S A 103(3):732–737. https://doi.org/10.1073/pnas.0506655103. PMID: 16407106; PMCID: PMC1334644

Borges S, Silva J, Teixeira P (2014) The role of lactobacilli and probiotics in maintaining vaginal health. Arch Gynecol Obstet 289:479–489. https://doi.org/10.1007/s00404-013-3064-9

Branda SS, Vik S, Friedman L, Kolter R (2005) Biofilms: the matrix revisited. Trends Microbiol 13:20–26

Brandwein M, Steinberg D, Meshner S (2016) Microbial biofilms and the human skin microbiome. NPJ Biofilms Microbiomes 2:3

Brennan CA, Garrett WS (2019) Fusobacterium nucleatum – symbiont, opportunist and oncobacterium. Nat Rev Microbiol 17(3):156–166

Buret AG, Motta JP, Allain T, Ferraz J, Wallace JL (2019) Pathobiont release from dysbiotic gut microbiota biofilms in intestinal inflammatory diseases: a role for iron? J Biomed Sci 26(1):1

Burmølle M, Ren D, Bjarnsholt T, Sørensen SJ (2014) Interactions in multispecies biofilms: do they actually matter? Trends Microbiol 22:84–91

Byrd AL, Belkaid Y, Segre JA (2018) The human skin microbiome. Nat Rev Microbiol 16 (3):143–155

Chandki R, Banthia P, Banthia R (2011) Biofilms: a microbial home. J Indian Soc Periodontol 15 (2):111–114

Chassaing B, Darfeuille-Michaud A (2013) The σE pathway is involved in biofilm formation by Crohn's disease-associated adherent-invasive Escherichia coli. J Bacteriol 195(1):76–84

Chiller K, Selkin BA, Murakawa GJ (2001) Skin microflora and bacterial infections of the skin. J Investig Dermatol Symp Proc 6(3):170–174

Choi YC, Morgenroth E (2003) Monitoring biofilm detachment under dynamic changes in shear stress using laser-based particle size analysis and mass fractionation. Water Sci Technol 47 (5):69–76

Collado MC, Rautava S, Aakko J, Isolauri E, Salminen S (2016) Human gut colonisation may be initiated in utero by distinct microbial communities in the placenta and amniotic fluid. Sci Rep 6:23129

Costello EK, Lauber CL, Hamady M, Fierer N, Gordon JI, Knight R (2009) Bacterial community variation in human body habitats across space and time. Science 326(5960):1694–1697. https://doi.org/10.1126/science.1177486

Costerton JW, Lewandowski Z, Caldwell DE, Korber DR, Lappin-Scott HM (1995) Microbial biofilms. Annu Rev Microbiol 49:711–745

Davey ME, O'Toole GA (2000) Microbial biofilm: from ecology to molecular genetics. Microbiol Mol Biol Rev 64:847–867

De Vos WM, de Vos EAJ (2012) Role of the intestinal microbiome in health and disease: from correlation to causation. Nutr Rev 70:S45–S56

Delcaru C, Alexandru I, Podgoreanu P (2016) Microbial biofilms in urinary tract infections and prostatitis: etiology, pathogenicity, and combating strategies. Pathogens 5(4):65

Dominguez-Bello MG, Costello EK, Contreras M, Magris M, Hidalgo G, Fierer N, Knight R (2010) Delivery mode shapes the acquisition and structure of the initial microbiota across multiple body habitats in newborns. Proc Natl Acad Sci U S A 107:11971–11975

Eckburg PB, Bik EM, Bernstein CN, Purdom E, Dethlefsen L, Sargent M, Gill SR, Nelson KE, Relman DA (2005) Diversity of the human intestinal microbial flora. Science 308:1635–1638

Edlund A, Yang Y, Hall AP, Guo L, Lux R, He X, Nelson KE, Nealson KH, Yooseph S, Shi W, McLean JS (2013) An in vitro biofilm model system maintaining a highly reproducible species and metabolic diversity approaching that of the human oral microbiome. Microbiome 1:25. https://doi.org/10.1186/2049-2618-1-25

Erb-Downward JR, Thompson DL, Han MK, Freeman CM, McCloskey L, Schmidt LA, Young VB, Toews GB, Curtis JL, Sundaram B, Martinez FJ, Huffnagle GB (2011) Analysis of the lung microbiome in the "healthy" smoker and in COPD. PLoS One 6(2):e16384

Fagö-Olsen H, Dines LM, Sørensen CH, Jensen A (2019) The adenoids but not the palatine tonsils serve as a reservoir for bacteria associated with secretory otitis media in small children. mSystems 4(1):e00169–e00118

Feuerstein R, Forde AJ, Lohrmann F, Kolter J, Ramirez NJ, Zimmermann J, Gomez de Agüero M, Henneke P (2020) Resident macrophages acquire innate immune memory in staphylococcal skin infection. elife 9:e55602

Fux CA, Wilson S, Stoodley P (2004) Detachment characteristics and oxacillin resistance of Staphylococcus aureus biofilm emboli in an in vitro catheter infection model. J Bacteriol 186 (14):4486–4491

Gannesen AV, Lesouhaitier O, Racine P-J, Barreau M, Netrusov AI, Plakunov VK, Feuilloley MGJ (2018) Regulation of monospecies and mixed biofilms formation of skin Staphylococcus aureus and Cutibacterium acnes by human natriuretic peptides. Front Microbiol 9:2912

Gnanadhas DP, Elango M, Janardhanraj S, Srinandan CS, Datey A, Strugnell RA, Gopalan J, Chakravortty D (2015) Successful treatment of biofilm infections using shock waves combined with antibiotic therapy. Sci Rep 5:17440

Grice EA, Kong HH, Conlan S, Deming CB, Davis J, Young AC, NISC Comparative Sequencing Program, Bouffard GG, Blakesley RW, Murray PR, Green ED, Turner ML, Segre JA (2009) Topographical and temporal diversity of the human skin microbiome. Science (New York, NY) 324(5931):1190–1192

Grigalauskienė R, Slabšinskienė E, Vasiliauskienė I (2015) Biological approach of dental caries management. Stomatologija 17(4):107–112

Hall-Stoodley L, Costerton JW, Stoodley P (2004) Bacterial biofilms: from the natural environment to infectious diseases. Nat Rev Microbiol 2(2):95–108. https://doi.org/10.1038/nrmicro821

Hannig C, Hannig M (2009) The oral cavity—a key system to understand substratum-dependent bioadhesion on solid surfaces in man. Clin Oral Investig 13(2):123–139

Hilty M, Burke C, Pedro H, Cardenas P, Bush A, Bossley C, Davies J, Ervine A, Poulter L, Pachter L, Moffatt MF, Cookson WO (2010) Disordered microbial communities in asthmatic airways. PLoS One 5(1):e8578

Hooper LV, Macpherson AJ (2010) Immune adaptations that maintain homeostasis with the intestinal microbiota. Nat Rev Immunol 10:159–169. https://doi.org/10.1038/nri2710

Horikoshi K, Grant WD (1998) Extremophiles: microbial life in extreme environments. Wiley-Liss, New York

Huang R, Li M, Gregory RL (2011) Bacterial interactions in dental biofilm. Virulence 2 (5):435–444

Huttenhower C, Gevers D, Knight R, Abubucker S, Badger JH, Chinwalla AT (2012) Structure, function and diversity of the healthy human microbiome. Nature 486:207–214

Kaiser D, Losick R (1993) How and why bacteria talk to each other. Cell 73:873–885

Kaplan CW, Ma X, Paranjpe A, Jewett A, Lux R, Kinder-Haake S, Shi W (2010) *Fusobacterium nucleatum* outer membrane proteins Fap2 and RadD induce cell death in human lymphocytes. Infect Immun 78(11):4773–4778

Karatan E, Watnick P (2009) Signals, regulatory networks, and materials that build and break bacterial biofilms. Microbiol Mol Biol Rev 73(2):310–347. https://doi.org/10.1128/MMBR.00041-08

Kilian M, Chapple IL, Hannig M, Marsh PD, Meuric V, Pedersen AM, Tonetti MS, Wade WG, Zaura E (2016) The oral microbiome—an update for oral healthcare professionals. Br Dent J 221(10):657–666

Kluytmans J, van Belkum A, Verbrugh H (1997) Nasal carriage of Staphylococcus aureus: epidemiology, underlying mechanisms, and associated risks. Clin Microbiol Rev 10:505–520

Kolenbrander PE, Andersen RN, Blehert DS, Egland PG, Foster JS, Palmer Jr RJ (2002) Communication among oral bacteria. Microbial Mol Biol Rev 66(3):486–505

Kolenbrander PE, Palmer RJ, Periasamy S, Jakubovics NS (2010) Oral multispecies biofilm development and the key role of cell-cell distance. Nat Rev Microbiol 8:471–480

Kriebel K, Hieke C, Müller-Hilke B, Nakata M, Kreikemeyer B (2018) Oral biofilms from symbiotic to pathogenic interactions and associated disease—Connection of periodontitis and rheumatic arthritis by peptidylarginine deiminase. Front Microbiol 9:53

Krzyściak W, Jurczak A, Piątkowski J (2016) The role of human oral microbiome in dental biofilm formation. In: Microbial biofilms—importance and applications. Intech Publishers, London, p 329

Lakshmi Narayanan L, Vaishnavi C (2010) Endodontic microbiology. J Conserv Dent 13 (4):233–239. https://doi.org/10.4103/0972-0707.73386
Lihoreau T, Agache P (2017) The skin surface ecosystem: a presentation. In: Humbert P et al (eds) Agache's measuring the skin. Springer, Cham. https://doi.org/10.1007/978-3-319-32383-1_4
Macfarlane S, Macfarlane GT (2006) Composition and metabolic activities of bacterial biofilms colonizing food residues in the human gut. Appl Environ Microbiol 72:6204–6211
Mah TF, O'Toole GA (2001) Mechanisms of biofilm resistance to antimicrobial agents. Trends Microbiol 9(1):34–39
Marsh PD, Zaura E (2017) Dental biofilm: ecological interactions in health and disease. J Clin Periodontol 44:S12–S22
Martin R, Makino H, Cetinyurek Yavuz A, Ben-Amor K, Roelofs M, Ishikawa E, Kubota H, Swinkels S, Sakai T, Oishi K, Kushiro A, Knol J (2016) Early-life events, including mode of delivery and type of feeding, siblings and gender, shape the developing gut microbiota. PLoS One 11(6):e0158498. https://doi.org/10.1371/journal.pone.0158498. PMID: 27362264; PMCID: PMC4928817
Mosaddad SA, Tahmasebi E, Yazdanian A, Rezvani MB, Seifalian A, Yazdanian M, Tebyanian H (2019) Oral microbial biofilms: an update. Eur J Clin Microbiol Infect Dis 38(11):2005–2019
Prescott SL, Larcombe DL, Logan AC, West C, Burks W, Caraballo L, Levin M, Etten EV, Horwitz P, Kozyrskyj A, Campbell DE (2017) The skin microbiome: impact of modern environments on skin ecology, barrier integrity, and systemic immune programming. World Allergy Organ J 10(1):29
Qin J, Li R, Raes J, Arumugam M, Burgdorf KS, Manichanh C, Nielsen T, Pons N, Levenez F, Yamada T, Mende DR, Li J, Xu J, Li S, Li D, Cao J, Wang B, Liang H, Zheng H, Xie Y, Tap J, Lepage P, Bertalan M, Batto JM, Hansen T, Le Paslier D, Linneberg A, Nielsen HB, Pelletier E, Renault P, Sicheritz-Ponten T, Turner K, Zhu H, Yu C, Li S, Jian M, Zhou Y, Li Y, Zhang X, Li S, Qin N, Yang H, Wang J, Brunak S, Doré J, Guarner F, Kristiansen K, Pedersen O, Parkhill J, Weissenbach J, Bork P, Ehrlich SD, Wang J, Qin J, Li R, Raes J, Arumugam M, Burgdorf KS, Manichanh C, MetaHIT Consortium (2010) A human gut microbial gene catalogue established by metagenomic sequencing. Nature 464:59–65
Ravel J, Gajer P, Abdo Z, Schneider GM, Koenig SS, McCulle SL, Karlebach S, Gorle R, Russell J, Tacket CO, Brotman RM, Davis CC, Ault K, Peralta L, Forney LJ (2011) Vaginal microbiome of reproductive-age women. Proc Natl Acad Sci U S A 108(Suppl 1):4680–4687. https://doi.org/10.1073/pnas.1002611107. PMID: 20534435; PMCID: PMC3063603
Rayner MG, Zhang Y, Gorry MC, Chen Y, Post JC, Ehrlich GD (1998) Evidence of bacterial metabolic activity in culture-negative otitis media with effusion. JAMA 279:296–299
Røder HL, Sørensen SJ, Burmølle M (2016) Studying bacterial multispecies biofilms: where to start? Trends Microbiol 24(6):503–513
Rotimi VO, Duerden BI (1981) The development of the bacterial flora in normal neonates. J Med Microbiol 14(1):51–62
SanMiguel A, Grice EA (2015) Interactions between host factors and the skin microbiome. Cell Mol Life Sci 72(8):1499–1515
Shapiro JA, Dworkin M (1997) Bacteria as multicellular organisms. Oxford University Press Inc., New York
Sintim HO, Gürsoy UK (2016) Biofilms as "connectors" for oral and systems medicine: a new opportunity for biomarkers, molecular targets, and bacterial eradication. OMICS 20(1):3–11
Sommer F, Bäckhed F (2013) The gut microbiota-masters of host development and physiology. Nat Rev Microbiol 11:227–238
Suntharalingam P, Cvitkovitch DG (2005) Quorum sensing in streptococcal biofilm formation. Trends Microbiol 13(1):3–6
Swidsinski A, Schlien P, Pernthaler A, Gottschalk U, Bärlehner E, Decker G, Swidsinski S, Strassburg J, Loening-Baucke V, Hoffmann U, Seehofer D, Hale LP, Lochs H (2005) Bacterial biofilm within diseased pancreatic and biliary tracts. Gut 54(3):388–395

Tagami H (2008) Location-related differences in structure and function of the stratum corneum with special emphasis on those of the facial skin. Int J Cosmet Sci 30(6):413–434

van de Wijgert JH, Borgdorff H, Verhelst R, Crucitti T, Francis S, Verstraelen H, Jespers V (2014) The vaginal microbiota: what have we learned after a decade of molecular characterization? PLoS One 9(8):e105998

Watanabe H, Goto S, Mori H, Higashi K, Hosomichi K, Aizawa N, Takahashi N, Tsuchida M, Suzuki Y, Yamada T, Horii A, Inoue I, Kurokawa K, Narita I (2017) Comprehensive microbiome analysis of tonsillar crypts in IgA nephropathy. Nephrol Dial Transplant 32 (12):2072–2079

Wessel AK, Arshad TA, Fitzpatrick M, Connell JL, Bonnecaze RT, Shear JB et al (2014) Oxygen limitation within a bacterial aggregate. MBio 5(2):e00992–e00914. https://doi.org/10.1128/mBio.00992-14

Wyss C (1992) Growth of Porphyromonas gingivalis, Treponema denticola, T. pectinovorum, T. socranskii, and T. vincentii in a chemically defined medium. J Clin Microbiol 30 (9):2225–2229

Ymele-Leki P, Ross JM (2007) Erosion from Staphylococcus aureus biofilms grown under physiologically relevant fluid shear forces yields bacterial cells with reduced avidity to collagen. Appl Environ Microbiol 73(6):1834–1841

Yoshida A, Kuramitsu HK (2002) Streptococcus mutans biofilm formation: utilization of a gtfB promoter-green fluorescent protein (PgtfB::gfp) construct to monitor development. Microbiology (Reading) 148(Pt 11):3385–3394

Zarco MF, Vess TJ, Ginsburg GS (2012) The oral microbiome in health and disease and the potential impact on personalized dental medicine. Oral Dis 18:109–1200

Zoetendal EG, von Wright A, Vilpponen-Salmela T, Ben-Amor K, Akkermans AD, de Vos WM (2002) Mucosa-associated bacteria in the human gastrointestinal tract are uniformly distributed along the colon and differ from the community recovered from feces. Appl Environ Microbiol 68:3401–3407

Zoetendal EG, Raes J, van den Bogert B, Arumugam M, Booijink CCGM, Troost FJ et al (2012) The human small intestinal microbiota is driven by rapid uptake and conversion of simple carbohydrates. ISME J 6:1415–1426. https://doi.org/10.1038/ismej.2011.212

Biofilms and Acute and Chronic Infections

4

Moupriya Nag, Dibyajit Lahiri, Dipro Mukherjee, Sayantani Garai, Ritwik Banerjee, and Rina Rani Ray

Abstract

Bacterial biofilm plays a major role in causing diseases as the indwelling bacteria within the biofilm are resilient to host immune system and various types of antibiotics. Various chronic infections of clinical origin such as cystic fibrosis (CF), urinary tract infections (UTI), pertussis, otitis media (OM), bacterial prostatitis (BP), mastitis, etc. are the types of biofilm-associated chronic infections. It is observed that the microorganisms involved in the abovementioned diseases can also cause acute infections in biological environments. Thus deciphering the transition from acute to chronic infections by understanding the bacterial strategy, mode of growth and spread within host will serve the purpose of solving the unanswered question: how the same microbe can cause acute infections in some environmental setting while chronic in others. This chapter tries to bridge the gap by understanding the host responses towards acute and chronic infections, recurrence of infections and development of various treatment strategies.

M. Nag · D. Lahiri · D. Mukherjee · S. Garai · R. Banerjee
Department of Biotechnology, University of Engineering & Management, Kolkata, West Bengal, India

R. R. Ray (✉)
Department of Biotechnology, Maulana Abul Kalam Azad University of Technology, West Bengal, Kolkata, West Bengal, India

R. R. Ray et al. (eds.), *Biofilm-Mediated Diseases: Causes and Controls*,
https://doi.org/10.1007/978-981-16-0745-5_4

4.1 Introduction

Biofilms are consortia of microbial communities adhered to various surfaces and are abundant in all environments. They possess positive impact in a wide variety of ecosystems (Geesey et al. 1977, 1978; Henrici 1933; Zobell and Allen 1935) and also play a major role in bioremediation (Mitra and Mukhopadhyay 2016) and heavy metal toxicity removal (Mishra and Malik 2014). On the other hand, they also possess negative impact such as biofilm-related infections in clinical settings (Vestby et al. 2020; Donlan and Costerton 2002). Microbial infections can be classified as acute and chronic type based on the time period from the onset of infection. Acute infection refers to the survival of microorganisms inside a host for a confined time period, such as few weeks or months, whereas long-time persisting microbes within a host lead to chronic infections (Furukawa et al. 2006). Acute infections predispose a person to the onset of chronic diseases if those microbial pathogens form a biofilm that will remain throughout their lives as in the case of cystic fibrosis caused by *P. aeruginosa* (Holsclaw 1980). Biofilm formation is the first step towards defence and survival mechanism by the indwelling microbial species against the antimicrobials. Inside biofilm, bacteria adapt to stress conditions such as restrained nutrient supply and extreme environmental conditions by exhibiting a modified metabolic activity, gene expression and protein production, resulting into a reduced rate of cell division and metabolism. Biofilm formation helps the microorganisms to remain hidden from the host immune system and may result into local tissue damage causing an acute infection at later instant of time. It has been observed that there exists a direct correlation between biofilm development and persistent infections in tissue-related and device-related cases (Lebeaux et al. 2014). Widespread use of noninvasive and invasive medical devices such as bandages, incision drapes, conductive gels intravascular catheters, contact lenses, pacemakers and dental and orthopaedic implants has led to the adhesion and colonization of microorganisms and results into biofilm-related infections. Biofilm formation takes place in four different steps including initial reversible attachment, irreversible attachment, maturation and dispersion into new places for spread of infection (Fig. 4.1).

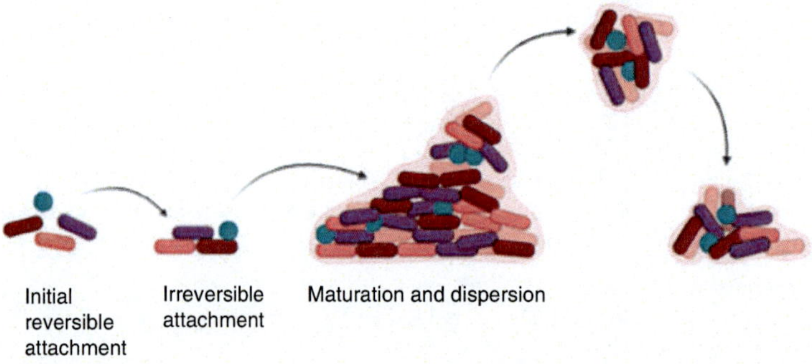

Fig. 4.1 Various steps in biofilm formation

Free-floating planktonic form of bacteria helps in spreading the infection by mixing with the bloodstream or around the source of the infection (Lewis 2007) (Lewis 2001). Planktonic bacteria can be easily eradicated by the use of antimicrobials and host immune responses or combined effect of both the processes, whereas bacteria surviving within the biofilm are highly tolerant against the host immune responses and antimicrobials frequently and can cause recurrence of infection. The only efficient way to eradicate a biofilm-related infection (Costerton et al. 1999) is to completely remove the colonized implanted device or surgical excision of infected tissue. This chapter sheds light on the transition of acute to chronic infections caused due to biofilm formation and the host responses towards the recurrence and persistent infections along with the reasons for treatment failures.

4.2 Biofilm in Acute Infections

Recent study points to the fact that distinct molecular mechanisms exist within the microorganisms while interacting with a host in order to cause acute or chronic infections. It has also been observed that a single type of microbe is capable of causing both chronic and acute infections via transition from planktonic to biofilm form (Furukawa et al. 2006) (Fig. 4.2).

For example, conglomeration of *P. aeruginosa* cells in the paranasal sinuses acts as reservoirs for iterative lung infections which eventually lead to chronic infections such as cystic fibrosis (CF) (Baltimore et al. 1989; Høiby et al. 2010; Lam et al. 1980; Singh et al. 2000). *P. aeruginosa* possesses type III secretion system (TTSS) and is responsible for production of a plethora of extracellular toxins in its planktonic state playing a pivotal role in acute infections such as pneumonia. Initial steps

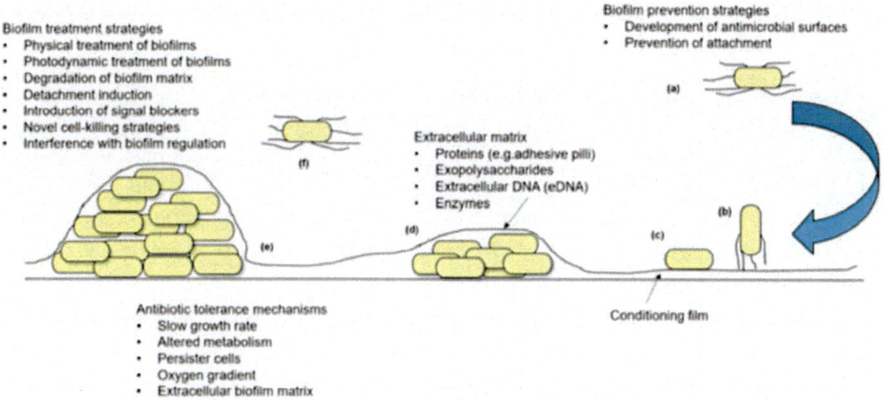

Fig. 4.2 Biofilm formation is behind the change of acute to chronic infection. The acute infection causing pathogen are mostly planktonic cells (**a**), reversible and (**b**) irreversible adherence to the substratum (**c**), microcolony formation via cell division and extracellular matrix production (**d**) and establishment of a mature three-dimensional biofilm structure with persisting cells (**e**) the cells actively disintegrating from the biofilm (**f**) for searching a new place in the body to form new biofilm

Table 4.1 Some examples of acute infections sites due to device implants that may lead to chronic infections if not cured

Device-related infection	Infection-causing microbe	Nature of species	References
Ventricular derivation	*S. epidermidis* *S. aureus*	Gram-positive bacteria	Ochieng' et al. (2015)
Contact lenses	*S. aureus*, *P. aeruginosa*, *S. marcescens*, *F. solani*, *Acanthamoeba*	Gram-positive and Gram-negative, both types of bacteria	Lagina (2016)
Endotracheal tubes	*E. faecalis*, *E. faecium*, *S. aureus*, *Klebsiella species*, *Stenotrophomonas maltophilia*, *P. aeruginosa*, *Haemophilus influenzae*	Gram-positive and Gram-negative, both types of bacteria	Thorarinsdottir et al. (2020)
Central vascular catheters	Coagulase-negative staphylococci (CoNS) or *Enterobacteriaceae*	Gram-negative bacteria	Gominet et al. (2017)
Peripheral vascular catheters	*Klebsiella pneumoniae* and *Candida* spp.	Gram-negative bacteria and fungi	Singhai et al. (2012)
Prosthetic cardiac valves, pacemakers, vascular grafts	*S. epidermidis*, *Propionibacterium acnes*	Gram-positive bacteria	Donlan and Costerton (2002); Okuda et al. (2018)
Breast implant	*Staphylococcus* spp.	Gram-positive bacteria	Hu et al. (2016)
Urinary catheters	*Klebsiella* sp. and *Enterococcus faecalis*, *E. coli*, *Pseudomonas aeruginosa*	Gram-positive and Gram-negative, both types of bacteria	Costerton et al. (1995)
Prosthetic joints and orthopaedic implants	*S. epidermis*, *S. aureus* and *Pseudomonas aeruginosa*	Gram-positive and Gram-negative, both types of bacteria	McConoughey et al. (2014)
Acute osteomyelitis	*Staphylococcus aureus*	Gram-positive bacteria	Brady et al. (2008)

include crossing the lung defence system followed by dispersion in bloodstream causing fatality within hours or days on first contact (Barbieri and Sun 2004; Ghosh 2004; Matsumoto 2004).

Once inside the host system, it is the expression of various factors that determine the type of infection, i.e., acute or persistent. For example, in case of an implant-based infection, insertion of an implant such as urinary catheters will drive the microbial response to specific host signals (host immune status, host nutrient status or tissue integrity) on entering the catheter insertion site. Some of the implant-based infections are listed in Table 4.1.

A highly contagious disease pertussis or better known as whooping cough is an acute infection caused by *Bordetella pertussis* or sometimes by *B. parapertussis*

(Melvin et al. 2014). Irie et al. (2004) studied the transition from acute to persistent infection in case of whooping cough caused by *Bordetella bronchiseptica*. The virulence and biofilm formation are governed by a two-component signal transduction system BvgAS (*Bordetella* virulence gene). It regulates three different growth phases of *B. bronchiseptica*: virulent (Bvg^+) phase, an intermediate (Bvg^i) phase and a nonvirulent (Bvg^-) phase (Locht et al. 2001; Mattoo et al. 2001). Bvg^+ and Bvg^i phase is dominated by the production of virulence factors such as fimbriae and filamentous haemagglutinin (FHA) (Cotter et al. 1998). Maximum biofilm is formed in the Bvg^i phase (Irie et al. 2004). Altogether it is seen that acute infections are followed by release of toxin molecules as opposed to biofilm formation in chronic infections. In another work by Resch et al. (2005), virulence factors such as proteases and toxins are found to be upregulated in planktonic cells of *S. aureus* for inducing acute infections, whereas virulence factors are downregulated in biofilm cells (Resch et al. 2005; Furukawa et al. 2006).

A common infection in children within the age group 3 months to 3 years (Auinger et al. 2003; DeAntonio et al. 2016; Schilder et al. 2016) is otitis media (OM) caused due to inflammation of the middle ear cavity by bacteria such as *H. influenzae*, *Moraxella catarrhalis*, S *pneumonia* and *Streptococcus pyogenes*. The infection can be subdivided into OM with effusion (OME), chronic supportive OM (CSOM) and acute OM (AOM) (Schilder et al. 2016). During AOM, it is proposed that bacteria from the nasopharynx can detach from its biofilms and move to the middle ear causing an acute infection (Coticchia et al. 2013).

In another infection of the prostate gland, namely, acute bacterial prostatitis (ABP) that causes urinary tract infection (UTI) in addition to pain in the pelvic and genital region. If remained untreated, it may lead to bladder infections, urosepsis, reduced fertility and death. ABP can also lead to chronic prostatitis (CP) or chronic pelvic pain syndrome (CPPS), if symptoms prevail for more than 3 months (Yoon et al. 2012, 2013). ABP is mainly caused by *Proteus mirabilis*, *E. faecalis*, *E. coli*, *Klebsiella* spp., *P. aeruginosa* and *Enterobacteriaceae*. Route of entry of bacteria causing UTIs involves the rectum, urethra, bladder, ureters and finally kidneys. Though our body possesses a clearance mechanism to remove UTI on its own within a few days of infection or following a short antibiotic course, its rate of recurrence is high (Liu et al. 2016; Olson and Hunstad 2016; Lakeman and Roovers 2016; Scott et al. 2015; Robino et al. 2013; Anderson et al. 2004).

Another example of acute infection is the foodborne disease typhoid. It is caused by *Salmonella typhi* with symptoms such as weakness, high fever, abdominal pain, headache and constipation.

Mastitis is an acute infection of the mammary gland resulting into its inflammation of lactiferous ducts or lumen of alveoli during lactation period and is primarily caused by *S. aureus*, *E. coli*, *S. agalactiae*, *Enterobacter* spp., *enterococci*, *Klebsiella* spp., *S. dysgalactiae*, *S. uberis*, *Citrobacter* spp. and *Pseudomonas* spp. (Hensen et al. 2000).

Biofilms of opportunistic pathogen *Pseudomonas aeruginosa* have also been observed to form within 8 h of infection in burn wounds of patients, suggesting bacterial colonization in acute infections (Pruitt et al. 1998; Tredget et al. 2004; Schaber et al. 2007). Post-burn infection by *P. aeruginosa* is found even after 24 h in the blood, liver and spleen of mice and observed that >90% of mice die within 48 h due to post-burn infections (Rumbaugh et al. 1999). Researchers also found the presence of bacterial colonies around blood vessels causing invasion and necrosis of blood vessels (Soave et al. 1978; Schaber et al. 2007).

Addition of a foreign material such as orthopaedic implants to a living world provides a golden opportunity for the adherence and colonization of microorganisms (Elek 1956; Zimmerli et al. 1982; Steckelberg and Osmon 2000; Saeed 2019). Moreover, regular movement of the implants within the body releases debris causing local inflammation of the tissues which provides a prosperous site for microbial infections (Arciola et al. 2005; Pulido et al. 2008; Darouiche 2004). For example, in the case of osteomyelitis, rate of microbial reinfection following orthopaedic surgery is found to be alarmingly high. Recent studies suggest reservoirs of bacterial biofilms of *S. abscess*, *Pseudomonas* sp., *Enterococcus* sp. and *Streptococcus* sp. in the local soft tissue and bone marrow and colonization of the osteocyte lacuno-canalicular network (OLCN) of cortical bone. Infection of the bone can be of two types: by endogenous seeding (haematogenous osteomyelitis) (Peltola et al. 2010) or by exogenous seeding (fracture site contamination during implantation). Osteomyelitis can be classified as subacute, acute (recent bone infection that causes systemic inflammation) or chronic stages based on disease severity (Kronig et al. 2015). Bone infection of longer duration comes under the category of chronic osteomyelitis, with minimal systemic symptoms and presence of pathological features, such as marrow fibrosis, lymphoplasmacytic in filtrate and reactive new bone formation (Masters et al. 2019).

4.3 Chronic Infections Persistency

A syntrophic consortium of microorganisms in which cells stick to each other and often to a surface, promoting bacterial persistence, is defined as a biofilm (Costerton et al. 1999). These multicellular communities in the biofilms develop a slippery coat or a matrix material that reduces the interaction between antimicrobial agents and the organisms, promoting resistance to drugs and hence become persistent (Fux et al. 2005). This polymeric material is composed primarily of polysaccharides, nucleic acids and/or amino acids (Flemming et al. 2007).

In the recent years, the development of vaccines and antimicrobial drugs has relieved the mankind from deadly epidemic diseases caused by free-moving bacterial pathogenic cells like *V. cholerae* and *Y. pestis*. Antibiotics play an important role in controlling and curing such infectious diseases in today's world, provided the microorganism is not drug-resistant or forms biofilms. In almost 50% of the cases, the cells that show commensalism or high frequency in our environment are responsible for the infections that affect the human body to mild extents (Costerton et al. 1999). Table 4.2 shows some of the major biofilm-forming pathogens causing infections in human. Microscopic studies on samples from human body, having

Table 4.2 Bacterial biofilm-associated chronic infections

Infection	Bacterial species	References
Persistence of infections within the urinary tract- and urinary catheter-associated infections. It also causes biliary tract infections	*Escherichia coli*	Römling and Balsalobre (2012); Lynch and Robertson (2008)
Cystic fibrosis pneumonia, Contact lens-associated infections, surface and other wound associated infections, urinary tract infections by the persistent organism developed upon the urinary tract, rhinosinusitis and development of chronic otitis media	*Pseudomonas aeruginosa*	Costerton et al. (1999); Høiby et al. (2010)
Orthopaedic implants, osteomyelitis, rhinosinusitis, endocarditis, otitis media	*Staphylococcus aureus*	Lynch and Robertson (2008)
Ventilator-assisted pneumonia	*K. pneumoniae, E. coli, P. aeruginosa*	Römling and Balsalobre (2012)
Central venous catheter, orthopaedic implants, osteomyelitis	*Staphylococcus epidermidis*	Lynch and Robertson (2008)
Congestion of the nasopharynx, chronic rhinosinusitis, chronic otitis media, dental carries	*Streptococcus pneumoniae*	Römling and Balsalobre (2012)
Periodontitis	*A. actinomycetemcomitans, P. gingivalis*	Römling and Balsalobre (2012)
Development of biofilm within the oral cavity, recurrent tonsillitis and within the nasopharynx	*Streptococcus pyogenes*	Römling and Balsalobre (2012)

persistent infections, have revealed the presence of biofilm-producing bacteria, covered with a layer of extracellular polymeric substance (EPS) (Khoury et al. 1992). The production of antibodies is triggered in our body by the accumulation of the sessile cells. However, these antibody molecules are not quite effective in killing the biofilm cells and may cause damage to the adjacent tissues or cells (Cochrane et al. 1988). Even in individuals having excellent humoral and cyto-immunity, infections involving biofilms are seldom resolved by host defence techniques and show recurring symptoms of antibiotic therapy, unless the biofilm-forming cells are removed from the host body completely (Costerton et al. 1999; Khoury et al. 1992). Thus, biofilms are breeding grounds of persistent infection unless the host defence system eliminates the planktonic cells secreted at any point during the disease (Dasgupta et al. 1988).

A study conducted at the National Institutes of Health (NIH), USA, shows the involvement of biofilm-causing microorganisms in human infections, up to 85% (Römling and Balsalobre 2012). Common infections like dental caries and periodontitis is caused by biofilm bacteria. Gram-positive pathogenic bacteria like *S. epididymis*, *S. aureus* and acidogenic *Streptococcus* sp. (Costerton et al. 1999) and Gram-negative *P. aeruginosa* and *Enterobacteriaceae* like *E. coli* are known to cause disease in human.

4.4 Biofilms and Drug Resistance

The biofilms developing in nature are usually resilient to the bacteriophages, amoeba and the industrial anti-biofoulants (Costerton et al. 1987). From the point of view of medicines, these sessile bacterial cells can resist host defence mechanisms and are less exposed to the antimicrobial drug molecules as compared to the planktonic cells of the same species (Nickel et al. 1985).

One of the ways how biofilms form resistance to the chemical molecules is by preventing the penetration of the molecules into them. These bactericides are unable to protrude the external slimy extra-polysaccharide layer to the full extent, since this polymeric layer is reported to reduce the rates of diffusion of the antibiotic molecules (Cheema et al. 1986). Other solutes, too, diffuse in lower rates in this EPS layer than they do in aqueous solution (Stewart 2002). Antimicrobial agents, like oxidants hypochlorite and hydrogen peroxides, are neutralized in the outer layers of the biofilm faster than it diffuses. These molecules are produced by the burst of phagocytotic cells by oxidation and the faulty uptake of reactive oxygen species, which causes the incapability of the phage cells to engulf the organisms in biofilms. Thus, various mathematical models are used by scientists to predict the strong penetration barrier that causes the drug inactivation (Kostakioti et al. 2013).

Scientists have come up with other theories of reduced biofilm susceptibility. The next hypothesis explains that reduced diffusion may be a reason of the quorum sensing property of the biofilms. In other words, at least few of the cells in the biofilms undergo nutrient restriction and grow in a slow or in a starved condition (Brown et al. 1988). The uneven distribution of various populations of same species within an area is also determined by micro-slicing and microscopic methods (Kinniment et al. 1996). This diverse population of microorganisms within a biofilm helps them to survive any host metabolic defence attacks, since the cells have varied metabolic rates.

Diagnosis of biofilm infections requires the adequate treatment strategies, since the conventional methods often fail in targeting the causative agents. Some of the criteria are devised by Parsek and Singh (Parsek 2015) for proper diagnosis of the biofilms. The devised conditions include the following: (a) resilient to antimicrobial molecules, (b) presence of a localized disease with the infection site having a collection of bacteria and (c) ineffective defence immune reactions in the host body. The problems in poor diagnosis and the correct detection of the biofilm pathogens can be overcome by proper preparation of sample and using latest molecular and microscopy techniques.

4.5 *Pseudomonas aeruginosa* Biofilms in Cystic Fibrosis Lung Infections

The loss of the cystic fibrosis transmembrane regulator (CFTR) chloride channel of epithelial cells is due to the genetic defect that leads to the bacterial infections to persist in the lungs. *P. aeruginosa* is responsible for the lung damage caused by the recurrent bacterial infection in patients, and they succumb to it, with an average life expectancy of approximately 30 years. Scientists have explained the pathogenesis of the lung in cases of cystic fibrosis in many ways (Joris et al. 1991), of which some are not mutually exclusive. One of the views is the blockage of the chloride ion channel that rises the salt content in the surface fluid passage and inhibits the activity of the bactericidal protein involved the passage immunity. This changes the balance of power within the cells in such a way that is just enough for the *P. aeruginosa* cells to colonize in the epithelium layer. The sessile cell communities secrete antigens as they thrive in the microcolonies in the lung. Studies have shown large concentration of the pathogenic cells circulating in the blood. The secreted antibodies react with specific antigens in the outer layer of the matrices and affect the microcolonies, but the defensive and bactericidal characters of these host molecules have been realised. Recently, the attempts to avoid the early aggregation by the pathogenic cell *P. aeruginosa* in young patients have been successful. In patients affected with the pathogen, antibiotic treatment provides relief from symptoms and pain but doesn't eliminate the biofilm itself (De Beer et al. 1994), and thus the persistent infection remains. Thus, we can say that antibiotics act on the planktonic cells that are shed by the biofilms. This alleviates the severe symptoms of the infection, but the drug-resistant biofilm communities are not eliminated.

Patients suffering from CF mostly associated with the biofilm formed by *P. aeruginosa* show various physiological symptoms and struggle. The organism associated with such disorder is pathogenic and a group of ubiquitous organism that initially develops acute infection followed by maturation and metastasis resulting in the development of chronic infections. The application of antimicrobial chemotherapy can be applied at the acute stage of infection.

The main problem in facing CF patients is that the antibiotics present in the market were developed for the planktonic-type pathogenic cells and the efficacy of the therapeutic agents are also based on the free-moving cells. The development of biofilm within the lungs can be well detected by direct observations and microscopic views. The development of biofilm results in the failures of various antimicrobial chemotherapies that are used for treating lungs-associated infections.

4.6 Biofilms on Chronic Wounds

The specimen obtained from the wounds when microscopically observed showed the presence of densely populated colonies of bacteria encompassed by an extracellular polymeric matrix (James et al. 2007). This observation in the morphology provides evidence of biofilms on the wounds. The biofilm observed in acute wound infections varies from the chronic wounds. Various tests have inferred that the biofilm is more predominant within chronic wounds in comparison to the acute ones. The biofilm forms consortia of polymicrobial organisms, and they are mostly anaerobic and are detected by various clinical analyses of wound-associated infections (Bowler et al. 2001). Various molecular techniques have been employed in detecting the anaerobic genera which are persistent in these chronic infections. The most common types of bacterial cells associated with the wounds are mostly Gram-positive in nature like *Staphylococcus* and *Enterococcus*. It has been observed that the predominance of *Staphylococcus aureus* causes delayed wound healing, but still there is a dearth of clear conception on the relation of *S. aureus* with the wound infections (Gardner and West 2004).

4.7 Endocarditis

The bacterial interaction with endothelium results in the development of endocarditis. The initial association of the bacterial cells is weak and the opportunistic bacteria forms a strong biofilm, damaging the heart valves, with any new development of wound (Kokare et al. 2009). The leaking of the heart valves is the first sign of manifestation of the infection, i.e., caused by bacterial population unsettling its normal functioning, causing the infections in blood, leading to persistent fever, severe inflammation and other complications (Kokare et al. 2009). In few cases the population is ruptured into several smaller parts, and they are carried by the circulatory system to various parts of the body like the brain, kidneys, etc. *Pneumococci*, *Staphylococcus*, *Streptococcus* and few other Gram-negative bacteria are some of the microorganisms associated with this infection. A protein that is secreted by the endothelial cells has the ability to bind to collagen, human cells and microorganisms (Banerjee et al. 2020). The formation of the biofilm within the heart valves and damaged tissues by *Staphylococcus* and *Streptococcus species* is associated with fibronectin protein.

4.8 Periodontitis

It is the infection of the soft tissues and bones in our mouth that supports the teeth. It is commonly known as "gum disease", caused by *Fusobacterium nucleatum* and *Porphyromonas gingivalis.* The colonization of the bacterial cells takes place upon the surface of the mucosa within the oral cavity (Kokare et al. 2009). The colonization of the bacterial cells results in the alteration of calcium flux, invasion of the mucosal cells and the release of various toxins. It has been further observed that the development of plaque occurs within a time span of 2–3 days resulting in the development of various bacterial infections. Although the saliva possesses bactericidal properties, it is unable to invade within the biofilm and results in the persistence of the infection.

4.9 Chronic Infections: Host Response

Biofilm infections are one of the key problems in the arena of modern medicine and health research. Bacterial biofilms that produce chronic infections are known to be the underlying cause of many modern-age deadly bacterial diseases, resulting in morbidity and increased mortality rates. Biofilms are consortia of more than one microbial community, and this heterogeneity in the population is known to aid its special properties compared to planktonic species, like development of resistance towards antibiotics. The transition of the bacterial cells from the planktonic to sessile forms occurs by the production of small molecules that protect them from the immunological responses of the host organism (Watters et al. 2016). Host systems, i.e., the organism affected by the pathogenic bacteria, have developed barriers and immunological defence mechanism in response to infection since ages. Even though specific immune responses towards biofilms have not been well studied, the failure of the host to completely clear the bacteria from the system is widely observed, which is believed to allow the biofilm infections to become chronic. Unlike the planktonic forms where the system can recognize antigens with relative ease, within the EPS (extracellular polymeric substances) forming communities, such ease of recognition is compromised as the immune cells from the host interact with the components of the EPS comprising diverse, hydrated mixture of extracellular DNA (bacterial and host), proteins, polysaccharides and lipids. Some compounds produced by biofilm-forming species of bacteria are reported to increase inflammation and induce cell death in host (Watters et al. 2016). This increases the ardent and immediate requirement to study the biofilm-associated host responses.

Even though as it seems from the medical standpoint that the host responses are directed against the biofilm as they cause chronic infections in the body, few bacterial species are considered important as they modulate and shape the immune system and help in the development of acquired immunity against other pathogenic strains (Russell et al. 2012). Probiotic biofilms colonize and protect the gut, cause a negligible immune response and, in states of inflammation, exhibit robust anti-inflammatory properties (Watters et al. 2016). These bacteria reside in privileged areas of immune tolerance, and the host defence system recognizes them as commensals (Tlaskalova-Hogenova et al. 2004).

The epithelial barrier, host microbiome and a variety of leukocytes serve as a potent barrier to the invading pathogens including the biofilm-forming infectious bacteria causing chronic illnesses (Watters et al. 2016). However, the host response to chronic biofilm infections is rather complex. The immunogenic response of the host varies greatly with the variation of the various types of biofilm formation and the organs associated with the chronic infections caused by the biofilm. Specific immune responses are generated for specific parts or components of a biofilm in reported cases. Usually, the body's response to any infection triggers any or all of the three components of the immune system, namely, the physical barriers like the skin and respiratory tract, the immune response and the inflammatory response. Among the three, immune response deals with the activation of the immune system, caused by antigens. It comprises both the innate and the acquired immunities which play a significant role on the course of biofilm-associated infections (Moser et al. 2017).

4.9.1 Immune Response in Chronic Biofilm Infections

4.9.1.1 Innate Immune Response in Chronic Biofilm Infections

Innate immunity has the ability to respond and epitomize the germline-associated cellular components and non-clonal components of the immune system that provides protection against non-specific pathogens influenced by repetitive encounters with infectious intruders (Kimbrell and Beutler 2001). The humoral component mainly consists of the cytokine barriers, whereas the cellular component withholds the epithelial cells, polymorphonuclear leukocytes or PMN, neutrophils, etc. Cellular components such as the neutrophils, monocytes and macrophages play a key role in the detection of infection. Endobronchial accumulation of active neutrophils in patients with cystic fibrosis (CF) chronic lung infection is known to be correlated with tissue damage and respiratory bursts (Kolpen et al. 2010; Houston et al. 2013). Neutrophils and various other components of the innate immune system are reported to be associated and activated in chronic biofilm infections (Table 4.3). The response of the neutrophils to *P. aeruginosa* biofilms has been extensively observed in the study of cystic fibrosis (CF) patients. Infected CF respiratory bronchioles house densely aggregated bacteria forming biofilms and a few planktonic bacteria, surrounded by numerous neutrophils (Bjarnsholt et al. 2009). The planktonic cells are readily phagocytosed by the surrounding neutrophils (Kolpen et al. 2010). An anaerobic condition is developed in the CF infected endobronchial mucus which is aided by the response of neutrophils to biofilms during chronic infection (Worlitzsch et al. 2002). In addition to this, the increased glucose uptake by neutrophils in CF lungs (Chen et al. 2006) and the high concentration of L-lactate in sputum from CF patients (Bensel et al. 2011) further substantiate the presence of active neutrophils during chronic infection, which is further proved by the fact that neutrophils that are active get their energy from the aerobic glycolysis and use large amount of oxygen at the time of respiratory burst for the production of large number of reactive oxygen species (ROS) or nitric oxide (Borregaard and Herlin 1982). The immune responses occuring by PMNs produces macrophages that help in phagocytising foreign material and microbes (Tables 4.3 and 4.4).

Table 4.3 Immunological components associated with biofilms

Immunological component	Pathogen	References
Neutrophils and monocytes	*Candida albicans*	Katragkou et al. (2010)
	Streptococcus mutans	Shapira et al. (2000)
	Staphylococcus aureus	Meyle et al. (2010)
Neutrophils and macrophages	*Pseudomonas aeruginosa*	Alhede et al. (2009)

Table 4.4 Components and their functions

Immunological component	Function	References
Polymorphonuclear neutrophils (PMNs)	Phagocyte microbes, release pro-inflammatory cytokines, IL-8, TNF-α and IL-1β, release antimicrobials like defensins, cathelicidins, lysozyme and lactoferrin	Witko-Sarsat et al. (2000)
Macrophages	Help in phagocytising microbes and pathogens entering into the body by the use of PMNs	Witko-Sarsat et al. (2000)

4.9.2 Acquired/Adapted Immune Response in Chronic Biofilm Infections

Unlike the innate immunity, the acquired immunity involves specific response which helps the host system to distinguish between commensals and pathogens. It helps develop the ability of the host to recognize re-exposures of the same pathogenic bacteria which is aided by the memory cells. There is a substantial clonal expansion of antigen-specific cells (100- to 1000-fold increase in numbers on subsequent exposures) and increased numbers of effector cells. This is the basis of development of acquired or adaptive immunity to infections. On subsequent exposures, the response becomes faster and more effective as compared to initial responses, where the innate immunity fails to distinguish between a first exposure and subsequent exposures to pathogens or antigenic substances.

The activation of the acquired immune system is mediated through the action of macrophages and dendritic cells (Banchereau and Steinman 1998). DCs or dendritic cells are pliable antigen-presenting cells which serve as a pivotal link between the innate and acquired immune response (Kalinski et al. 1999). DCs have the unique ability to prime naive T cells into a Th1, Th2 or Th17 response (Banchereau et al. 2000) and have been correlated with the cytokine response to the development of effector functions in cystic fibrosis with chronic *P. aeruginosa* lung infection (Moser et al. 2005). T cells recognize processed antigens presented by such antigen-presenting cells like DCs and macrophages. However, even a highly active and efficient immune system is not able to relieve the host from most chronic biofilm infections. This is mainly because of the release of reactive oxidative radicals and enzymes from inflammatory cells. The host cell release of elastases, proteases and other exoenzymes causes substantial degradation of essential cell surface proteins, which further contribute to the inhibition or loss of effectiveness of most of the immune cells (Horvat and Parmely 1988). In fact, the negative impact of the host on its own anti-biofilm responses by such tissue damage is probably the major cause of the biofilm-related pathology and continues to be the most threatening cause that leads to the development of chronic infections (Doring and Hoiby 1983).

4.9.3 Host Immune Responses to Biofilm Constituents

4.9.3.1 The Immune Response to Exopolysaccharides

The major component that forms the architecture of EPS is exopolysaccharides or extracellular polysaccharides that provide mechanical stability, help in attachment of the cells and absorption of various organic and inorganic compounds as well as prevent the penetration of drugs and antibiotics, thus rendering resistance to the biofilm (Flemming and Wingender 2010). It serves as a potent carbon source in times of nutrient depletion which occurs majorly in areas of chronic infection due to manifestation of glucose by neutrophils as mentioned earlier. Studies have shown that certain bacterial extracellular polysaccharides can inhibit the immune response in host systems. Exopolysaccharides present in the EPS of *Lactobacillus plantarum* biofilms have been reported to decrease the production of the pro-inflammatory cytokines like MCP-1, IL-6 and IL-8 (monocyte chemotactic protein-1 also known as CCL2), by increasing the negative regulators of toll-like receptor-4 (TLR4), an important pattern recognition receptor (PRR) for the detection of pathogen-associated molecular patterns (PAMPs) (Murofushi et al. 2015). Capsular exopolysaccharides from *Salmonella typhi* can also reduce the expression of IL-8 by human intestinal epithelial cells (Raffatellu et al. 2005). In *Burkholderia cepacia* the polysaccharides were reported to inhibit neutrophil chemotaxis and production of reactive oxygen species (ROS). Studies have also shown that alginates being produced from the EPS of the biofilm formed by *P. aeruginosa* in CF possess the ability of killing macrophages and inhibiting the complement activation system, thus enhancing the resistance in *P. aeruginosa* against phagocytosis (Pier et al. 2001) (Fig. 4.3).

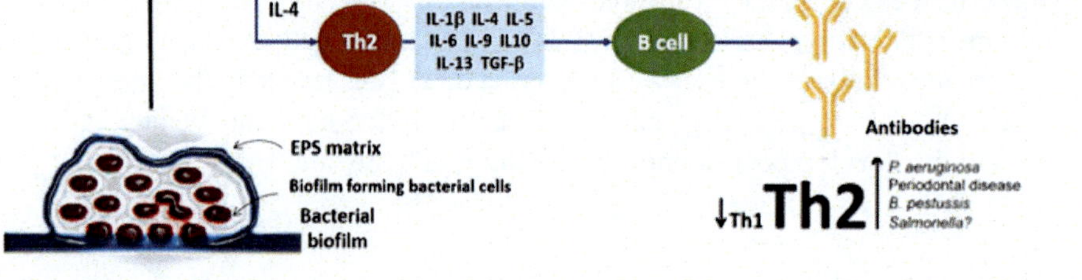

Fig. 4.3 Immunogenic response to the formation of the biofilm

4.9.3.2 Immune Response to Extracellular DNA

Extracellular DNA (eDNA) is known to facilitate horizontal gene transfer (HGT) between microbes which is believed to confer the biofilm community's additional resistance to host immune responses (Okshevsky and Meyer 2015). The bacterial DNA in the EPS is known to be capable of triggering both innate and adaptive immune responses. Oligodeoxynucleotides containing unmethylated 5-′-cytosinephosphate-guanine-3′ (CpG) motifs, present in bacterial DNA, are recognized as PAMPs by TLR9 on the surface of innate immune cells and B cells (Krieg 2002). In vitro experiments in *P. aeruginosa* biofilms showed reduction in the release of the neutrophil pro-inflammatory cytokines, IL-8 and IL-1b on degradation of the eDNA with DNAse1 (Fuxman Bass et al. 2010). The *P. aeruginosa* eDNA has also been observed to have immunomodulatory effects in terms of creating a cation-limited environment by chelation that induces the bacterial expression of resistance genes to cationic antimicrobial peptides (Mulcahy et al. 2008).

4.9.3.3 Immune Response to Exoproteins

Extracellular proteins or exoproteins maintain and modify the EPS by performing enzymatic degradation of structural elements and also acquire nutrition by performing enzymatic digestion of certain macromolecules (Zhang and Bishop 2003). DNA-binding exoproteins play significant roles in conferring physical stability to the biofilm components. Exoproteins are also believed to aid cell-to-cell communication by acting as signalling molecules, and therefore, targeting these proteins holds a potential to hinder biofilm persistence. Proteome study of *S. aureus* biofilms revealed more than 30 different proteins, with various functions (Table 4.5) (Kaplan 2010); more than 20 immunogenic proteins were found in *S. aureus* osteomyelitis biofilms (Brady et al. 2006).

4.9.3.4 Immune Responses to Biofilm-Associated Small Molecules

Small molecules produced by the bacterial biofilms like secondary metabolites, redox-active secondary messengers, antibiotics and cell-to-cell signalling or quorum sensing (QS) molecules (Table 4.6) can potentially affect the host immune response (Enany et al. 2013). N-(3-oxododecanoyl)-homoserine lactone (3-oxo-C12-HSL), a QS molecule made by *P. aeruginosa*, is known to activate the intracellular LasR transcriptional regulator promoting the expression of many different bacterial genes

Table 4.5 Exoproteins and their functions in cell response

Exoproteins	Function	References
Leukocidin and haemolysin	Pathogenesis	Kaplan (2010); Brady et al. (2006)
IgG-binding protein, immunodominant antigens, lipoprotein and protein A	Immunomodulation	
Glyceraldehyde-3-phosphate dehydrogenase, fructose bisphosphate aldolase, alcohol dehydrogenase, phosphoglycerate mutase and enolase	Carbohydrate metabolism	
DNA-binding protein, superoxide dismutase, nuclease and foldase	DNA metabolism	

Table 4.6 QS molecules and their function in host responses

QS and other small molecules	Function in host response	References
N-(3-oxododecanoyl)-homoserine lactone	Attract neutrophils and stimulating the production of the chemokine IL-8; cause apoptosis in fibroblasts, vascular endothelial cells, macrophages and neutrophils, airway epithelial cells and human mesenchymal stem cells	Smith et al. (2001)
Cyclic diguanylate monophosphate (c-di-GMP)	Variation in the concentration of c-di-GMP alters the motility of the bacterial cells; helps in the adhesion of cells, formation of biofilm and control of the cell cycle; helps in metabolism of the cell wall; repairs the expression of genes; regulates various types of virulence factors	Holban et al. (2014)
Cyclic adenosine monophosphate (c-di-AMP)	Help in the mechanism of recognition of PAMPs that in turn help in enhancing the interaction between the host and interferon type I	McWhirter et al. (2009)
c-di-GMP		
Pyocyanin	It possesses the ability of killing the host cells including the immunogenic cells and the cells which are competing with the microbial cells.	Jayaseelan et al. (2014)

and also causes a variety of effects in different types in mammalian cells (Massion et al. 1994). In addition to HSLs, cyclic dinucleotide (c-di-NMPs) and other small molecules like a redox-active, membrane-permeable pigment produced by *P. aeruginosa* have been shown to be involved in various processes.

4.10 Treatments for Biofilm-Associated Diseases

From the conclusions drawn by many studies; it has become crystal clear that almost 65% of all bacterial infections are related with the bacterial biofilms. The infections are mainly of two types: device- and non-device-associated infections (Carek et al. 2001). Device-related infections affect over 2% of the joint replacements, almost around 4% for heart valves, 10% for ventricular shunts and 40% for ventricular-assisted devices.

4.10.1 Current Treatment Approaches

The diseases and infections which are caused by the biofilms are usually progressive in nature (Carek et al. 2001). These may be diagnosed with a combined effect of antibacterial therapies and tissue debridement. In some specific conditions of implant-related infections, diagnosis is different which involves the delivery of high-dose antibiotics in the severity of infection. The persistence of the specified symptoms related to the infection may result in surgical replacement. The development of resistance against the antibiotics by the biofilm occurs due to the usage of higher antibiotic dosages. This also results in the reduction of bacterial mortality.

4.10.2 Reasons Behind Treatment Failure in Device-Related Osteomyelitis

The formation of biofilm on the medical or implanted devices is caused by the virulent feature of the pathogen such as *Staphylococcus epidermidis* and is regarded as an opportunistic pathogen in orthopaedic device-related infection or in ODRI (Berendt and Byren 2004).

Although it is widely known, no definite clinical proof is available (Berendt and Byren 2004). This lack of evidence is linked with the output which is further bonded with the tendency of formation of biofilm by *S. epidermidis* in orthopaedic device-related infection. In several studies, these observations are confirmed.

For example, in some studies, *S. epidermidis* isolates were collected and cultured from patients with orthopaedic device-related infection. Antibiotic resistance patterns and biofilm-forming ability were detected. The medical history and other relevant information about the infected patients were collected, and depending on that, the treatment options were determined (Berendt and Byren 2004). This determination procedure took as long as 26 months. The initial result of the treatment had an univariate logistic regression model that was measured for the determination of the effectiveness of biofilm formation and antibiotic resistance. In several studies, it was found that on every 100 number of infected patients, a majority that is almost 80 of those patients were found to be infected at the lower extremity).

A tendency or a pattern was observed in the lower extremity, and in those sectors, a very low cure rate was found as the biofilm formation capacity of the isolates rises up in the stipulated time span. It was seen that almost 84% cure rate was found in the patients where biofilm did not form, whereas the percentage decreased to 76% and 60% in patients with weak and strong biofilm formation, respectively. The antibiotic resistance however did not influence the cure rate. A tendency of immunosuppression was found with the highly statistical and noticeable low cure rate. The tendency of the high biofilm formation ability which results in decreased cure rate in *Staphylococcus epidermidis* orthopaedic device-related infection further results in the ability of biofilm formation in infecting pathogens.

4.10.3 Reasons Behind Treatment Failure and Recurrence in Vaginal Infections

Biofilm in the vagina is one of the most important biofilm-related problems in women. The microbial cells within biofilm are capable of decreasing and also increasing the susceptibility of the antimicrobial agents. The biofilms which are related to the vagina-oriented bacterial and fungal microbes have a very significant role in infectious diseases, specifically depending on the parameters such as persistence and recurrence. In the last few years, biofilms are also located during the treatment failures of vaginal infections, specifically vulvovaginal candidiasis (VVC) and bacterial vaginosis (BV), particularly in the setting of treatment failure and recurrence. The impact of biofilms on the management and treatment of BV and recurrent VVC is extremely high, and there is also an extreme urgency for more research and development with the target of developing newer therapeutics against newer pathogenic vaginal biofilms. It is well known that BV occurs from dysbiosis of normal *Lactobacillus*-predominant vaginal flora which is associated with a synergistic relationship with a large number of microbes consisting of *Gardnerella vaginalis* and other anaerobes or the BV-associated bacteria (BVAB), etc. The study of epidemiology of BV reveals that it is acquired generally through sexual transmission as a polymicrobial consortium. Moreover, according to some recent data, BV is related to the development of adherent polymicrobial biofilm which possesses a lot of *G. vaginalis* and smaller numbers of BVAB. This further includes *Atopobium vaginae* in the surface of the vaginal epithelial cells. This is confirmed by the fluorescent in situ hybridization report of vaginal biopsy which was collected from a patient with BV. This is again accompanied with the non-Quomation of the selected specific number of cells covered with biofilms related to bacterial species, and that leads to the synthesis of "clue cells". These cells can be observed on the saline microscopy of vaginal secretions. Again, the *G. vaginalis* species are found to possess a larger potential of virulence which is definitely having greater percentage of adherence, cytotoxicity and the capability for the forming biofilm (De Jonghe and Glaesener 1995). All of these are relative to BVAB which helps in the hypothesis which predicts that *G. vaginalis* biofilm formation can be a starting phase in the pathogenesis of BV.

4.10.4 Reasons Behind Treatment Failure and Recurrence in Otitis Media

A specific type of *Haemophilus influenzae* is one of the major reasons of acute otitis media or AOM. This consists of chronic and recurrent otitis in individuals less than 10–15 years. Through many research works, it has been tried to find out if the non-typeable *H. influenzae* culture isolations cause these infections by forming biofilms and carry out resistance mechanisms to β-lactams. In many studies, it has been confirmed that *H. influenzae* isolates result in the formation of tympanocentesis. All the isolates were observed and checked for the presence of *blaTEM* genes, amino acid changes in the PBP3 or the transpeptidase domain of penicillin-binding protein 3 and biofilm formation in microtiter plates. It was seen that in almost 89.6% of the isolates, a specific stage of the available three observed stages was seen (Berendt and Byren 2004). Those were the biofilm formation which accounted for almost 83.3% and the resistance mechanisms to β-lactams which accounted for almost 33.3% resulting in the conversion of transpeptidase domain of PBP3 which is of greater significance and accounts for almost 22.9%. After this, the β-lactamase production takes place in a percentage of almost 10.4%. This is followed by 27.1% determination of isolates that possessed two or more of these three features. These observations prove that the successful diagnosis of non-typeable *H. influenzae* which causes chronic and recurrent AOM in young children can be altered by the higher formation of biofilm capacity of the isolates which takes place by the presence of β-lactam resistance mechanisms, PBP3 mutations to be precise.

4.10.5 Reasons Behind Treatment Failure and Recurrence in Orthopaedic Infections

The small-colony variants play an important role in failure in diagnosis and treatment of biofilm-related infections in orthopaedics. The process of infection during replacing the joint may give rise to a diagnostic challenge which is especially responsible for the late arrival of the infection. The extreme pain in the patient's body can be a result of the experience gained from the other normally obtained complicacies occurring during the process of surgery (Calhoun et al. 1988). The performed preoperative tests, which are spontaneous and consistent and also which are sensitive and specified for some specific infections taking place in patients requires revision surgery. The specific methods of diagnosis which depend only on the historical and physical parameters result in ineffective accuracy rate. But, the patients with earlier arrival of the infection symptoms suffer from pain and redness. It is associated with swelling at the joints, several wound drainages, etc. However, the patients who have the infections may recover with time, with a complaint of an early arrival of joint pain. This is not related to the symptoms of frequent fevers with joint infections. In this case, it can be said that the signs and symptoms are very less (Espersen et al. 1991). Haematological testing, radiographs and bone scans give positive results in both the aseptic loosening and infective cases, which make the indicators ineffective for long time of usage. Again, in the measurement of sensitivity, preoperative aspiration of the infected joint does not cross 70% of infection probability.

The detection of these sorts of infections includes parametric tests that involve blood testing, radiography and microbiological screening. The infections that are suspected to be associated with various prosthetic joints can be inferred by proper checking and observation. This creates a serious orthopaedic problem, with significant therapeutic implications. Hence it is of utmost importance to distinguish between aseptic loosening and infection as the treatment for these two is different from each other (Table 4.7).

Table 4.7 Description of the three outcome classifications used

Parametric conditions associated with the disorder	Probable outcomes observed		
Surgical parameters	Cured	Not cured	References
Specimen was obtained from two consecutive negative cultures	Yes		Hellmark et al. (2013)
Cessation of the systemic antibiotic treatment	Yes		Høiby et al. (2015)
Functional parameters	Persistence	Absence	Hellmark et al. (2013)
Signs of infection at the time of follow-up examinations	Positive	Negative	Høiby et al. (2015)

4.10.6 Reasons Behind Treatment Failure and Recurrence in Coccidioidal Meningitis

The recurrence of coccidioidal meningitis appears by the formation of fungal biofilm over the topmost point of the ventriculo-peritoneal shunt tubing, used to relieve the fluid pressure from the brain. It is the main reason behind a minimum of 4 years of persistence of *Coccidioides immitis* in spite of the continuous intake of maximum doze of fluconazole by the patient. The biofilms associated with fungi are regarded as a reason for the treatment failure and fungal persistence. This occurs specifically where artificial prostheses or indwelling catheters are present. Biofilms are a collection of one or more types of microorganisms, generally including bacteria, fungi and protists, that grow on many different surfaces.

Biofilm can grow on both abiotic surfaces and biotic surfaces like plant tissues and animal tissues and on implanted medical devices such as catheters and pacemakers. But the most common and striking similarity in all surfaces of attachment is that they must be wet. These surfaces are always periodically or continuously in contact with any fluid, especially water in this case—they are able to form chronic infections especially where they are formed on the surfaces of medical appliances, for example, in the catheters and prostheses (Carek et al. 2001). However, on this topic very few research works have been performed; some of the important clinically significant fungal biofilms also synthesize and do possess some related features like the antimicrobial resistance and attachment to indwelling medical appliances. Also, several studies were reported where *Coccidioides immitis* forms a biofilm on a medical device and that in persistent infectious conditions. *Candida* spp. and *Cryptococcus neoformans* have been seen to synthesize biofilms on catheters. Also, it is seen that *C. immitis* can also form biofilm colonies on the indwelling catheters (Carek et al. 2001). The fungal biofilms can be regarded as a major reason in the failure of diagnosis of many structural fungal infections, such as formed on the catheters or other artificial prostheses, which aggravate the situation of the affected individuals.

4.10.7 Reasons Behind Treatment Failure and Recurrence of Osteomyelitis

If a bone is injured and is exposed, the microbes can reach that injured bone from a place nearby through blood vessels and cause infection. This is referred to as osteomyelitis. It can be of two types: haematogenous osteomyelitis, produced by direct seeding of bacteria from the bloodstream, and contiguous focus osteomyelitis, where microbes are directly inoculated to the bone.

In the process of acute infections, if correct antibiotics are applied at the right time, then the infection will generally be irradicated after a couple of weeks of its action (Carek et al. 2001). But the process of diagnosis against these specific infections in the starting or early phase is way harder. Acute infections generally do not show any sorts of radiographic changes till a couple of weeks after the occurrence of the infection. MRI or magnetic resonance imaging is a good treatment option in the diagnosis of acute infections when the medical implants are not present. However, there are options available for a lag time after the first surgery or infection is done. But in case of chronic osteomyelitis, generally it is associated with huge areas of devitalized cortical and cancellous bone inside the wound. This is due to the antibiotics which don't penetrate well into devitalized bone but comply the dead areas to be totally debrided. This consists of cortex, marrow and devitalized scar tissue. The soft tissue which covers the portion of bone should be revived; otherwise the infection already present there will continue to grow, resulting in the reappearance of a fresh infection.

Almost all the patients with contiguous focus osteomyelitis possess diabetes mellitus and range from 35 to 70 years of age. Due to an enormous amount of increase in the diabetic population, osteomyelitis in the diabetic foot is to be taken into account. The shorter bones present in the feet including bones like the talus, distal fibula, calcaneus and tibia are linked with the infection. Sometimes, the infection takes place by infected nail beds, trophic skin ulceration, minor trauma in the feet or cellulitis. Neuropathy in the infected patients takes place by pairing with the proper function of the foot which is also associated to protective pain responses (Carek et al. 2001). This then rises to a progression of soft tissue infection inside the surface of the bone. Osteomyelitis in patients having dislocated or infected vasculature may be hard to treat. The patients can be infected in several positions such as a perforating foot ulcer, ingrown toe nail or a deep space infection or cellulitis. The staphylococci are commonly very infectious and serve as an agent in osteomyelitis. This is also said to start the empirical therapy on the infected individuals with a special condition which demands an anti-staphylococcal agent. After undergoing a correct process of treatment, the specified antibiotic therapy is applied. The treatment is done for more than consecutive 4 weeks, and it is administered intravenously).

4.11 Conclusion

Biofilm is the myriad of wide varieties of sessile communities that remain protected within self-secreted polymeric substances that not only provide nourishment but protect the bacteria from environmental resistance. The polymeric substance also helps in the better adherence of the sessile communities upon a solid surface. In this chapter it was well demarcated that infections start in its acute form and with the dispersal and metastasis of the sessile cells to distinct part of the body result in the development of chronic infections. The chapter also critically reviewed the interaction of the host organism having its influence on the development of biofilm within the body. The chapter also showed the failure of the conventional treatments, and newer researches have to be performed for developing alternate therapies in treating such biofilm-associated chronic and acute infections of the body.

References

Alhede M, Bjarnsholt T, Jensen PO, Phipps RK, Moser C, Christophersen L (2009) *Pseudomonas aeruginosa* recognizes and responds aggressively to the presence of polymorphonuclear leukocytes. Microbiology 155:3500–3508

Anderson GG, Dodson KW, Hooton TM, Hultgren SJ (2004) Intracellular bacterial communities of uropathogenic Escherichia coli in urinary tract pathogenesis. Trends Microbiol 12:424–430

Arciola CR, An Y, Campoccia D, Donati M, Montanaro L (2005) Etiology of implant orthopedic infections: a survey on 1027 clinical isolates. Int J Artif Organs 28:1091–1100

Auinger P, Lanphear BP, Kalkwarf HJ, Mansour ME (2003) Trends in otitis media among children in the United States. Pediatrics 112:514–520

Baltimore RS, Christie CDC, Smith GJW (1989) Immunohistopathologic localization of pseudomonas aeruginosa in lungs from patients with cystic fibrosis: implications for the pathogenesis of progressive lung deterioration. Am Rev Respir Dis 140:1650–1661

Banchereau J, Steinman RM (1998) Dendritic cells and the control of immunity. Nature 392:245–252

Banchereau J, Briere F, Caux C, Davoust J, Lebecque S, Liu YJ (2000) Immunobiology of dendritic cells. Annu Rev Immunol 18:767–811

Banerjee D, Popoola J, Shah S, Ster IC, Quan V, Phanish M (2020) COVID-19 infection in kidney transplant recipients. Kidney Int 97(6):1076–1082. https://doi.org/10.1016/j.kint.2020.03.018

Barbieri JT, Sun J (2004) Pseudomonas aeruginosa ExoS and ExoT. Rev Physiol Biochem Pharmacol 152:79–92

Bensel T, Stotz M, Borneff-Lipp M, Wollschlager B, Wienke A, Taccetti G (2011) Lactate in cystic fibrosis sputum. J Cyst Fibros 10:37–44

Berendt T, Byren I (2004) Bone and joint infection. Clin Med 4:510–518

Bjarnsholt T, Jensen PO, Fiandaca MJ, Pedersen J, Hansen CR, Andersen CB (2009) Pseudomonas aeruginosa biofilms in the respiratory tract of cystic fibrosis patients. Pediatr Pulmonol 44:547–558

Borregaard N, Herlin T (1982) Energy metabolism of human neutrophils during phagocytosis. J Clin Investig 70:550–557

Bowler PG, Duerden BI, Armstrong DG (2001) Wound microbiology and associated approaches to wound management. Clin Microbiol Rev 14(2):244–269

Brady RA, Leid JG, Camper AK, Costerton JW, Shirtliff ME (2006) Identification of Staphylococcus aureus proteins recognized by the antibody-mediated immune response to a biofilm infection. Infect Immun 74(6):3415–3426

Brady RA, Leid JG, Calhoun JH, Costerton JW, Shirtliff ME (2008) Osteomyelitis and the role of biofilms in chronic infection. FEMS Immunol Med Microbiol 52(**1**):13–22
Brown MR, Allison DG, Gilbert P (1988) Resistance of bacterial biofilms to antibiotics: a growth-rate related effect? J Antimicrob Chemother 22(6):777–780
Calhoun JH, Cantrell J, Cobos J, Lacy J, Valdez RR, Hokanson J, Mader JT (1988) Treatment of diabetic foot infections: Wagner classification, therapy, and outcome. Foot Ankle 9:101–106
Carek PJ, Dickerson LM, Sack JL (2001) Diagnosis and management of osteomyelitis. Am Fam Physician 63:2413–2420
Cheema MS, Rassing JE, Marriot C (1986) The diffusion characteristics of antibiotics in mucus glycoprotein gels. J Pharm Pharmacol 38:53–10
Chen DL, Ferkol TW, Mintun MA, Pittman JE, Rosenbluth DB, Schuster DP (2006) Quantifying pulmonary inflammation in cystic fibrosis with positron emission tomography. Am J Respir Crit Care Med 173:1363–1369
Cochrane DMG, Brown MRW, Anwar H, Weller PH, Lam K, Costerton JW (1988) Antibody response to Pseudomonas aeruginosa surface protein antigens in a rat model of chronic lung infection. J Med Microbiol 27:255–261
Costerton JW, Cheng KJ, Geesey GG, Ladd TI, Nickel JC, Dasgupta M, Marrie TJ (1987) Bacterial biofilms in nature and disease. Annu Rev Microbiol 41:435–464
Costerton JW, Lewandowski Z, Caldwell DE, Korber DR, Lappin-Scott HM (1995) Microbial biofilms. Annu Rev Microbiol 49:711–745
Costerton JW, Stewart PS, Greenberg EP (1999) Bacterial biofilms: a common cause of persistent infections. Science 284:1318–1322
Coticchia JM, Chen M, Sachdeva L, Mutchnick S (2013) New paradigms in the pathogenesis of otitis media in children. Front Pediatr 1:52
Cotter PA, Yuk MH, Mattoo S, Akerley BJ, Boschwitz J, Relman DA, Miller JF (1998) Filamentous hemagglutinin of Bordetella bronchiseptica is required for efficient establishment of tracheal colonization. Infect Immun 66:5921–5929
Darouiche RO (2004) Treatment of infections associated with surgical implants. N Engl J Med 350:1422–1429
Dasgupta MK, Lam K, Ulan RA, Bettcher KB, Burns V, Tyrrell DL, Dossetor JB, Costerton JW (1988) An extracorporeal model of biofilm-adherent bacterial microcolony colonization for the study of peritonitis in continuous ambulatory peritoneal dialysis. Am J Nephrol 8(2):118–122
De Beer D, Srinivasan R, Stewart PS (1994) Direct measurement of chlorine penetration into biofilms during disinfection. Appl Environ Microbiol 60:4339–4344
De Jonghe M, Glaesener G (1995) Type B Haemophilus influenzae infections. Experience at the pediatric hospital of Luxembourg. [French]. Bull Soc Sci Med Grand-Duche de Luxembourg 132:17–20
DeAntonio R, Yarzabal JP, Cruz JP, Schmidt JE, Kleijnen J (2016) Epidemiology of otitis media in children from developing countries: a systematic review. Int J Pediatr Otorhinolaryngol 85:65–74
Donlan RM, Costerton JW (2002) Biofilms: survival mechanisms of clinically relevant microorganisms. Clin Microbiol Rev 15:167–193
Doring G, Hoiby N (1983) Longitudinal study of immune € response to Pseudomonas aeruginosa antigens in cystic fibrosis. Infect Immun 42:197–201
Elek SD (1956) Experimental staphylococcal infections in the skin of man. Ann N Y Acad Sci 65:85–90
Enany S, Yoshida Y, Magdeldin S (2013) Two dimensional electrophoresis of the exoproteome produced from community acquired methicillin resistant Staphylococcus aureus belonging to clonal complex 80. Microbiol Res 168(8):504–511
Espersen F, Frimodt-Moller N, Thamdrup RV, Skinhoj P, Bentzon MW (1991) Changing pattern of bone and joint infections due to Staphylococcus aureus: study of cases of bacteremia in Denmark, 1959–1988. Rev Infect Dis 13:347–358
Flemming HC, Wingender J (2010) The biofilm matrix. Nat Rev Microbiol 8(9):623–633

Flemming HC, Neu TR, Wozniak DJ (2007) The EPS matrix: the "house of biofilm cells". J Bacteriol 189(22):7945–7947. https://doi.org/10.1128/JB.00858-07. PMID: 17675377; PMCID: PMC2168682.
Furukawa S, Kuchma SL, O'Toole GA (2006) Keeping their options open: acute versus persistent infections. J Bacteriol 188(4):1211–1217
Fux CA, Costerton JW, Stewart PS, Stoodley P (2005) Survival strategies of infectious biofilms. Trends Microbiol 13(1):34–40
Fuxman Bass JI, Russo DM, Gabelloni ML (2010) Extracellular DNA: a major proinflammatory component of Pseudomonas aeruginosa biofilms. J Immunol 184(11):6386–6395
Gardner A, West SA (2004) Spite and the scale of competition. J Evol Biol 17(6):1195–1203
Geesey GG, Richardson WT, Yeomans HG, Irvin RT, Costerton JW (1977) Microscopic examination of natural sessile bacterial populations from an alpine stream. Can J Microbiol 23:1733–1736
Geesey GG, Mutch R, Costertin JW, Green RB (1978) Sessile bacteria: an important component of the microbial population in small mountain streams. Limnol. Oceanography 23:1214–1223
Ghosh P (2004) Process of protein transport by the type III secretion system. Microbiol Mol Biol Rev 68:771–795
Gominet M, Compain F, Beloin C, Lebeaux D (2017) Central venous catheters and biofilms: where do we stand in 2017? Biofilm Infect 125:4
Hellmark B, Berglund C, Nilsdotter-Augustinsson Å, Unemo M, Söderquist B (2013) Staphylococcal cassette chromosome mec (SCCmec) and arginine catabolic mobile element (ACME) in Staphylococcus epidermidis isolated from prosthetic joint infections. Eur J Clin Microbiol Infect Dis 32(5):691–697
Henrici AT (1933) Studies of freshwater bacteria. I. A direct microscopic technique. J Bacteriol 25:277–287
Hensen SM, Pavicic MJAM, Lohuis JACM, de Hoog JAM, Poutrel B (2000) Location of *Staphylococcus aureus* within the experimentally infected bovine udder and the expression of capsular polysaccharide type 5 in situ. J Dairy Sci 83:1966–1975
Høiby N, Bjarnsholt T, Givskov M, Molin S, Ciofu O (2010) Antibiotic resistance of bacterial biofilms. Int J Antimicrob Agents 35(4):322–332
Høiby N, Bjarnsholt T, Moser C, Bassi GL, Coenye T, Donelli G, Stoodley LH, Holá V, Imbert C, Kirketerp-Møller K, Lebeaux D, Oliver A, Ullmann AJ, Williams C, Zimmerli W (2015) ESCMID guideline for the diagnosis and treatment of biofilm infections 2014. Clin Microbiol Infect 21(1):S1–S25
Holban AM, Bleotu C, Chifiriuc MC, Bezirtzoglou E, Lazar V (2014) Role of *Pseudomonas aeruginosa* quorum sensing (QS) molecules on the viability and cytokine profile of human mesenchymal stem cells. Virulence 5(2):303–310
Holsclaw DS (1980) Cystic fibrosis: overview and pulmonary aspects in young adults. Clin. Chest Med 1:407–421
Horvat RT, Parmely MJ (1988) Pseudomonas aeruginosa alkaline protease degrades human gamma interferon and inhibits its bioactivity. Infect Immun 56:2925–2932
Houston N, Stewart N, Smith DS, Bell SC, Champion AC, Reid DW (2013) Sputum neutrophils in cystic fibrosis patients display a reduced respiratory burst. J Cyst Fibros 12:352–362
Hu H, Johani K, Almatroudi A, Vickery K, Natta BV, Kadin ME, Brody G, Clemens M, Cheah CY, Lade S, Joshi PA, Prince HM, Deva AK (2016) Bacterial biofilm infection detected in breast implant-associated anaplastic large-cell lymphoma. Plast Reconstr Surg 137(6):1659–1669
Irie Y, Mattoo S, Yuk MH (2004) The Bvg virulence control system regulates biofilm formation in Bordetella bronchiseptica. J Bacteriol 186:5692–5698
James GA, Swogger E, Wolcott R, Pulcin E, Secor P, Sestrich J, Costerton JW, Stewart PS (2007) Biofilms in chronic wounds. Wound Repair Regen 16:37–44
Jayaseelan S, Ramaswamy D, Dharmaraj S (2014) Pyocyanin: production, applications, challenges and new insights. World J Microbiol Biotechnol 30(4):1159–1168

Joris B, Ledent P, Dideberg O, Fonze E, Lamotte-Brasseur J, Kelly JA (1991) Comparison of the sequences of class A beta-lactamases and of the secondary structure elements of penicillin-recognizing proteins. Antimicrob Agents Chemother 35:2294–2301

Kalinski P, Hilkens CM, Wierenga EA, Kapsenberg ML (1999) T-cell priming by type-1 and type-2 polarized dendritic cells: the concept of a third signal. Immunol Today 20:561–567

Kaplan JB (2010) Biofilm dispersal: mechanisms, clinical implications, and potential therapeutic uses. J Dent Res 89(3):205–218

Katragkou A, Kruhlak MJ, Simitsopoulou M, Chatzimoschou A, Taparkou A, Cotten CJ (2010) Interactions between human phagocytes and Candida albicans biofilms alone and in combination with antifungal agents. J Infect Dis 201:1941–1949

Khoury AE, Lam K, Ellis B, Costerton JW (1992) Prevention and control of bacterial infections associated with medical devices. ASAIO J 38(3):M174–M178

Kimbrell DA, Beutler B (2001) The evolution and genetics of innate immunity. Nat Rev Genet 2:256–267

Kinniment SL, Wimpenny JWT, Adams D (1996) Development of a steady-state oral microbial biofilm community using the constant-depth film fermenter. Microbiology 142:631–638

Kokare CR, Chakraborty S, Khopade AN et al (2009) Biofilm: importance and applications. Indian J Biotechnol 8:159–168

Kolpen M, Hansen CR, Bjarnsholt T, Moser C, Christensen LD, van Gennip M (2010) Polymorphonuclear leucocytes consume oxygen in sputum from chronic Pseudomonas aeruginosa pneumonia in cystic fibrosis. Thorax 65:57–62

Kostakioti M, Hadjifrangiskou M, Hultgren SJ (2013) Bacterial biofilms: development, dispersal, and therapeutic strategies in the dawn of the postantibiotic era. Cold Spring Harbor Perspect Med 3(4):a010306. https://doi.org/10.1101/cshperspect.a010306

Krieg AM (2002) CpG motifs in bacterial DNA and their immune effects. Annu Rev Immunol 20:709–760

Kronig I, Vaudaux P, Suvà D, Lew D, Uçkay I (2015) Acute and chronic osteomyelitis. In: Clinical infectious disease, 2nd edn. Temple University, Philadelphia, pp 448–453

Lagina A (2016) Beware the Biofilm. Review of cornea and contact lenses

Lakeman MM, Roovers JPW (2016) Urinary tract infections in women with urogynaecological symptoms. Curr Opin Infect Dis 29:1–97

Lam J, Chan R, Lam K, Costerton JW (1980) Production of mucoid microcolonies by Pseudomonas aeruginosa within infected lungs in cystic fibrosis. Infect Immun 28:546–556

Lebeaux D, Ghigo JM, Beloin C (2014) Biofilm-related infections: bridging the gap between clinical management and fundamental aspects of recalcitrance toward antibiotics. Microbiol Mol Biol Rev 78(3):510–543

Lewis K (2001) Riddle of biofilm resistance. Antimicrob Agents Chemother 45:999–1007

Lewis K (2007) Persister cells, dormancy and infectious disease. Nat Rev Microbiol 5:48–56

Liu SC, Han XM, Shi M, Pang ZL (2016) Persistence of uropathogenic Escherichia Coli in the bladders of female patients with sterile urine after antibiotic therapies. Acta Acad Med Wuhan 36:710–715

Locht C, Antoine R, Jacob-Dubuisson F (2001) Bordetella pertussis, molecular pathogenesis under multiple aspects. Curr Opin Microbiol 4:82–89

Lynch AS, Robertson GT (2008) Bacterial and fungal biofilm infections. Annu Rev Med 59:415–428

Massion PP, Inoue H, Richman-Eisenstat J et al (1994) Novel Pseudomonas product stimulates interleukin-8 production in airway epithelial cells in vitro. J Clin Invest 93(1):26–32

Masters EA, Trombetta RP, de Mesy Bentley KL, Boyce BF, Gill AL, Gill SR, Nishitani K, Ishikawa M, Morita Y, Ito H, Bello-Irizarry SN, Ninomiya M, Brodell JD Jr, Lee CC, Hao SP, Oh I, Xie C, Awad HA, Daiss JL, Owen JR, Kates SL, Schwarz EM, Muthukrishnan G (2019) Evolving concepts in bone infection: redefining "biofilm", "acute vs. chronic osteomyelitis", "the immune proteome" and "local antibiotic therapy". Bone Research 7:20

Matsumoto K (2004) Role of bacterial proteases in pseudomonal and serratial keratitis. Biol Chem 385:1007–1016

Mattoo S, Foreman-Wykert AK, Cotter PA, Miller JF (2001) Mechanisms of Bordetella pathogenesis. Front Biosci 6:E168–E186

McConoughey SJ, Howlin R, Granger JF, Manring MM, Calhoun JH, Shirtlif M, Kathju S, Stoodley P (2014) Biofilms in periprosthetic orthopedic infections. Future Microbiol 9 (8):987–1007

McWhirter SM, Barbalat R, Monroe KM (2009) A host type I interferon response is induced by cytosolic sensing of the bacterial second messenger cyclic-di-GMP. J Exp Med 206 (9):1899–1911

Melvin JA, Scheller EV, Miller JF, Cotter PA (2014) Bordetella pertussis pathogenesis: current and future challenges. Nat Rev Genet 12:274–288

Meyle E, Stroh P, Gunther F, Hoppy-Tichy T, Wagner C, Hansch GM (2010) Destruction of bacterial biofilms by polymorphonuclear neutrophils: relative contribution of phagocytosis, DNA release, and degranulation. Int J Artif Organs 33:608–620

Mishra A, Malik A (2014) Novel fungal consortium for bioremediation of metals and dyes from mixed waste stream. Bioresour Technol 171:217–226

Mitra A, Mukhopadhyay S (2016) Biofilm mediated decontamination of pollutants from the environment. AIMS Bioeng 3(1):44–59

Moser C, Jensen PO, Pressler T, Frederiksen B, Lanng S, Kharazmi A (2005) Serum concentrations of GM-CSF and G-CSF correlate with the Th1/Th2 cytokine response in cystic fibrosis patients with chronic Pseudomonas aeruginosa lung infection. APMIS 113:400–409

Moser C, Pedersen HT, Lerche CJ, Kolpen M, Line L, Thomsen K, Høiby N, Jensen PØ (2017) Biofilms and host response–helpful or harmful. APMIS 125:320–338

Mulcahy H, Charron-Mazenod L, Lewenza S (2008) Extracellular DNA chelates cations and induces antibiotic resistance in Pseudomonas aeruginosa biofilms. PLoS Pathog 4(11): e1000213

Murofushi Y, Villena J, Morie K (2015) The toll-like receptor family protein RP105/MD1 complex is involved in the immunoregulatory effect of exopolysaccharides from Lactobacillus plantarum N14. Mol Immunol 64(1):63–75

Nickel JC, Ruseska I, Wright JB, Costerton JW (1985) Tobramycin resistance of *Pseudomonas aeruginosa* cells growing as a biofilm on urinary catheter material. Antimicrob Agents Chemother 27(4):619–624

Ochieng' N, Okechi H, Ferson S, Albright AL (2015) Bacteria causing ventriculoperitoneal shunt infections in a Kenyan population. J Neurosurg Pediatr 15(2):150–155

Okshevsky M, Meyer RL (2015) The role of extracellular DNA in the establishment, maintenance and perpetuation of bacterial biofilms. Crit Rev Microbiol 41(3):341–352

Okuda K, Nagahori R, Yamada S, Sugimoto S, Sato C, Sato M, Iwase T, Hashimoto K, Mizunoe Y (2018) The composition and structure of biofilms developed by propionibacterium acnes isolated from cardiac pacemaker devices. Front Microbiol 9:182

Olson PD, Hunstad DA (2016) Subversion of host innate immunity by uropathogenic Escherichia coli. Pathogens 5:2

Parsek MR (2015) Controlling the connections of cells to the biofilm matrix. J Bacteriol 198 (1):12–14

Peltola H, Pääkkönen M, Kallio P, Kallio MJ, Group OSAS (2010) Short-versus long-term antimicrobial treatment for acute hematogenous osteomyelitis of childhood: prospective, randomized trial on 131 culture-positive cases. Pediatr Infect Dis J 29:1123–1128

Pier GB, Coleman F, Grout M, Franklin M, Ohman DE (2001) Role of alginate O acetylation in resistance of mucoid Pseudomonas aeruginosa to opsonic phagocytosis. Infect Immun 69 (3):1895–1901

Pruitt BA, McManus AT, Kim SH, Goodwin CW (1998) Burn wound infections: current status. World J Surg 22:135–145

Pulido L, Ghanem E, Joshi A, Purtill JJ, Parvizi J (2008) Periprosthetic joint infection: the incidence, timing, and predisposing factors. Clin Orthop Relat Res 466:1710–1715

Raffatellu M, Chessa D, Wilson RP, Dusold R, Rubino S, Baumler AJ (2005) The Vi capsular antigen of Salmonella enterica serotype Typhi reduces Toll-like receptor-dependent interleukin-8 expression in the intestinal mucosa. Infect Immun 73(6):3367–3374

Resch A, Rosenstein R, Nerz C, Gotz F (2005) Differential gene expression profiling of Staphylococcus aureus cultivated under biofilm and planktonic conditions. Appl Environ Microbiol 71:2663–2676

Robino L, Scavone P, Araújo L, Algorta G, Zunino P, Vignoli R (2013) Detection of intracellular bacterial communities in a child with Escherichia coli recurrent urinary tract infections. Pathog Dis 68:78–81

Römling U, Balsalobre C (2012) Biofilm infections, their resilience to therapy and innovative treatment strategies. J Intern Med 272:541–561

Rumbaugh KP, Griswold JA, Iglewski BH, Hamood AN (1999) Contribution of quorum sensing to the virulence of Pseudomonas aeruginosa in burn wound infections. Infect Immun 67:5854–5862

Russell SL, Gold MJ, Hartmann M (2012) Early life antibiotic-driven changes in microbiota enhance susceptibility to allergic asthma. EMBO Rep 13:440–447

Saeed K (2019) The 2018 international consensus meeting on musculoskeletal infection: summary from the biofilm workgroup and consensus on biofilm related musculoskeletal infections. J Orthop Res 37:1007–1017

Schaber JA, Triffo WJ, Suh SJ, Oliver JW, Hastert MC, Griswold JA, Auer M, Hamood AN, Rumbaugh KP (2007) Pseudomonas aeruginosa forms biofilms in acute infection independent of cell-to-cell signaling. Infect Immun 75(8):3715–3721

Schilder AG, Chonmaitree T, Cripps AW, Rosenfeld RM, Casselbrant ML, Haggard MP, Venekamp RP (2016) Otitis media. Nat Rev Dis Primers 2(1):16063

Scott VCS, Haake DA, Churchill BM, Justice SS, Kim JH (2015) Intracellular bacterial communities: a potential etiology for chronic lower urinary tract symptoms. Urology 86:425–431

Shapira L, Tepper P, Steinberg D (2000) The interactions of human neutrophils with the constituents of an experimental dental biofilm. J Dent Res 79:1802–1807

Singh PK, Schaefer AL, Parsek MR, Moninger TO, Welsh MJ, Greenberg EP (2000) Quorum-sensing signals indicate that cystic fibrosis lungs are infected with bacterial biofilms. Nature 407:762–764

Singhai M, Malik A, Shahid M, Malik A, Rawat V (2012) Colonization of peripheral intravascular catheters with biofilm producing microbes: evaluation of risk factors. Niger Med J 53(1):37–41

Smith RS, Fedyk ER, Springer TA, Mukaida N, Iglewski BH, Phipps RP (2001) IL-8 production in human lung fibroblasts and epithelial cells activated by the Pseudomonas autoinducer N-3-oxododecanoyl homoserine lactone is transcriptionally regulated by NF-kappa B and activator protein-2. J Immunol 167(1):366–374

Soave R, Murray HW, Litrenta MM (1978) Bacterial invasion of pulmonary vessels. Pseudomonas bacteremia mimicking pulmonary thromboembolism with infarction. Am J Med 65:864–867

Steckelberg JM, Osmon DR (2000) Prosthetic joint infections. In: Infections associated with indwelling medical devices, 3rd edn. American Society of Microbiology, Washington, pp 173–209

Stewart PS (2002) Mechanisms of antibiotic resistance in bacterial biofilms. Int J Med Microbiol 292(2):107–113. https://doi.org/10.1078/1438-4221-00196

Thorarinsdottir HR, Kander T, Holmberg A, Petronis S, Klarin B (2020) Biofilm formation on three different endotracheal tubes: a prospective clinical trial. Crit Care 24:382

Tlaskalova-Hogenova H, Stepankova R, Hudcovic T (2004) Commensal bacteria (normal microflora), mucosal immunity and chronic inflammatory and autoimmune diseases. Immunol Lett 93:97–108

Tredget EE, Shankowsky HA, Rennie R, Burrell RE, Logsetty S (2004) Pseudomonas infections in the thermally injured patient. Burns 30:3–26
Vestby LK, Grønseth T, Simm R, Nesse LL (2020) Bacterial biofilm and its role in the pathogenesis of disease. Antibiotics 9:59
Watters C, Fleming D, Bishop D, Rumbaugh KP (2016) Host responses to biofilm. Prog Mol Biol Transl Sci 142:193–239
Witko-Sarsat V, Rieu P, Descamps-Latscha B, Lesavre P, Halbwachs-Mecarelli L (2000) Neutrophils: molecules, functions and pathophysiological aspects. Lab Investig 80(5):617–653
Worlitzsch D, Tarran R, Ulrich M, Schwab U, Cekici A, Meyer KC (2002) Effects of reduced mucus oxygen concentration in airway Pseudomonas infections of cystic fibrosis patients. J Clin Investig 109:317–325
Yoon BI, Kim S, Han DS, Ha US, Lee SJ, Kim HW, Han CH, Cho YH (2012) Acute bacterial prostatitis: how to prevent and manage chronic infection? J Infect Chemother 18:444–450
Yoon BI, Han DS, Ha US, Lee SJ, Sohn DW, Kim HW, Han CH, Cho YH (2013) Clinical courses following acute bacterial prostatitis. Prostate Int 1:89–93
Zhang X, Bishop PL (2003) Biodegradability of biofilm extracellular polymeric substances. Chemosphere 50(1):63–69
Zimmerli W, Waldvogel FA, Vaudaux P, Nydegger UE (1982) Pathogenesis of foreign body infection: description and characteristics of an animal model. J Infect Dis 146:487–497
Zobell CE, Allen EC (1935) The significance of marine bacteria in the fouling of submerged surfaces. J Bacteriol 29:239–251

5 Bacteria and Biofilms in Chronic Infections and Nosocomial Diseases

Smaranika Pattnaik

Abstract

In the present clinical scenario, the prolonged therapeutic approaches against chronic infectious diseases are primarily due to the formation of biofilm. Although the bacterial strains express a cascade of resistant mechanisms, biofilm is considered as the most important drug-resistant strategy as opted by the candidate bacteria. Considering cystic fibrosis, chronic wound infections, endophthalmitis, and keratitis as models, it is observed that the causal organisms, namely, *Pseudomonas aeruginosa*, *Staphylococcus aureus/epidermidis*, *Escherichia coli*, etc. are highly biofilm producers. Therefore, various alternative and novel approaches were suggested in the past. Combinatorial approach, combining both antibiofilm agent and antibiotic doses, was proved to be most effective.

5.1 Introduction

Bacterial adhesion on the host tissue is the primary requisite for colonization. The establishment of nonspecific interactions between host cell receptors and adhesins (polysaccharides, proteins, fimbriae, etc.) enables bacteria to adhere (Sarowska et al. 2019). The bacteria like *Escherichia coli* used its fimbriae to adhere to surfaces. Figure 5.1 shows the typical Gram-negative cells protruded with fimbriae. Specific bacterial strains adhere to specific host tissue, which is known as "bacterial tissue tropism" (Klemm and Schembri 2000).

S. Pattnaik (✉)
Department of Biotechnology and Bioinformatics, Sambalpur University, Burla, Odisha, India
e-mail: smaranika2010@suniv.ac.in

R. R. Ray et al. (eds.), *Biofilm-Mediated Diseases: Causes and Controls*,
https://doi.org/10.1007/978-981-16-0745-5_5

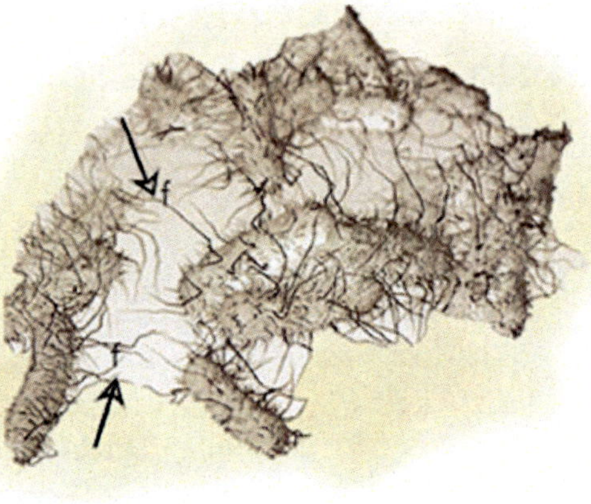

Fig. 5.1 Typical Gram-negative bacteria are protruded with fimbriae (indicated by arrows) to adhere to the tissue surfaces

Hence, colonization is the first stage of microbial infection by establishing the virulent bacteria at the appropriate portal of entry (Dani 2014). Antibiotics, the wonder drugs, are prescribed to kill the tissue-colonized virulent bacteria. The antibiotics hit the targets (a) DNA replication DNA synthesis and DNA gyrase, (b) RNA synthesis, (c) protein synthesis (50S or 30S subunit inhibitors), (d) cell wall biosynthesis, (e) cell membrane biosynthesis, and (f) fatty acid synthesis pathways of a sensitive bacterial cell (O'Rourke et al. 2020). However, due to many resistance mechanisms conferred by colonizing bacteria toward conventional antibiotics (Kapoor et al. 2017), they seem to be a critical situation in the clinical setup.

The emergence of multidrug-resistant bacteria and the dominance of extended-spectrum of drug-resistant (XDR) bacteria have created havoc in public health (Wang et al. 2017). In the twenty-first century, AMR (antimicrobial resistance) is considered as one of the ten most important public health threats (Munita and Arias 2016), worldwide.

The resistant bacteria may express one or more types of resistant mechanisms to evade the antibiotic action. The bacteria may go for the expression of antibiotic-modifying or antibiotic-degrading enzymes. The bacterial group showing resistance toward aminoglycosides express modification enzymes, including *N*-acetyl transferases, *O*-phosphotransferases, and *O*-adenyltransferases that acetylate, phosphorylate, or adenylylate (Peterson and Kaur 2018). Besides, the production of extended-spectrum β-lactamases (ESBLs), by the *Enterobacteriaceae* members, confers resistance toward beta-lactam antibiotics, including penicillins, cephalosporins, and the monobactam aztreonam (Shaikh et al. 2015). Moreover, metallo-β-lactamases, which hydrolyze all the β-lactam antibiotics and plasmid-mediated serine carbapenemases expressed by *Klebsiella pneumonia* (Cp-Kpn strains), represent a challenge for physicians.

As well, bacteria are much expert in implementing "drug efflux action." Various groups of bacteria including *Pseudomonas aeruginosa* avail "multidrug efflux pump mechanisms" to throwing away of the drug molecules. The efflux pumps are protein channels, which are intercalated in the cell wall of bacteria. If the drug molecules get a chance to enter inside the bacterial wall, the pertinent gene expression makes the protein pumps active to push the drug molecules out of its cell. The efflux protein pumps are of five types which are classified based upon their sequences, target specificity, and energy assistance (Issa et al. 2018). These are (a) the ATP-binding cassette (ABC), the small multidrug resistance (SMR), the major facilitator superfamily (MFS), the resistance-nodulation-cell division (RND), and the multidrug and toxic compound extrusion (MATE).

In addition to the abovementioned mechanisms of drug resistance, other mechanisms are also functional, but the citing of all the mechanisms is out of the scope of this chapter. As this chapter is focused on biofilms only, it is a mandate to skew toward the role of biofilm in developing pathogenesis, specifically in a clinical setup. Synthesis of biofilm in situ is regarded as the most effective method of drug resistance and is convened by bacterial strains.

Biofilm, a network of heterogeneous molecules, is composed of exopolysaccharide (EPS), proteins, lipids, and e-DNA, which are webbed in an organized manner. Both the Gram-positives and Gram-negatives of clinical importance use biofilm as a "molecular barricade" and resist the action of antibiotics. Under the pressure of antibiotics, the bacteria provisionally halt the lifestyle of free-living and settle on a solid or tissue surface. By activating "quorum sensing" molecular signaling tools, the whole invading bacterial populations come closer and start to adhere. The bacterial cells accumulate all the extracellular and intracellular biomolecules and start to synthesize biofilm. Then, the bacterial cells slow down their metabolic activities but remain alive inside the biofilm. When the pressure of antibiotics diminishes, the bacterial cells regain their metabolic activity, and the biofilm matrix starts to disintegrate. And the bacterial cells recuperate their free-living activities and escape from the disintegrating biofilm and search for a new spot to form a new biofilm matrix.

It is understood that the bacteria remain hidden inside the biofilm and create a suitable niche inside the host tissue or surface of external implants. The bacterial strains express pertinent proteins, by activating either upstream or downstream gene operon system to make a biofilm matrix. Among the notorious bacterial strains, *Pseudomonas aeruginosa*, *Escherichia coli*, *Proteus mirabilis*, and *Klebsiella* spp. express fimbrial appendages to anchor suitable surfaces. *Staphylococcus aureus* and its related species like *Staphylococcus epidermidis* adopt different mechanism in organizing biofilm. It is reported that (Otto 2008) there is a manifestation of MSCRAMMs (microbial surface components recognizing adhesive matrix molecules), which adhere to the host matrix proteins. The poly-*N*-acetylglucosamine (PNAG) is considered the main component of staphylococcal biofilms. Teichoic acids, specifically, lipoteichoic acid, and wall teichoic acids have also critical roles in forming biofilms (Naclerio et al. 2020). This bacterium expresses both proteinaceous and nonproteinaceous adhesins. AtlA protein is meant for adhesion as well as autolysis. The autolysis of cells is required to liberate extracellular DNA (e-DNA) to the environment, to make a biofilm. Also, PIA (polysaccharide intracellular adhesion) synthesized by *ica* gene (Paharik and Horswill 2016) take part in biofilm organization.

Moreover, biofilm is considered to be a bacterial community output. Before the presence of antibiotic molecules or any signaling molecules in the environment, the bacterial cells communicate with each other and settle on the available surfaces. As the communication transmits in a quorum, through quorum sensing molecules, the phenomenon is "quorum sensing" mechanism, community-acquired strategy.

Now it is clear that biofilms have an enormous impact on the success or failure of medicaments prescribed by clinicians. The reason behind this is due to (a) failure of antibiotic molecules to penetrate a biofilm, (b) slow growth and stress response, (c) heterogeneity in RNA/protein content (not DNA content), (d) general stress response, (e) quorum sensing, and (f) biofilm phenotype. Working with strains of *Pseudomonas aeruginosa*, in the presence of various concentrations of β-lactams, Rojo-Molinero et al. (2019) had observed that in biofilms, both drug-resistant and sensitive populations were present. The drug-resistant strains produced more amounts of β-lactamase and efflux pump proteins in contrast to drug-sensitive strains. It implies that there is a correlation between drug resistance and biofilm production and there may be involvement of novel genes, as inferred from transcriptomics (Penesyan et al. 2019).

Henceforth, it is understood that the bacterial strains with biofilm-forming ability maintain a higher MIC (minimum inhibitory concentration) relative to their free-floating counterparts (Sun et al. 2013). This leads to a prolonged period of antibiotic therapy in chronic infections and the possibility of fatality increases. Besides, the development of biofilm on the surfaces of tissue implanted devices is imposing major hurdles in the treatment of patients. In the clinical scenario, the bacteria enter into the circulation system (Singh et al. 2017), resulting in manifestation of bacteremia in blood and leading to systemic infections.

Pseudomonas aeruginosa in cystic fibrosis pneumonia, *Escherichia coli* in urinary tract infections, and *Mycobacterium tuberculosis* in human tuberculosis cause important biofilm-associated chronic diseases, due to which the internal organs of the body are affected. The bacteria can survive inside the body, being cryptic inside the network of biofilms, without causing excessive damage to host cells, hence making the infection persistent and escaping itself from the attack of macrophages. This is to add here that *P. aeruginosa* and *E. coli* are much adroit to cause both symptomatic acute and chronic infections (Grant and Hung 2013).

In addition to immunological factor, the bacteria can opt for biofilm-associated infections due to (1) induction of stress, (2) SOS checkpoint, (3) quorum sensing, (4) ppGDP signaling, (5) toxin, (6) antitoxin module, etc.

The bacterial cells colonizing in persistent or chronic infections are of three types: (a) respondents, (b) tolerant, and (c) persisters, which pose indifferent to antibiotic treatments (Trastoy et al. 2018). The respondents respond to antibiotic treatment, maybe sensitive or resistant, while the second group of bacteria tolerates the antibiotic doses but does not activate the drug-resistant mechanism. However, the persisters are neither respondents nor tolerants; simply they are indifferent toward antibiotic action. It is reported that most of the persisters remain present in the core of

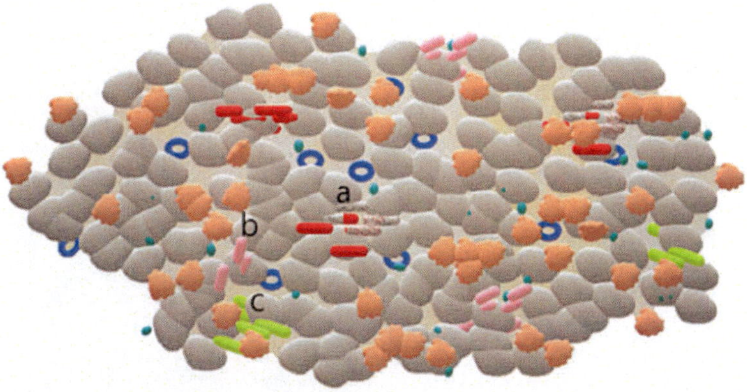

Fig. 5.2 A typical tissue showing bacterial colonization intracellularly. In response to these bacterial invaders, host nonspecific immune system is operative, recruiting macrophages (brown lobes structures), cytokinins (cyan balls), and lymphocytes (blue rings). Three types of bacteria are seen: (**a**) red cells are Responding cells, (**b**) green cells are tolerant, and (**c**) pink cells are persisters

biofilms. The invading bacteria try to defend the action of antibiotics as well as nonspecific immunocytes, which are present in placebo. The putative scenario is depicted in Fig. 5.2.

According to Abe and Suzuki (2020), biofilms are "best suited" place for horizontal gene transfer (HGT) of antibiotic resistance genes (ARGs). ARGs spreading via HGT have been shown to depend on the environment, bacterial communities, and mobile genetic elements. Classically, HGT mechanisms include conjugation, transformation, and transduction. Plasmids, harbored with drug-resistant (DR) genes, are extra-chromosomal circular DNAs shuttle among bacterial strains, horizontally. The transposons and IS (insertion sequence) elements flanked with DR genes also move to the recipient bacteria. Further, if the recipient is a *hfr* strain, then it donates the ARGs to another sensitive recipient. And the process goes on. As a result of which, the whole community of bacteria becomes a carrier of drug-resistant genes.

Figure 5.3 demonstrates the transfer of plasmids or antibiotic-resistant genes from a donor cell (putative antibiotic-resistant cell) to a recipient cell (putative antibiotic-sensitive cell), through a conjugation bridge, inside the biofilm. The donor bacterial cells, confined inside the biofilm, come closer and establish contact with recipient cell, making use of fimbriae to put a bridge, so that the plasmid as a whole or in part will move. As plasmids are carrier of drug-resistant genes, a sensitive cell is mutated into a resistant cell, after receiving the drug-resistant gene(s). Therefore, biofilm is the key factor for horizontal drug-resistant gene transfer among the bacterial cells. This is the underlying mechanism, for the spread of drug resistance in clinical as well as nosocomial bacteria.

Nosocomial infections are also known as hospital-acquired/hospital-associated infections, caused by certain opportunistic bacterial strains. The most frequent types of infections include central line-associated bloodstream infections (CLABSI),

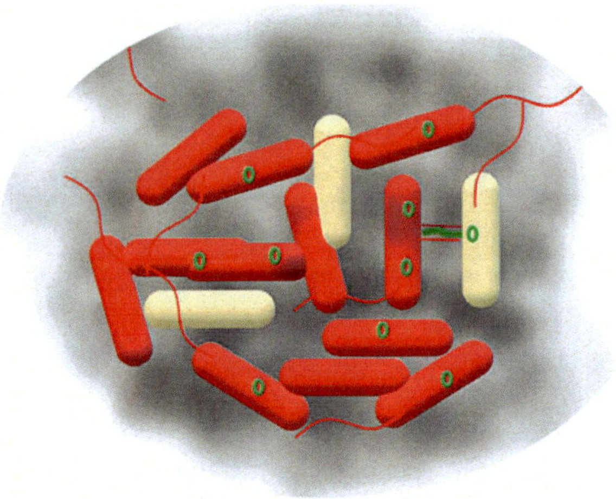

Fig. 5.3 This artwork depicts horizontal gene transfer events while inside a biofilm. The plasmid harbored in drug-resistant bacteria (indicated in red color rods) is transferring genes via the process of conjugation to a recipient cell, which happens to be a drug-sensitive cell (indicated by light yellow color)

catheter-associated urinary tract infections, surgical site infections (SSIs), and ventilator-associated pneumonia (Khan and Aziz 2017). The term CLABSI is released by the US Centers for Disease Control and Preventions (CDC). In fact, CLABSI is an infection in the central line due to intravascular catheters, while UTI (urinary tract infections) are due to the use of catheters in the urinary tract. As biofilm ultimately develops on the catheter or its associated parts, there is development of bacteriuria. SSIs are initiated during the surgical operations. Sometimes, the infection is superficial, but it may lead to persistent infection. Infections are classified as either incisional or organ/space infections (Owens and Stoessel 2008). Ventilator-associated pneumonia (VAP) is defined as pneumonia occurring after 48–72 h of intubation and ventilation. Evidences are reported with the development of endotracheal biofilm in VAP (Gil-Perotin et al. 2012).

In addition to the biofilm development in nosocomial infections, chronic infections are also exaggerated with biofilm growth and therefore the leading causes of death and disability worldwide (WHO report, www.who.int). Among chronic diseases, cystic fibrosis (CF), chronic wound infections, and endophthalmitis are serious threats to mankind because of drug resistance conferred by biofilm growth.

Recognizing the importance of effective therapeutic approaches, the chapter highlights the features and etiology of certain diseases like cystic fibrosis, chronic wounds, endophthalmitis, and keratitis, which are associated with biofilm growth. As well, there is a simultaneous call for eradicating biofilm producers in the clinical situation through pertinent approaches.

5.2 Biofilm and Lung Cystic Fibrosis (CF)

Cystic fibrosis (CF) is caused due to mutation in CFTR (cystic fibrosis transmembrane conductance regulator) gene, which imparts protein coagulation and the formation of thick layers of mucus in the lower respiratory tract. This disease is characterized by excessive mucus production that covers the epithelium (Perez et al. 2011).

This abnormal mucus can clog the airways, leading to severe problems with breathing and bacterial infections in the lungs. Over time, infections result in permanent lung damage, including the formation of scar tissue (fibrosis) and cysts in the lungs (Fig. 5.4).

The CF-associated persistent infections are caused primarily by *Pseudomonas aeruginosa*, a Gram-negative ubiquitous bacterium. The bacterium is notorious for its multidrug resistance by expressing a cascade of drug-resistant and virulent genes. Hence, there is a delay in treatment with the possibility of high fatality (Harrington et al. 2020). Although other bacterial strains, namely, *Burkholderia multivorans*, *Achromobacter xylosoxidans*, *Stenotrophomonas maltophilia*, *Dyella* species, and *Mycobacterium abscessus* have also been shown with the production of biofilms in CF airways (Høiby et al. 2017), *P. aeruginosa* has a prominent role in CF biofilms with drug resistance and persistence. The unique feature of bacterial persistence is due to hyper-mutability, downstream gene expression, and production of β-lactamases (Fernández-Barat et al. (2017).

5.2.1 Origin and Demonstration of Biofilm in CF of Lungs

It is evidenced that (Yoon et al. 2002) different phenotypic forms of *P. aeruginosa* in an anaerobic environment of the lungs are present, which are different from test tube-grown phenotypes. Vandeplassche et al. (2019) have added that in CF lungs, the phenotypes of *P. aeruginosa* cells are AC (acute phenotype) and CP (chronic

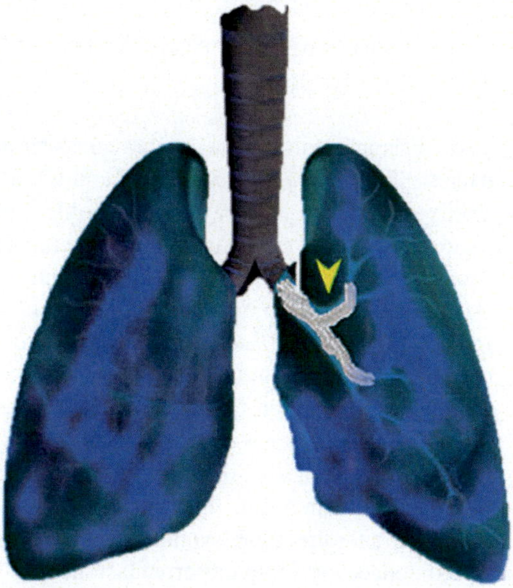

Fig. 5.4 Figure is depicting a typical Cystic fibrosis in the lungs' airways. The wrinkled gray matter (yellow arrow) seen in the airway is indicative of Cystic fibrosis lesions

phenotype) types. AC phenotypes are motile, protease producing, and associated with type III secretion, while CP are involved in biofilm production, drug resistance, but decreased expression of virulence factor.

The CP is differently called mucoid phenotype (MP) due to the overproduction of the viscous exopolysaccharide, called alginate. It is reported that alginate surrounds the bacteria and creates a material that seems quite like mucus, for which the MP can organize biofilm and persists under low-oxygen environments (Lopes et al. 2014). Schick and Kassen (2018) also studied on the CF biofilms and stated that mucin is responsible to increase the viscosity of thick mucus layer, which is developed in CF airways.

Likewise, Kovach et al. (2017) had suggested another extracellular polysaccharide, Psl, which increases the stiffness of biofilm. Further, it was proposed that CdrA, a protein, binds to mannose groups on Psl and is a likely cross-linker for the Psl components of the biofilm matrix.

From the past research, it was observed that the mucoid bacteria being "human serum sensitive" (Fegan et al. 1990) were devoid of rough cell wall as part of lipopolysaccharide. Some of the MP also execute "human serum resistance," with smooth wall, laden with more polysaccharide. It was also added that serum-sensitive MP strains do not activate complement cascade system as effectively as serum-resistant strains, hence conferring a possible selective advantage in the host lung. However, strains of *P. aeruginosa* invading and colonizing the lungs are different from persister cells.

Further, an anaerobic environment in the lungs is prevailed due to the growth of thick biofilm (Panda and Pattnaik 2020), making conditions more favorable for bacterial persistence (Behera and Pattnaik 2019) and, of course, difficult to eradicate (Maurice et al. 2018).

5.2.2 Mechanism Underlying the Biofilm Organization in Lungs in Chronic CF

The CF lung environment is reported to be a specialized niche, for opportunistic bacteria like *P. aeruginosa* (Wagner and Iglewski 2008). This bacterium has enough ability to produce an array of metabolites (elastase, proteases, phospholipase C, hydrogen cyanide, exotoxin A, and exoenzyme S, as well as cell-associated factors, such as lipopolysaccharide, flagella, and pili). These factors more or less are equally contributed to biofilm organizations in the lungs. Additionally, the quorum sensing molecules (QSM) of *P. aeruginosa* have also a major role in regulating the biofilm organization in CF lungs. Barr et al. (2015) had detected, 2-nonyl-4-hydroxy-quinoline (NHQ) as a QSM, which they considered as a biomarker for detecting the degree of CF-associated biofilm development. Further, the role of acyl-homoserine lactones (AHLs) as a QSM cannot be denied for biofilm organization in CF lungs. Smith and Iglewski (2003) had suggested that LasR transcriptional regulator and the LasI synthase genes express the AHL signal molecule, *N*-(3-oxododecanoyl)-L-homoserine lactone (3O-C12-HSL). The airway surface liquid

(ASL), due to its hyper-absorption activity, can produce a concentrated airway mucus that interacts with *P. aeruginosa* for stimulation of biofilm growth (Ong et al. 2017). Further, *P. aeruginosa* being equipped with TFP (type IV pili) can form biofilm very efficiently in CF lungs. Besides, CUP (chaperone-usher pathway) fimbriae also engaged in biofilm production (Varga et al. 2015). While Lozano et al. (2018) had made an extensive study on biofilm-producing *P. aeruginosa* cells and concluded that SCV (small-colony variants) were fast-growing with effective biofilm formation in comparison to mucoid variants (MFs).

Alginate has substantial input in developing a biofilm, while the oxygen radicals induced by polymorphonuclear leukocytes (PMN) generate alginate. As bacteria are inside the host, PMNs reach at the site of infection in response to infection sensitized inflammation (Tan et al. 2018).

This is interesting to note here that *P. aeruginosa* express an intracellular signaling molecule cyclic-di-GMP (c-di-GMP) and gear up early biofilm formation by regulating (a) bacterial motility and (2) EPS (exopolysaccharide) production (Kuchma et al. 2007).

Fig. 5.5 illustrates that the airway surface liquid (ASL) present in airways of CF lungs is dehydrated with floating pockets of "biofilm rafts" (Smith et al. 2013). The typical biofilm rafts are the "interwoven biofilm mats" composed of exopolysaccharide (EPS), residential protein (R-Protein), and extracellular DNA (e-DNA). Both phenotypically and genotypically different *P. aeruginosa* cells encrypt inside the biofilm mats.

5.3 Biofilm and Chronic Wounds

Wounds may be superficial or deep injured and are subjected to healing with the application of chemotherapeutics. During the process of healing, the bacterial intruders may cause infection (Sattar et al. 2019). If bacteria enter deep into the wound and colonize both intra- and intercellularly, then the process of wound healing is delayed. Of late, bacteria may start to produce biofilm, thus resisting the action of drugs. Most chronic wounds contain biofilm that perpetuates the inflammatory phase of wound healing (Attinger and Wolcott 2012). And the wound is defined as a chronic wound (CW) with a delayed healing procedure. The bacterial biofilm is attached to the wound bed and fused with the extracellular matrix (ECM) secreted by the biofilm itself. The oxygen stress, prevailing in chronic wound, also helps the biofilm development (Kim et al. 2019). Although biomolecules, namely, proteins, lipids, and e-DNA, contribute biofilm formation in CWs, EPS is the founder molecule of biofilms. The EPS contained in biofilm acts as a mechanical barrier (Mendoza et al. 2019) to macrophages and dendritic cells, thus enabling bacteria to escape the host's immunological events destined to eradicate intruders.

The type of EPS is dependent on the genera of invading bacterial strains, because the CF is colonized by a multitude of bacterial species, belonging to the *Enterobacteriaceae* family, *Staphylococcaceae* family, or *Pseudomonadaceae*

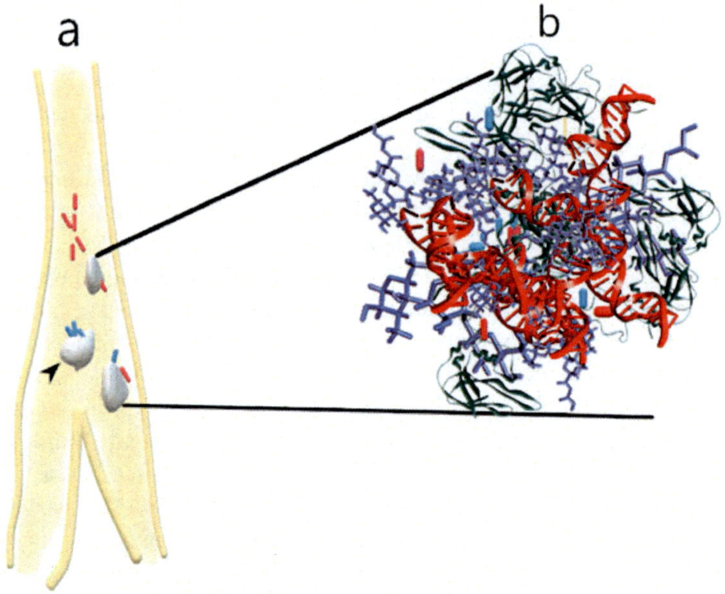

Fig. 5.5 (**a**) Biofilm rafts floating in the mucus of airways of CF lung; (**b**) magnified view of a biofilm raft composed of interwoven exopolysaccharide (EPS), extracellular DNA (e-DNA), and residential protein (after Smith et al. 2013)

family. For example, *S. aureus* and *S. epidermidis* produce exopolysaccharide named polysaccharide intercellular adhesin (PIA), which is poly-*N*-acetyl glucosamine (Joo and Otto 2012).

Chronic wounds (CWs) include diabetic foot ulcers, pressure or decubitus ulcers, venous leg ulcers, and nonhealing surgical site infections (Clinton and Carter 2015). Chronic lower extremity ulcers are those that do not progress through the healing process promptly and have become a major challenge to healthcare systems worldwide (Frykberg and Banks 2015; Wu et al. 2018).

Morgan et al. (2019) had predicted a model on the initiation of biofilm in chronic wounds. It was suggested that chronic infection involves a two-step process, namely, (a) initial phase and (b) established phase of infection. The initial phase is the infection phase with active expression of virulence, while the establishment phase is dominated by biofilm formation and diminishing of virulence. However, biofilm formation itself is a virulence factor as it helps in the retention of infection inside the body.

A typical chronic wound is demonstrated with vasodilation at an early stage followed by clotting in a later phase. The biofilms are seen as gray/dark patches intercalated among the healthy and wound tissues (Fig. 5.6).

Figure 5.7 depicts the invasion of bacterial cells in the ruptured tissue of a typical wound and gradually the development of biofilm in the presence of antibiotics. In chronic wounds, there is a continuous cycle of settlement of planktonic cells on

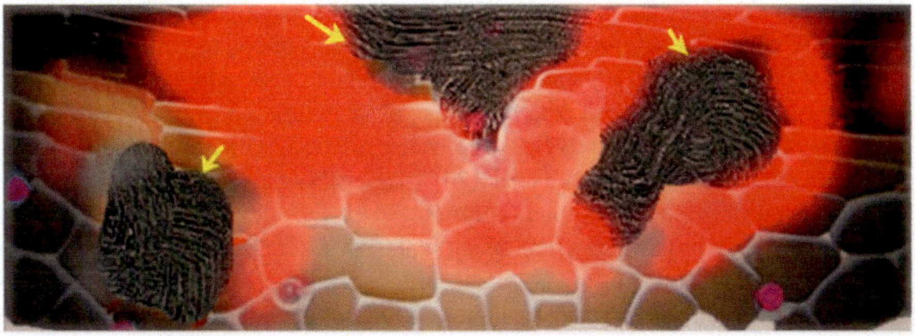

Fig. 5.6 Figure is demonstrating a typical wound with vasodilation and development of biofilm, by encrypting the infecting bacteria

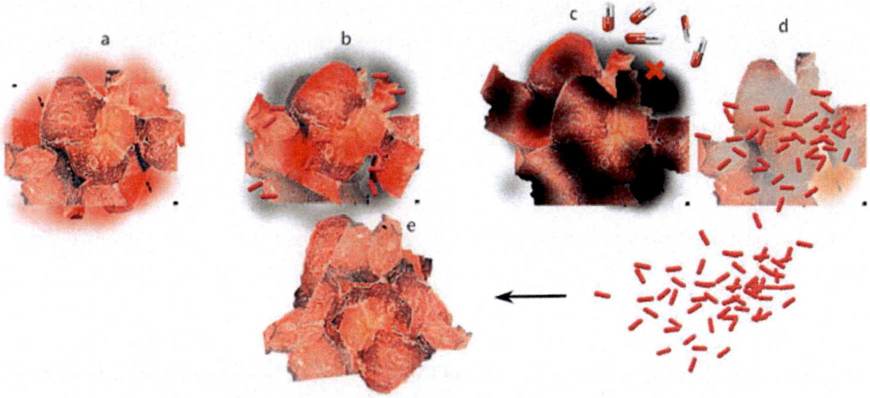

Fig. 5.7 The invasion of bacterial cells in the ruptured tissue of a typical wound and gradually the development of biofilm in the presence of antibiotics. Further, the planktonic cells released from biofilm after damaging the invaded wound tissue sufficiently. Then the bacterial cells search for new tissue to invade

tissue surface with the formation of biofilm and release of sessile bacteria from mature biofilm to invade nearby tissue to initiating the process of biofilm organization.

5.3.1 Signaling Events Leading to Biofilm Formation in Chronic Wounds

In response to bacterial colonization, there is a massive influx of immune cells such as neutrophils and macrophages to the site of a wound, marked by excessive toll-like receptor (TLR) signaling (Shang et al. 2019). To evade the action of immune cells, the bacteria tend to express various defensive mechanisms including biofilm formation. This makes simulation of the inflow of cytokines, chemokines, and growth factors. And there is the creation of a highly proteolytic environment, due to the release of matrix metalloproteinases (MMPs), which are engaged in wound repair (Nguyen et al. 2016). In a parallel event, the immune cells and defective cells are worn out. Accumulation and deficient removal of infiltrating immune cells lead to excessive production of reactive oxygen species (ROS), which creates high oxidative stress in the wound bed. Wu et al. (2018) agreeably had explained the depletion of oxygen consumption, which contributes to the development of hypoxic conditions near a biofilm infection, using mathematical derivations. The prolonged proliferative phase of biofilms in wound beds is also fuelled by the generation of hyperactive alkaline environment which maintained pH values > pH 11.0 (Charles et al. 2019).

5.4 Biofilm and Endophthalmitis

Endophthalmitis is a purulent inflammation of the intraocular fluids, due to bacterial invasion. The vitreous cavity, the anterior chamber of the eye, and adjacent ocular tissues such as the choroid or retina, sclera, or cornea (Maalouf et al. 2012) can also be infected. The traumatic injury, post-surgical tissue, and previously infected corona are the prime entry points of the bacteria. The infection may continue inside the vitreous cavity transiently while amplifying the bacterial population in blood circulation resulting in bacteremia. During this eclipse period, the disease remains asymptomatic. Finally, the infection becomes systemic and leads to fatal consequences.

Among the causal organisms, coagulase-negative staphylococci (CoNS) are responsible for about 70% of post-cataract surgery endophthalmitis, followed by *Staphylococcus aureus* and viridans group streptococci. Besides, enterococci, the oblong-shaped Gram-positive, catalase-negative bacteria, are also notorious to cause endophthalmitis (Callegan et al. 2002). However, vancomycin-resistant and biofilm-producing strains of *Staphylococcus epidermidis* from patients with conjunctivitis, corneal ulcers, and endophthalmitis were reported by Juárez-Verdayes et al. (2006). In the recent past, Priya et al. (2015) had reported about isolation and identification of type III secretion system (T3SS) genotypic strains of *P. aeruginosa*, from the patients, complaints of endophthalmitis after cataract surgery. Notably, the presence of gut microbiota, *Enterococcus faecalis*, is mentioned in a case report (Aderman et al. 2018) of a bilateral recurrent endophthalmitis post-cataract surgery also. However, *Staphylococcus epidermidis* is documented as the most common bacteria to cause post-cataract endophthalmitis infection.

5.4.1 Endophthalmitis and Biofilm

As biofilm is a prime resistant mechanism, adapted by MDR (multidrug-resistant) bacteria in chronic infections, bacteria associated with endophthalmitis also opt for biofilm organization to defend the action of antibacterial agents. In most of the cases, the biofilm growth is observed with the biological implant, namely, the intraocular lenses (IOLs). The surface of IOLs is best suited for the growth of biofilm, because the planktonic bacteria want to settle on a surface only. As *Staphylococcus aureus* possesses several surface-associated adhesins for biofilm formation, it can easily settle on the surfaces of lenses (Leid et al. 2002). The strains of other Gram-positive cocci are also involved in establishing biofilms on IOL surfaces (Suzuki et al. 2005). Shivaji et al. (2019) had mentioned about the involvement of *P. aeruginosa*, *Staphylococcus aureus*, *S. epidermidis*, *Micrococcus luteus*, *S. marcescens*, *Neisseria* spp., *Moraxella* spp., *Bacillus* spp., *E. coli*, *Proteus mirabilis*, *Enterobacter agglomerans*, and *Klebsiella* spp. in organizing biofilm to resist the action of antibiotics. Postoperative endophthalmitis was reported (Park et al. 2014) with IOL biofilms (Fig. 5.8) structured by *Staphylococcus epidermidis*, which is a normal microbiota of human beings. However, being opportunistic, it may be infectious and drug-resistant by forming biofilms. The presence of *S. epidermidis* was substantiated by electron microscopic studies made by Das et al. (2019). Miller et al. (2011) had stated that IOL was a vector for recurrent endophthalmitis, evidenced from DNA analyses.

Not only IOLs but also other devices like reused-reprocessed cassette of the phacoemulsifier machine could be a potent source of the endophthalmitis outbreak (Elnour et al. 2019), caused by *Pseudomonas aeruginosa*.

In a study made by Fazly Bazzaz et al. (2014), it was reported that the *S. epidermidis* prefers to attach itself the hydrophilic acrylic IOLs than hydrophobic IOLs because biofilm is layered on a substrate in water (Tanaka et al. 2019). Therefore, the degree of "wettability" is an important factor for biofilm growth on IOLs, leading to endophthalmitis.

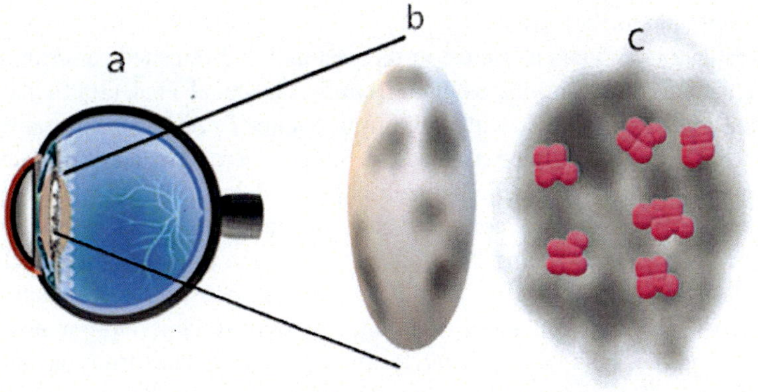

Fig. 5.8 Growth of biofilm on the surface of intraocular lenses (contact lens). (**a**) The CS of a typical eye, where the implantation of IOL is magnified. (**b**) The magnified view of IOL, with patches of biofilm. (**c**) The cluster of staphylococcal cells inside the biofilm

5.5 Biofilm and Keratitis

Keratitis is an inflammation in the cornea of the eye due to infection caused by invasive bacteria. The corneal epithelium, stroma, and endothelium are the sensitive sites, where the bacteria can colonize. Nevertheless, the source of infection is due to the use of contact lenses (McLaughlin-Borlace et al. 1998) by the immunocompromised patients. The pseudomonas keratitis is common in the clinical setup. The other causative organisms like *Streptococcus pneumoniae* spp., *Moraxella*, *Staphylococcus* spp. (Srinivasan et al. 2008), and nontuberculous *Mycobacteria* sp. have also substantial contribution to disease development. Certain predisposing factors, like thermal injury, trauma, UV irradiation, etc., can instigate the bacterial infection. For acute keratitis, topical antibacterial agents are prescribed, but the delay in curing is apprehensive with the development of biofilm. Saraswathi and Beuerman (2015) had suggested that "crystalline keratopathy," an indolent corneal infection caused by *Streptococcus viridans*, *Staphylococcus*, *Pseudomonas*, or *Enterococcus*, could be due to the accumulation of bacterial EPS. It was reported that (Zegans et al. 2016) alginate produced by *Pseudomonas aeruginosa* is the basic compound to form the biofilm matrix on the surface of the contact lens-originated keratitis. Quite some time back, Choy et al. (2008) had studied on strong biofilm formation phenotype of *P. aeruginosa*, and it was inferred that the biofilm formation was regulated by *exoU* gene. It was also published (Holland et al. 2000) that the other types of keratitis, (1) which was associated with the laser beam in a tissue called as in situ keratomileusis and (2) diffuse lamellar keratitis (DLK), were also manifested with biofilm growth and release of endotoxins.

In a nutshell, biofilm is the universal cause of drug resistance as well as delaying the process of therapy in cystic fibrosis, chronic wounds, endophthalmitis, and keratitis. The biological implants like catheters, IOLs, or contact lenses are the contributing factors for the development of biofilm in patients with weak immunological responses. It was found that "drug resistance" is a common phenomenon among the causative agents, specifically *Pseudomonas aeruginosa* and *Staphylococcus aureus/epidermidis* strains.

Investigators had taken interest in the past and had screened for drug resistance toward the prescribed antibiotics. For this purpose, Kamali et al. (2020) had made an "antibiotic sensitivity" test with biofilm-producing *P. aeruginosa* strains isolated from CF patients and inferred that the strains were resistant to all antibiotics including amikacin, piperacillin, tazobactam, and levofloxacin. It was proposed that the multidrug resistance conferred by the bacteria could be due to the interaction with the antibiotics, inducing DNA damage response (the SOS response), and the oxidative stress responses, because these stress responses cause mutation. Singh et al. (2000) had reported about oxacillin, vancomycin, and fluoroquinolone resistance among biofilm-forming *Staphylococcus aureus* isolates from ulcerative keratitis infections. It is observed that the antibiotic resistance is species-specific (Cepas et al. 2019), like gentamicin and ceftazidime resistance that was related to biofilm formation in *Escherichia coli*, piperacillin/tazobactam and colistin in *Klebsiella pneumoniae*, and ciprofloxacin in *Pseudomonas aeruginosa*. The plausible explanation is that the bacterial strains have different types of biofilm constituents used in making each type of biofilm. This is pertinent to mention that the adhesins and extrapolymeric substances used in organizing biofilms are different from each other. But with this comprehensive study, it may be inferred that the biofilm-forming bacterial strains were resistant toward almost all available antibiotics; thus the phenomenon of "pandrug resistance" (Padhan and Pattnaik 2019) may be assigned.

Henceforth, there is a need for novel antibiofilm agents and/or methodologies to combat biofilm-producing bacteria. While pondering about the novel or alternate strategies, it is important to address two important points: (a) disassemblement of biofilm molecules and (b) hooking the cryptic bacterial cells. Once the biofilm is disrupted, the bacteria may be targeted by conventional antibiotics. With these predicaments in mind, the authors had taken interest in the past and were successful to eradicate the biofilm as well as the deep colonizing bacteria in chronic diseases. And so, these agents may be considered as "prospective antibiofilm agents."

5.6 Prospective Antibiofilm Agents

Among the prospective antibiofilm agents, specific enzymes for degradation of EPS, proteins, or e-DNA had been considered by some authors.

5.6.1 EPS-Degrading Enzymes

As EPS contributes a major role in the biofilm matrix, the EPS-degrading enzymes were critically studied. It is reported that the topical application of α-mannosidases could reduce bacterial biofilms along with killing the bacteria infecting the corneas (Kugadas et al. 2019). Keeping aside the possibility of destruction of neutrophils at the site of chronic infection, this enzyme may be tagged with traditional drug moieties, and reformulated antibiotics may be prescribed to manage biofilm-associated infections. Figure 5.9 is a rational model showing the putative mechanism of action of α-mannosidase to thinning of biofilm, thus provoking the bacterial cells coming out of the biofilm matrix and targeting the free planktonic form of bacteria with conventional antibiotics.

In this context, LuTheryn et al. (2019) came forward with the concept of combinatorial approach contemplated with "encapsulation" of biologically active nitric oxide as the gaseous core of a shelled microbubble and consequently agitating mechanically by ultrasonic sound to disturb the biofilm from the surface. The combined mechanical action of the oscillating microbubble and nitric oxide had induced the process of dispersal and elimination of biofilm.

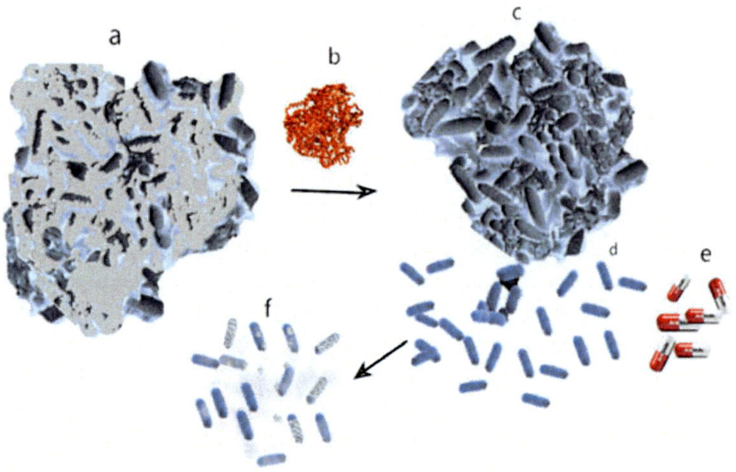

Fig. 5.9 (**a**) A hypothetical model demonstrating the mechanism of action of alpha mannosidase to kill the encrypted bacteria inside biofilms. The biofilms may be thinned out using (**b**) α-mannosidase, the key enzyme to dismantle the (**c**) biofilm matrix, thus releasing the (**d**) cryptic bacteria outside (**e**) Prescribed antibiotic and (**f**) killing the planktonic bacterial cells

5.6.2 Targeting EPS (Extrapolymeric Network) by Alginate Oligosaccharides

Powell et al. (2018) had demonstrated that a low molecular weight antimicrobial agent, OligoG CF-5/20 (an alginate oligosaccharide derived from the marine algae *Laminaria hyperborea* with a guanosine content >85%), could disrupt the biofilm and enable colistin to permeate through the biofilm and kill the disguised Gram-negative non-mucoid *P. aeruginosa* strain. Besides, they had shown a disruption of "e-DNA-Ca^{2+} e-DNA" bridge in the biofilm matrix, rather forming "alginate-Ca^{2+} bridge."

5.6.3 Targeting e-DNA by DNase

Tetz et al. (2009) had pointed out that the extracellular DNA (e-DNA) could be destroyed by DNase I leading to a decrease in the extracellular matrix/EPS, and as a result, the test antibacterial agents could act more effectively to reduce the biofilm biomass and the numbers of CFU (colony-forming unit). However, Aung et al. (2017) had attempted a combinatorial approach (amikacin + gatifloxacin + DNase), for the destruction of the biofilm matrix component, e-DNA, for treating mycobacterial infection in vitro. Chen et al. (2016) had constructed a DNase-mimetic artificial enzyme (DMAE) for antibiofilm by confining passivated gold nanoparticles with multiple cerium (IV) complexes on the surface of colloidal magnetic Fe_3O_4/SiO_2 core/shell particles. Biofilm formation was inhibited for prolonged periods. In the recent past, Swartjes et al. (2013) had applied a DNase I enzyme coating to polymethylmethacrylate, using dopamine as an intermediate. It was asserted that the DNase I coating strongly reduced the adhesion of *Staphylococcus aureus* and *Pseudomonas aeruginosa* and prevented biofilm formation without affecting host cell adhesion and proliferation.

5.6.4 Targeting Quorum Sensing Molecules

Quorum sensing (QS) is considered to be a promising target for the identification and development of antibiofilm agents since this intercellular communication system positively controls the growth and development of biofilms. The *pqs* QS system expressing 2-heptyl-3-hydroxy-4-quinolone of *P. aeruginosa* was inhibited by an FDA-approved drug named clofoctol.

5.6.5 Imposing Negative Pressure to Distract Biofilm

Imposing negative pressure wound therapy (NPWT) has also been widely used to reduce biofilm. Figure 5.10a, b demonstrates negative pressure wound therapy (NPWT) in staphylococcal infected chronic wounds with biofilm. It is illustrated that normal saline is administered into the infected wounds, which creates a negative pressure and imposes the biofilm to be disorganized. When suction pressure is applied, the biofilm-freed bacteria are pulled outside (live/or dead) and are collected in disposed test tubes.

A study made by Li et al. (2020) had summarized that *S. aureus* responds to NPWT by regulating gene expression, manifesting a decrease in biofilm formation, but increasing bacterial adhesion. The NPWTs installed in hospital systems have already added potential benefits (Aycart et al. 2018).

Studies have shown that AMPs (antimicrobial peptides) and their combinations with antibiotics exhibit antibiofilm activities (Chung and Khanum 2017; Batoni et al. 2016). AMPs are naturally occurring polypeptide sequences comprising cationic and hydrophobic short-chain amino acids and have gained the term "HDPs" (host defense proteins). This is relevant to mention that bacteria striving in amino acid starvation, fatty acid limitation, iron limitation, heat shock, etc. trigger the upregulation of the two signaling nucleotides (1) guanosine tetraphosphate (ppGpp) and pentaphosphate (pppGpp) to divert nutrients from growth and division processes to promote survival, ultimately resulting in biofilm formation.

The HDPs bind with these signaling molecules, thus restricting further biofilm growth (Haney et al. 2019). Approximately 106 human-AMPs have been identified to date and are considered as potential antibiofilm agents. Chen et al. (2016) had made an attempt to eradicate of endophthalmitis biofilms using synergistic chemotherapy and photodynamic therapy mediated by zeolitic imidazolate framework-based drug delivery systems.

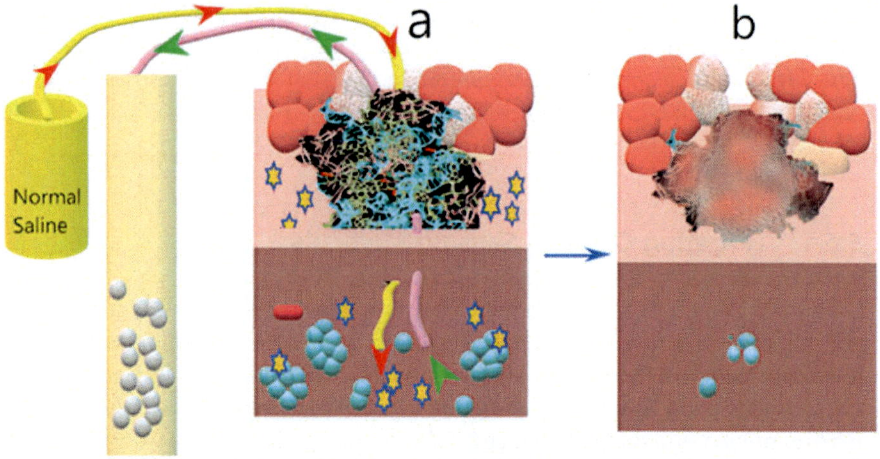

Fig. 5.10 Negative pressure wound therapy (NPWT) in staphylococcal infected chronic wound with biofilm. (**a**) Normal saline is administered as indicated in red arrows into the infected wound, which creates a negative pressure (indicated by stars) imposing the biofilms be disorganized; due to the suction pressure, the biofilm-freed bacteria are pulled outside (live/or dead), indicated by green arrows, which are collected in disposed test tubes. (**b**) The disorganized biofilm in the vicinity of the wound and presence of a few no. of bacterial cells

5.6.6 Biofilm Gene Knockout Strategies

"Gene knockout" strategies are observed to be most contemporary in the context of elimination of biofilms in clinical systems. Noirot-Gros et al. (2019) had adapted the CRISPRi (clustered regularly interspaced short palindromic repeats interference) "gene knockout approach" against biofilm-associated gene (*gacS*) present in a strain of *P. fluorescens.*

The Cas9 (CRISPR-associated protein 9) nuclease complexed with a synthetic guide RNA (gRNA) was delivered into a bacterial cell's genome, pre-nicked at a desired sequence. The integration was made near short DNA repeats with the help of Cas9 proteins. As a result, an array of repeat sequences was created in the host chromosome interspersed by short, unique spacers. The spacer precursors (proto-spacers) from the invading nucleic acids were selected by unique proto-spacer adjacent motifs (PAMs), as depicted in Fig. 5.11.

5.6.7 Nanoparticle-Based Technology

Nanoparticle-based approaches are spotted as a promising technology to reduce the incidence of biofilm growth in chronic wound infections. Kalishwaralal et al. (2010) had studied about biologenetically synthesized silver nanoparticles, which had exhibited a potential antibiofilm activity against biofilms formed by *Pseudomonas aeruginosa* and *Staphylococcus epidermidis* the causal organisms of microbial

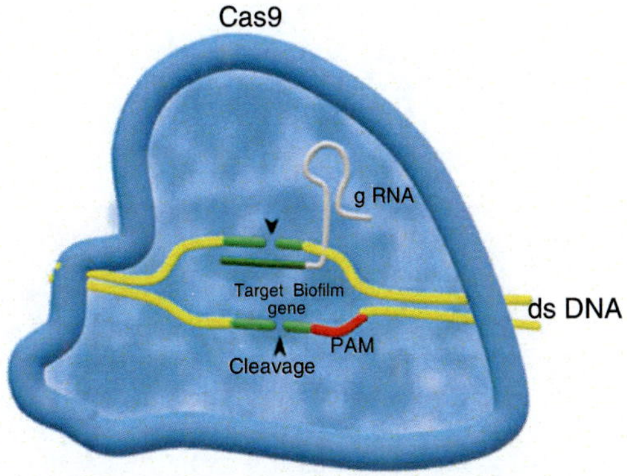

Fig. 5.11 The CRISPRi system is practiced to knock out biofilm genes. The Cas9 nuclease complexed with guide RNA (gRNA) into a bacterial cell genome, nick near a specific sequence. The oligonucleotides are integrated into bacterial genome near short DNA repeats with the help Cas9 protein. Repeat sequences are created in the host chromosome interspersed by short, unique spacers. The selection of spacer precursors (proto-spacers) from the invading nucleic acids is achieved by unique proto-spacer adjacent motifs (PAMs), indicated by the red patch (after Choudhury et al. 2016) and upon induction of Cas9 expression, test *Pseudomonas fluorescence* cells expressing the *gac* short nucleotide guide were impaired for swarming motility, while disorganization biofilm formation was demonstrated upon the silencing of *gacS* gene

keratitis. Figure 5.12 shows an accepted model of the application of nanoparticles in reducing the biofilm burden of infected tissues or biological implants. Nanoparticles are packed with anti-adhesion agents, thus restricting the attachment of bacterial cells on the surface. Further, nanoparticles imbued with ROS-producing agents, which can be released near the biofilms and release super oxides (O_2^-), hydroxyl (HO), peroxyl (ROO), perhydroxyl radicals (HO_2), and singlet oxygen ($^1O_2$), or conventional drug-impregnated nanoparticles may be administered to enter inside the biofilm and consequently kill the hidden bacteria. Likewise, nanoparticles sensitive to internal stimuli such as varying pH, concentrations of specific enzymes, or chemicals deliver the drug near the targets only.

5.7 Conclusion and Prospective

From this brief analysis of the biofilm-associated nosocomial as well as chronic infections, it was concluded that the biofilm is pervasive in all types of infections. The bacterial strains causing either clinical or nosocomial infections can build a biofilm matrix to resist the action of prescribed antibiotics. Biofilm also gives a suitable platform for the transfer of antibiotic resistance genes (ARGs) horizontally, by facilitating close contact between conjugation compatible bacteria, thus impeding the spread of drug resistance. The most important feature of biofilm is it coats the persisters, which are obviously the "genotypic turmoil" of a bacterial strain. The slow recovery rate of cystic fibrosis, chronic wound infections, and endophthalmitis

Fig. 5.12 The application of nanoparticles impregnated with biofilm target specific antibiofilm agents. (**a**) Nanoparticles are packed with anti-adhesion agents, thus restricting the attachment of bacterial cells on the surface. (**b**) Nanoparticles are imbued with ROS-producing agents, which are released near the biofilms and releasing super oxides (O_2^-), hydroxyl (HO), peroxyl (ROO), perhydroxyl radicals (HO_2), and singlet oxygen ($^1O_2$). (**c**) Conventional drug-impregnated nanoparticles are administered, which can penetrate inside the biofilm and consequently kill the hidden bacteria. (**d**) Nanoparticles are responding to the internal stimuli such as varying pH, concentrations of specific enzymes, or chemicals, which are associated with pathological conditions of infection and inflammation, thus releasing drugs in controlled delivery manner

models described in this chapter gives an insight into the significance of biofilm in the clinical system. As the prescribed antibiotics act as "biofilm-triggering" agents, it is a mandate to opt for alternative strategies to dismantle the biofilm and target the isolated bacterial cells by conventional antibiotics. Therefore, researchers have taken interest and experimented with novel approaches to dismantle the biofilm as well as to kill the invading bacteria. Concisely, most of the authors have given importance to the combinatorial approach encompassing both the antibiofilm agent and antibacterial target in one bucket. Nevertheless, this fact cannot be denied that biofilm is a serious threat to mankind.

References

Abe KNN, Suzuki S (2020) Biofilms: hot spots of horizontal gene transfer (HGT) in aquatic environments, with a focus on a new HGT mechanism. FEMS Microbiol Ecol 96(5):fiaa031

Aderman CM, Chao DL, Obeid A, Schwartz DM, Bhisitkul RB, Chiu CS (2018) Bilateral *Enterococcus faecalis* endophthalmitis with multiple recurrences. Retin Cases Brief Rep 15 (1):38–42

Attinger C, Wolcott R (2012) Clinically addressing biofilm in chronic wounds. Adv Wound Care 1 (3):127–132

Aung TT, Chor WHJ, Yam JKH, Givskov M, Yang L, Beuerman RW (2017) Discovery of novel antimycobacterial drug therapy in biofilm of pathogenic nontuberculous Mycobacterial keratitis. Ocular Surf 15(4):770–783

Aycart MA, Eble DJ, Ross KM, Orgill DP (2018) Mechanisms of action of instillation and dwell negative pressure wound therapy with case reports of clinical applications. Cureus 10(9):e3377

Barr HL, Halliday N, Cámara M, Barrett DA, Williams P, Forrester DL, Simms R, Smyth AR, Honeybourne D, Whitehouse JL, Nash EF (2015) *Pseudomonas aeruginosa* quorum sensing molecules correlate with clinical status in cystic fibrosis. Eur Respir J 46(4):1046–1054

Batoni G, Maisetta G, Esin S (2016) Antimicrobial peptides and their interaction with biofilms of medically relevant bacteria. Biochim Biophys Acta (BBA)-Biomembr 1858(5):1044–1060

Bazzaz BSF, Khameneh B, Jalili-Behabadi MM, Malaekeh-Nikouei B, Mohajeri SA (2014) Preparation, characterization and antimicrobial study of a hydrogel (soft contact lens) material impregnated with silver nanoparticles. Contact Lens Anter Eye 37(3):149–152

Behera S, Pattnaik S (2019) Persister cell development among Enterobacteriaceae, Pseudomonadaceae, Mycobacteriaceae and Staphylococcaceae biotypes: a review. Biocatal Agricult Biotechnol 22:101401

Callegan MC, Engelbert M, Parke DW, Jett BD, Gilmore MS (2002) Bacterial endophthalmitis: epidemiology, therapeutics, and bacterium-host interactions. Clin Microbiol Rev 15(1):111–124

Cepas V, López Y, Muñoz E, Rolo D, Ardanuy C, Martí S, Xercavins M, Horcajada JP, Bosch J, Soto SM (2019) Relationship between biofilm formation and antimicrobial resistance in Gram-negative Bacteria. Microb Drug Resist 25(1):72–79

Charles CJ, Rout SP, Wormald R, Laws AP, Jackson BR, Boxall SA, Humphreys PN (2019) *In-situ* biofilm formation in hyper alkaline environments. Geomicrobiol J 36(5):405–411

Chen Z, Ji H, Liu C, Bing W, Wang Z, Qu X (2016) A multinuclear metal complex based DNase-mimetic artificial enzyme: matrix cleavage for combating bacterial biofilms. Angew Chem 128 (36):10890–10894

Choudhury SR, Cui Y, Lubecka K, Stefanska B, Irudayaraj J (2016) CRISPR-d Cas9 mediated TET1 targeting for selective DNA demethylation at BRCA1 promoter. Oncotarget 7(29):46545

Choy MH, Stapleton F, Willcox MD, Zhu H (2008) Comparison of virulence factors in *Pseudomonas aeruginosa* strains isolated from contact lens-and non-contact lens-related keratitis. J Med Microbiol 57(12):1539–1546

Chung PY, Khanum R (2017) Antimicrobial peptides as potential anti-biofilm agents against multidrug-resistant bacteria. J Microbiol Immunol Infect 50(4):405–410

Clinton A, Carter T (2015) Chronic wound biofilms: pathogenesis and potential therapies. Lab Med 46(4):277–284

Dani A (2014) Colonization and infection. Central Eur J Urol 67(1):86

Das D, Bhattacharjee H, Gogoi K, Das J, Misra P, Dhir P, Deka A (2019) Intraocular lens biofilm formation supported by scanning electron microscopy imaging. Indian J Ophthalmol 67 (10):1708–1709

Elnour AA, Negm S, Ismail A, Elhusseiny MA (2019) An outbreak of post cataract *Pseudomonas aeruginosa* acute endophthalmitis in Egypt. Bull Natl Res Centre 43(1):13

Fegan MARK, Francis PAUL, Hayward AC, Davis GH, Fuerst JA (1990) Phenotypic conversion of *Pseudomonas aeruginosa* in cystic fibrosis. J Clin Microbiol 28(6):1143–1146

Fernández-Barat L, Ciofu O, Kragh KN, Pressler T, Johansen U, Motos A, Torres A, Hoiby N (2017) Phenotypic shift in *Pseudomonas aeruginosa* populations from cystic fibrosis lungs after 2-week antipseudomonal treatment. J Cyst Fibros 16(2):222–229

Frykberg RG, Banks J (2015) Challenges in the treatment of chronic wounds. Adv Wound Care 4 (9):560–582

Gil-Perotin S, Ramirez P, Marti V, Sahuquillo JM, Gonzalez E, Calleja I, Menendez R, Bonastre J (2012) Implications of endotracheal tube biofilm in ventilator-associated pneumonia response: a state of concept. Crit Care 16(3):1–9

Grant SS, Hung DT (2013) Persistent bacterial infections, antibiotic tolerance, and the oxidative stress response. Virulence 4(4):273–283

Haney EF, Barbosa SC, Baquir B, Hancock RE (2019) Influence of non-natural cationic amino acids on the biological activity profile of innate defense regulator peptides. J Med Chem 62 (22):10294–10304

Harrington NE, Sweeney E, Harrison F (2020) Building a better biofilm-formation of *in vivo*-like biofilm structures by *Pseudomonas aeruginosa* in a porcine model of cystic fibrosis lung infection. Biofilms 2020:100024

Høiby N, Bjarnsholt T, Moser C, Jensen PØ, Kolpen M, Qvist T, Aanaes K, Pressler T, Skov M, Ciofu O (2017) Diagnosis of biofilm infections in cystic fibrosis patients. APMIS 125 (4):339–343

Holland SP, Mathias RG, Morc DW, Chiu J, Slade SG (2000) Diffuse lamellar keratitis related to endotoxins released from sterilizer reservoir biofilms. Ophthalmology 107(7):1227–1233

Issa HB, Phan G, Broutin I (2018) Functional mechanism of the efflux pumps transcription regulators from *Pseudomonas aeruginosa* based on 3D structures. Front Mol Biosci 5:57

Joo HS, Otto M (2012) Molecular basis of in vivo biofilm formation by bacterial pathogens. Chem Biol 19(12):1503–1513

Juárez-Verdayes MA, Reyes-López MÁ, Cancino-Díaz ME, Muñoz-Salas S, Rodríguez-Martínez S, De la Serna FJ, Hernández-Rodríguez CH, Cancino-Díaz JC (2006) Isolation, vancomycin resistance and biofilm production of *Staphylococcus epidermidis* from patients with conjunctivitis, corneal ulcers, and endophthalmitis. Rev Latinoam Microbiol 48 (3–4):238–246

Kalishwaralal K, BarathManiKanth S, Pandian SR, Deepak V, Gurunathan S (2010) Silver nanoparticles impede the biofilm formation by Pseudomonas aeruginosa and Staphylococcus epidermidis. Colloids Surf B Biointerfaces 79:340–344

Kamali E, Jamali A, Ardebili A, Ezadi F, Mohebbi A (2020) Evaluation of antimicrobial resistance, biofilm forming potential, and the presence of biofilm-related genes among clinical isolates of *Pseudomonas aeruginosa*. BMC Res Notes 13(1):27

Kapoor G, Saigal S, Elongavan A (2017) Action and resistance mechanisms of antibiotics: a guide for clinicians. J Anaesthesiol Clin Pharmacol 33(3):300

Khan RA, Aziz Z (2017) A retrospective study of antibiotic de-escalation in patients with ventilator-associated pneumonia in Malaysia. Int J Clin Pharm 39(4):906–912

Kim JH, Yang B, Tedesco A, Lebig EGD, Ruegger PM, Xu K, Borneman J, Martins-Green M (2019) High levels of oxidative stress and skin microbiome are critical for initiation and development of chronic wounds in diabetic mice. Sci Rep 9(1):1–16

Klemm P, Schembri MA (2000) Bacterial adhesins: function and structure. Int J Med Microbiol 290 (1):27–35

Kovach K, Davis-Fields M, Irie Y, Jain K, Doorwar S, Vuong K, Dhamani N, Mohanty K, Touhami A, Gordon VD (2017) Evolutionary adaptations of biofilms infecting cystic fibrosis lungs promote mechanical toughness by adjusting polysaccharide production. npj Biofilm Microb 3(1):1–9

Kuchma SL, Brothers KM, Merritt JH, Liberati NT, Ausubel FM, O'Toole GA (2007) BifA, a cyclic-Di-GMP phosphodiesterase, inversely regulates biofilm formation and swarming motility by *Pseudomonas aeruginosa* PA14. J Bacteriol 189(22):8165–8178

Kugadas A, Geddes-McAlister J, Guy E, Di Giandomenico A, Sykes DB, Mansour MK, Mirchev R, Gadjeva M (2019) Frontline science: employing enzymatic treatment options for management of ocular biofilm-based infections. J Leukoc Biol 105(6):1099–1110

Leid JG, Costerton JW, Shirtliff ME, Gilmore MS, Engelbert M (2002) Immunology of Staphylococcal biofilm infections in the eye: new tools to study biofilm endophthalmitis. DNA Cell Biol 21(5–6):405–413

Li T, Wang G, Yin P, Li Z, Zhang L, Tang P (2020) Adaptive expression of biofilm regulators and adhesion factors of *Staphylococcus aureus* during acute wound infection under the treatment of negative pressure wound therapy in vivo. Exp Therap Med 20(1):512–520

Lopes SP, Azevedo NF, Pereira MO (2014) Emergent bacteria in cystic fibrosis: in vitro biofilm formation and resilience under variable oxygen conditions. Biomed Res Int 2014:678301

Lozano C, Azcona-Gutiérrez JM, Van Bambeke F, Sáenz Y (2018) Great phenotypic and genetic variation among successive chronic Pseudomonas aeruginosa from a cystic fibrosis patient. PLoS One 13(9):e0204167

LuTheryn G, Hind C, Glynne-Jones P, Webb JS, Sutton JM, Carugo D (2019) Ultrasound-responsive bioactive-microbubbles for the eradication of biofilms. In 24th Eur. Symp. Ultrasound Contrast Imaging

Maalouf F, Abdulaal M, Hamam RN (2012) Chronic postoperative endophthalmitis: a review of clinical characteristics, microbiology, treatment strategies, and outcomes. Int J Inflamm 2012:313248. https://doi.org/10.1155/2012/313248

Maurice NM, Bedi B, Sadikot RT (2018) *Pseudomonas aeruginosa* biofilms: host response and clinical implications in lung infections. Am J Respir Cell Mol Biol 58(4):428–439

McLaughlin-Borlace L, Stapleton F, Matheson M, Dart JKG (1998) Bacterial biofilm on contact lenses and lens storage cases in wearers with microbial keratitis. J Appl Microbiol 84 (5):827–838

Mendoza RA, Hsieh JC, Galiano RD (2019) The impact of biofilm formation on wound healing. Wound Healing-Curr Perspect. https://doi.org/10.5772/intechopen.85020

Miller KV, Eisley KM, Shanks RM, Lahr RM, Lathrop KL, Kowalski RP, Noecker RJ (2011) Recurrent enterococcal endophthalmitis seeded by an intraocular lens biofilm. J Cataract Refract Surg 37(7):1355–1359

Morgan SJ, Lippman SI, Bautista GE, Harrison JJ, Harding CL, Gallagher LA, Cheng AC, Siehnel R, Ravishankar S, Usui ML, Olerud JE (2019) Bacterial fitness in chronic wounds appears to be mediated by the capacity for high-density growth, not virulence or biofilm functions. PLoS Pathog 15(3):e1007511

Munita JM, Arias CA (2016) Mechanisms of antibiotic resistance. Virulence mechanisms of bacterial pathogens. Microbiol Spectr 4:481–511

Naclerio GA, Onyedibe KI, Sintim HO (2020) Lipoteichoic Acid Biosynthesis Inhibitors as Potent Inhibitors of *S. aureus* and *E. faecalis* Growth and Biofilm Formation. Molecules 25(10):2277

Nguyen TT, Mobashery S, Chang M (2016) Roles of matrix metalloproteinases in cutaneous wound healing. InTech, London, pp 37–71. https://doi.org/10.5772/intechopen.85020

Noirot-Gros MF, Forrester S, Malato G, Larsen PE, Noirot P (2019) CRISPR interference to interrogate genes that control biofilm formation in Pseudomonas fluorescens. Sci Rep 9(1):1–14

O'Rourke A, Beyhan S, Choi Y, Morales P, Chan AP, Espinoza JL, Dupont CL, Meyer KJ, Spoering A, Lewis K, Nierman WC (2020) Mechanism-of-action classification of antibiotics by global transcriptome profiling. Antimicrob Agent Chemother 64(3):e01207–e01219

Ong T, Marshall SG, Karczeski BA, Sternen DL, Cheng E, Cutting GR (2017) Cystic fibrosis and congenital absence of the vas deferens. In GeneReviews. University of Washington, Seattle

Otto M (2008) Targeted immunotherapy for staphylococcal infections. BioDrugs 22(1):27–36

Owens CD, Stoessel K (2008) Surgical site infections: epidemiology, microbiology and prevention. J Hosp Infect 70:3–10

Padhan M, Pattnaik S (2019) Biotyping and Phylogenetic analysis of bacterial isolates prevalent in a local hospital environment. Res J Biotechnol 14(11):52–61

Paharik AE, Horswill AR (2016) The staphylococcal biofilm: adhesins, regulation, and host response. Virulen Mechan Bact Pathog 4:529–566

Panda SP, Pattnaik S (2020) Effect of environmental factors on biofilm production of the bacterial strains isolated from the patient with nosocomial infection. Res J Biotechnol 15(6):67–75

Park JC, Ramasamy B, Shaw S, Ling RH, Prasad S (2014) A prospective and nationwide study investigating endophthalmitis following pars plana vitrectomy: clinical presentation, microbiology, management and outcome. Br J Ophthalmol 98(8):1080–1086

Penesyan A, Nagy SS, Kjelleberg S, Gillings MR, Paulsen IT (2019) Rapid microevolution of biofilm cells in response to antibiotics. NPJ Biofilm Microb 5(1):1–14

Perez LRR, Costa MCN, Freitas ALPD, Barth AL (2011) Evaluation of biofilm production by *Pseudomonas aeruginosa* isolates recovered from cystic fibrosis and non-cystic fibrosis patients. Braz J Microbiol 42(2):476–479

Peterson E, Kaur P (2018) Antibiotic resistance mechanisms in bacteria: relationships between resistance determinants of antibiotic producers, environmental bacteria, and clinical pathogens. Front Microbiol 9:2928

Powell LC, Pritchard MF, Ferguson EL, Powell KA, Patel SU, Rye PD, Sakellakou SM, Buurma NJ, Brilliant CD, Copping JM, Menzies GE (2018) Targeted disruption of the extracellular polymeric network of *Pseudomonas aeruginosa* biofilms by alginate oligosaccharides. NPJ Biofilm Microb 4(1):1–10

Priya JL, Prajna L, Mohankumar V (2015) Genotypic and phenotypic characterization of *Pseudomonas aeruginosa* isolates from post-cataract endophthalmitis patients. Microb Pathog 78:67–73

Rojo-Molinero E, Macià MD, Oliver A (2019) Social behavior of antibiotic resistant mutants within *Pseudomonas aeruginosa* biofilm communities. Front Microbiol 10:570

Saraswathi P, Beuerman RW (2015) Corneal biofilms: from planktonic to microcolony formation in an experimental keratitis infection with *Pseudomonas aeruginosa*. Ocular Surf 13(4):331–345

Sarowska J, Futoma-Koloch B, Jama-Kmiecik A, Frej-Madrzak M, Ksiazczyk M, Bugla-Ploskonska G, Choroszy-Krol I (2019) Virulence factors, prevalence and potential transmission of extraintestinal pathogenic *Escherichia coli* isolated from different sources: recent reports. Gut Pathog 11(1):10

Sattar F, Sattar Z, Mohsin Zaman SA (2019) Frequency of post-operative surgical site infections in a Tertiary care hospital in Abbottabad, Pakistan. Cureus 11(3):e4243

Schick A, Kassen R (2018) Rapid diversification of *Pseudomonas aeruginosa* in cystic fibrosis lung-like conditions. Proc Natl Acad Sci 115(42):10714–10719

Shaikh S, Fatima J, Shakil S, Rizvi SMD, Kamal MA (2015) Antibiotic resistance and extended spectrum beta-lactamases: types, epidemiology and treatment. Saudi J Biol Sci 22(1):90–101

Shang L, Deng D, Buskermolen JK, Roffel S, Janus MM, Krom BP, Crielaard W, Gibbs S (2019) Commensal and pathogenic biofilms alter toll-like receptor signaling in reconstructed human gingiva. Front Cell Infect Microbiol 9:282

Shivaji S, Konduri R, Ramchiary J, Jogadhenu PS, Arunasri K, Savitri S (2019) Gene targets in ocular pathogenic *Escherichia coli* for mitigation of biofilm formation to overcome antibiotic resistance. Front Microbiol 10:1308

Singh PK, Schaefer AL, Parsek MR, Moninger TO, Welsh MJ, Greenberg EP (2000) Quorum-sensing signals indicate that cystic fibrosis lungs are infected with bacterial biofilms. Nature 407 (6805):762–764

Singh S, Singh SK, Chowdhury I, Singh R (2017) Understanding the mechanism of bacterial biofilms resistance to antimicrobial agents. Open Microbiol J 11:53

Smith RS, Iglewski BH (2003) *P. aeruginosa* quorum-sensing systems and virulence. Curr Opin Microbiol 6(1):56–60

Smith DJ, Lamont IL, Anderson GJ, Reid DW (2013) Targeting iron uptake to control *Pseudomonas aeruginosa* infections in cystic fibrosis. Eur Respir J 42(6):1723–1736

Srinivasan M, Mascarenhas J, Prashanth CN (2008) Distinguishing infective versus noninfective keratitis. Indian J Ophthalmol 56(3):203

Sun F, Qu F, Ling Y, Mao P, Xia P, Chen H, Zhou D (2013) Biofilm-associated infections: antibiotic resistance and novel therapeutic strategies. Future Microbiol 8(7):877–886

Suzuki T, Uno T, Kawamura Y, Joko T, Ohashi Y (2005) Postoperative low-grade endophthalmitis caused by biofilm-producing coccus bacteria attached to posterior surface of intraocular lens. J Cataract Refract Surg 31(10):2019–2020

Swartjes JJ, Das T, Sharifi S, Subbiahdoss G, Sharma PK, Krom BP, Busscher HJ, van der Mei HC (2013) A functional DNase I coating to prevent adhesion of bacteria and the formation of biofilm. Adv Funct Mater 23(22):2843–2849

Tan Q, Ai Q, Xu Q, Li F, Yu J (2018) Polymorphonuclear leukocytes or hydrogen peroxide enhance biofilm development of mucoid *Pseudomonas aeruginosa*. Mediat Inflamm 2018:8151362. https://doi.org/10.1155/2018/8151362

Tanaka N, Kogo T, Hirai N, Ogawa A, Kanematsu H, Takahara J, Awazu A, Fujita N, Haruzono Y, Ichida S, Tanaka Y (2019) *In-situ* detection based on the biofilm hydrophilicity for environmental biofilm formation. Sci Rep 9(1):1–11

Tetz GV, Artemenko NK, Tetz VV (2009) Effect of DNase and antibiotics on biofilm characteristics. Antimicrob Agent Chemother 53(3):1204–1209

Trastoy R, Manso T, Fernandez-Garcia L, Blasco L, Ambroa A, Del Molino MP, Bou G, Garcia-Contreras R, Wood TK, Tomas M (2018) Mechanisms of bacterial tolerance and persistence in the gastrointestinal and respiratory environments. Clin Microbiol Rev 31(4):e00023–e00018

Vandeplassche E, Sass A, Lemarcq A, Dandekar AA, Coenye T, Crabbé A (2019) *In vitro* evolution of *Pseudomonas aeruginosa* AA2 biofilms in the presence of cystic fibrosis lung microbiome members. Sci Rep 9(1):1–14

Varga JJ, Barbier M, Mulet X, Bielecki P, Bartell JA, Owings JP, Martinez-Ramos I, Hittle LE, Davis MR, Damron FH, Liechti GW (2015) Genotypic and phenotypic analyses of a *Pseudomonas aeruginosa* chronic bronchiectasis isolate reveal differences from cystic fibrosis and laboratory strains. BMC Genomics 16(1):883

Wagner VE, Iglewski BH (2008) *P. aeruginosa* biofilms in CF infection. Clin Rev Allergy Immunol 35(3):124–134

Wang J, Foxman B, Mody L, Snitkin ES (2017) Network of microbial and antibiotic interactions drive colonization and infection with multidrug-resistant organisms. Proc Natl Acad Sci 114 (39):10467–10472

Wu Y, Klapper I, Stewart PS (2018) Hypoxia arising from concerted oxygen consumption by neutrophils and microorganisms in biofilms. Pathog Dis 76(4):fty043

Wu YK, Cheng NC, Cheng CM (2019) Biofilms in chronic wounds: pathogenesis and diagnosis. Trends Biotechnol 37(5):505–517

Yoon SS, Hennigan RF, Hilliard GM, Ochsner UA, Parvatiyar K, Kamani MC, Allen H, DeKievit TR, Gardner PR, Schwab U, Rowe JJ (2002) *Pseudomonas aeruginosa* anaerobic respiration in biofilms: relationships to cystic fibrosis pathogenesis. Dev Cell 3(4):593–603

Yu D, Zhao L, Xue T, Sun B (2012) *Staphylococcus aureus* autoinducer-2 quorum sensing decreases biofilm formation in an icaR-dependent manner. BMC Microbiol 12(1):288

Zegans ME, DiGiandomenico A, Ray K, Naimie A, Keller AE, Stover CK, Lalitha P, Srinivasan M, Acharya NR, Lietman TM (2016) Association of biofilm formation, *Psl* exopolysaccharide expression, and clinical outcomes in *Pseudomonas aeruginosa* keratitis: analysis of isolates in the steroids for corneal ulcers trial. JAMA Ophthalmol 134(4):383–389

Immune Response to Biofilm

6

Sreejita Ghosh and Rina Rani Ray

Abstract

Bacterial biofilms are considered as one of the most potent and destructive reasons for various persistent and chronic infections. Bacterial biofilm networks are intricately designed for each species of bacteria so that they can efficiently evade the host immune reactions (both innate and adaptive immune responses) and continue to survive within the host. The biofilms mainly target their defense mechanisms against polymorphonuclear cells (neutrophils) which are the main cells to invade sites of infection. In this chapter, we go on to describe how the host immune cells respond to biofilms and the mechanisms employed by the biofilms to counter the host immunity. These biofilms may release different toxic products that are responsible for overcoming the antimicrobial activity of neutrophil extracellular traps (NET) in the host, or some of the biofilms may increase their network and become concentrated to evade the host complement system, and thereby efficient biofilm clearance cannot be achieved. Moreover, the biofilms tend to skew the host T-cell response leading to a standstill between pathogen and host, and thus the infection becomes chronic.

6.1 Introduction

Bacteria after penetrating the normal barriers of the host body adhere to receptors. They survive several filters of the defense system of the body to settle in the host body and cause disease.

S. Ghosh · R. R. Ray (✉)
Department of Biotechnology, Maulana Abul Kalam Azad University of Technology, West Bengal, Kolkata, West Bengal, India

R. R. Ray et al. (eds.), *Biofilm-Mediated Diseases: Causes and Controls*,
https://doi.org/10.1007/978-981-16-0745-5_6

The members of the immune system (such as antibodies, polymorphonuclear neutrophils (granulocytes), mononuclear phagocytes from the lung, spleen and from other organs) arrive, multiply, and recruit immune cells at the infection site to eliminate the bacteria from the host body. This reaction may be local, i.e., *in statu nascendi*, or systemic in nature. The vibrant interaction between invaded bacteria and the immune reaction of the host leads to the establishment of infection ranging from acute to chronic diseases, with reduced immune responses.

In spite of a well-functioning immune system, sometimes host body can not completely eliminate the bacteria resulting in chronic or permanent infections. A typical reason for the weakened abatement of the microorganism is due to the formation of biofilms by pathogenic bacteria. It is difficult to eradicate biofilms as they can withstand antibiotics and immune response of human body resulting in chronic infections. Biofilm-forming microorganisms can grow as both free-flowing (planktonic cells) or embedded (sessile cells) and are responsible for acute and chronic infections. The course of the infection (acute or chronic) is mostly independent of the type of the microorganisms infecting.

Biofilm-forming infections can be of two types: native biofilm infections directly on host tissues and bacterial infections on medical devices and implants (Sun et al. 2013).

Bacterial virulence and impaired immune system are responsible for the establishment of an acute infection. On the other hand, a variety of other host factors can predispose to biofilm-related infections like intra- or extravascular devices like orthopedic implant, pacemakers, heart valves, vascular grafts, and stents which cause manyfold increase in the danger of developing biofilm-related infection. Again, presence of damaged cells, necrotic tissues due to impaired vascularization, inefficient ciliary movement of mucosal membrane, and insufficient activities of polymorphonuclear neutrophils (PMNs) at the site of infection lead to the generation of different chronic infections. The immunological response against biofilm-forming bacteria is intricate and involves the cumulative effect of innate as well as adaptive immune systems (Maurice et al. 2018).

On the other hand, microbes of human microbiome remain as multispecies biofilm (de Vos 2015) in specific niches. The host shows immunotolerance to these microbes to maintain a kind of immunostasy.

The present chapter focuses on the immune response exerted by the host body following invasion of biofilm-forming bacteria, which may be local reaction or systemic response and the strategies adopted by opportunistic pathogens residing in biofilm to overcome host immune responses. The immunotolerance exhibited by the host body toward the commensal colonizing bacteria will also be highlighted.

6.2 Immune Response for Pathogens and Pathobionts

6.2.1 Local Immune Response

Biofilm-forming infections can be of two types: native biofilm infections directly on host tissues and bacterial infections on medical devices and implants (Sun et al. 2013).

6.2.1.1 Immune Response from Host Tissues

The immune system of our body starts responding just after the invasion of bacteria in normal germ-free locations of the human body. It involves a complex interaction of signals with the cell populations and is generally been divided into the innate response and adaptive (cell mediated and humoral) immune responses. Innate immunity consists of common epithelial barrier, complement, microbiome of the host, and a wide variety of leukocytes comprising non-phagocytic (natural killer) and phagocytic (neutrophils and macrophages) cells. The host body employs the soldiers of its innate immune system to remove the invading pathogen. It is accomplished by the recruitment of classical and alternative complement pathway, mannose binding pathway, opsonization, and phagocytosis.

After entry in the host body, bacteria face the first line of defense as physical barrier at epithelial tissues. The skin is always exposed to symbiotic microorganisms. Antimicrobial peptides (AMPs) produced by symbiotic microbiota of the skin and also the presence of keratinocytes actively take part in providing protection against invasive bacterial infections (Herman and Herman 2018). Phospholipid head groups carrying negative charge on the bacterial cell wall are targeted by positively charged domains of AMP (Zasloff 2002). Although, the complete mechanism for the function of AMP is still not totally known, yet it is found that functions of AMPs are triggered by TNF, LPS, and IL-1β. All these functions are enhanced by inflammation due to microbial attack and are the main ways for downstream modulation of the immune system (Zasloff 2002). In addition to this, electrolyte concentration at the luminal side is controlled by ion pumps, thereby maintaining optimum state for AMP to function (Zasloff 2002).

Neutrophils, after sensing a localized infection, migrate actively from blood vessels to the site of infection. After arrival at the infection site, they engulf bacteria by phagocytosis with their collection of cytotoxic substances. The reactive oxygen species (ROS) plays the most noteworthy role to kill bacteria extracellularly or intracellularly (Wagner et al. 2004).

Actually, neutrophil recognition does not occur in any antigen-specific manner but by employing certain receptors like pathogen-associated molecular patterns (PAMPs) and microbe-associated molecular patterns (MAMPs).

Processed antigen after phagocytosis is presented via class II major histocompatibility complex molecules (MHC class II). These presented fragments are recognized by circulation helper T cells (Th1 and Th2). They produce different cytokines like interferon g (IFN-g) by Th1 cells and interleukin 4 (IL-4) by Th2 cells. These

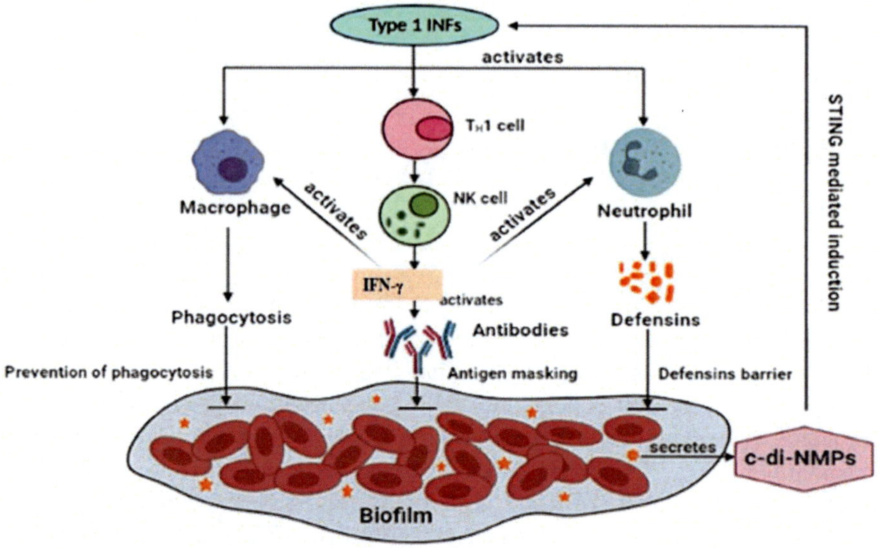

Fig. 6.1 Immune evasion by bacterial cells in biofilm

cytokines in turn promote cell-mediated (IFN-g) and humoral (IL-4) immune mechanisms (Fig. 6.1).

Planktonic bacteria are phagocytosed by the process of "opsonization" in which the bacteria get surrounded by complement (C3b/C3bi) and antibodies. Immunoglobulin G receptors like CD32, CD16, and CD64 are also known to take part in phagocytosis and intracellular killing along with the complement receptors like CR1 and CR3.

Since biofilm, as such, is not recognized through opsonization, there must be some activating factor/protein(s) present in the slimy material of EPS. As for example, in *Staphylococcus epidermidis*, the EPS of the biofilm activate the neutrophils, thereby releasing lactoferrin and leading to the upregulation of activation-associated adherence proteins CD11b/CD18 and CD66 (Meyle et al. 2012). For the sake of identifying the key protein, researchers firstly separated lipoteichoic acid, poly-*N*-acetyl (1–6) β-glucosamine, and LPS which act as contaminant from biofilm EPS. They found the presence of **GroEL, a chaperon** (a bacterial heat shock protein) susceptible to trypsin, to be responsible for raising immune reaction in the host body (Dapunt et al. 2016).

Another way by which the neutrophils fight with the bacteria is by releasing **NET** (neutrophil extracellular traps), a DNA-based trap, which is also known as "NETosis," i.e., the invading bacteria are trapped within the DNA strands from host neutrophils and get killed by cathepsin G, neutrophil elastase, and a variety of bacterial enzymes present within NET (Brinkmann and Zychlinsky 2012). Since biofilm is actually an abundant clump of live bacteria, their cumulative metabolism may change the pH and oxygen levels of the host microenvironment and elicits immune response and weaving of NET.

Another innate local immune response is **complement killing** that takes place by any of the three pathways which are classical, alternative, or lectin pathway. The classical pathway is triggered by the formation of antigen-antibody complex and associated to adaptive immune response. Pattern recognition receptors (PRRs) are used by cells of the innate immune system in order to recognize widely conserved molecules that are specific to microbial components like peptidoglycan and LPS. These commencing host cells play a critical role in killing the pathogenic bacterial cells, thereby switching on other features of the immune system by the production of cytokines and chemokines.

However, the alternative and lectin pathways depend on pathogen-associated molecular patterns (PAMPs). The MBL pathway gets activated by PRR lectin which binds to mannose residues present in bacterial carbohydrates, whereas the alternative pathway depends on opsonization of C3b on the surfaces of microbes. These three pathways converge at C3 cleavage and lytic pathway activation. C5a which acts as a chemoattractant for neutrophils and macrophages is produced by the lytic pathway (Killick et al. 2017). There are many intermediate proteins present in the complement cascade which function as signals to activate and differentiate lymphocytes for AI (Killick et al. 2017). Opsonization of complement leads to the production of C3a/C5a locally at the APC-T-cell interface, improvement of antigen uptake by dendritic cells (DCs), and regulation of lymphocyte activation. In addition to this, B cells get activated by C3d-opsonized antigens through the cofunctioning of CD21 and BCR at concentrations ranging from 25% to 50% of concentrations needed to activate antigen only.

The host response against biofilm-forming gram-negative bacteria *P. aeruginosa* biofilms is typically complicated (Kaya et al. 2020), and it is known to involve an amalgamated activity of a series of cell types from the innate as well as the adaptive immune systems (Maurice et al. 2018). Further, in vitro studies revealed the functions of neutrophils and monocytes against *P. aeruginosa* biofilms (Ciornei et al. 2010); however, very little is known about the interactions between biofilms and peripheral blood mononuclear cells (PBMCs) in humans (Kaya et al. 2020).

The interplay between human macrophages and neutrophils isolated freshly and biofilms formed by *P. aeruginosa* results in the significant immune reactions like accumulation, respiratory burst, penetration, phagocytosis, and elimination of biofilm-forming bacteria (Moser et al. 2017). Bacterial extracellular DNA present in the biofilm matrix can also activate neutrophils by means of TLR9-independent mechanism that in turn leads to increased production of IL-8 and intracellular signaling (Fuxman Bass et al. 2008; Alvarez et al. 2006).

Alginate is the main matrix component present in the mucoid of *P. aeruginosa* biofilm and is recognized as the major virulence factor in patients suffering from chronic CF. Neutrophils counter alginate by triggering monocytes to generate cytokines in vitro and enhance respiratory burst. TLR4 and TLR2 have been utilized in such activating mechanisms, but the receptors used for activating neutrophils are yet to be known. Apart from alginate, some extrapolysaccharide constituents like Pel and Psl are also present in the *P. aeruginosa* biofilm matrix (Ryder et al. 2007). These constituents can possibly initiate an innate immune response particularly against the biofilm. However, an innate immune response against proteins is unlikely to have been initiated because questions are being raised as to whether biofilm-specific proteomes truly exist (Vilain et al. 2004).

In vitro biofilms produced by *Staphylococcus aureus*, *Candida albicans*, and *Streptococcus mutans* have also been found to have been targeted by the antimicrobial activity produced by human monocytes and neutrophils (Moser et al. 2017).

But, in general, host immune responses cannot exert effective immune response against a biofilm infection, mostly resulting in chronic disease. There are different mechanisms including neutrophils or reprogramming of immunological response, myeloid-derived suppressor cells (MDSCs), and straightaway leukocyte killing macrophages (MΦs) (Yamada and Kielian 2019) are being employed.

Therefore, T cells retrieved from contaminated parts are found to be terminally differentiated and produce IFN-γ, which is a cytokine that can increase functioning of phagocytic cells indicating that invading T cells can assist local immune defense (Kotsougiani et al. 2010). T cells, especially CD8+ cells, are known to get activated in the neighboring injury lavage.

Apart from the T-cell response, a huge amount of invading activated neutrophils have been found at the contaminated site shown by the stimulation of adhesion proteins (CD18, CD11b), MHC class II molecules, chemokine receptors CXCR6, or Fc receptors. Also, neutrophils show reduced chemotactic activity and a modified functional response of enhanced generation of oxygen radicals.

Although within first 5 days of infection, local IL-1b inflammatory response against biofilm formed by *P. aeruginosa* was found to increase in mice strain BALB/c in comparison to that of C3H/HeN strain (Trøstrup et al. 2013), yet, in all experimental murine models, a reduced local neutrophil response against *P. aeruginosa* biofilm wounds was noted (Moser et al. 2017).

6.2.1.2 Immune Response due to Implant

Infections due to implants include interplay between host immune response with that of biomaterial and the pathogen. In the absence of a foreign substance, infections in the tissues caused by invading pathogens are continually cleared by the host immune system. On the contrary, due to infections caused by implants, a local tissue response like foreign body reaction, chronic and acute inflammation, fibrous encapsulation, and generation of granulation tissue gets activated. This leads to the formation of a niche immune depression (Arciola et al. 2012), which is described as a locus minoris resistentiae that causes predisposition of the implant to establishment of microbial colony and finally infection (Schierholz and Beuth 2001). Additionally, the biomaterial serves as the substrate for bacterial cells to adhere causing formation of biofilms, thereby leading to microbial colonization on the implant. In such sessile state, pathogens form microcolonies giving rise to biofilm growth which makes the pathogens resistive against all immunological attacks by the hostile host environment.

Cationic antimicrobial peptides (CAMPs) consisting of cathelicidins, thrombocidins, and defensins which are present partly in neutrophils and in epithelial cells cluster to provide local defense. However, their bactericidal potentialities

Table 6.1 Cells of host immune system and their functions in biofilm clearance

Cells of the immune system	Functions
PMN	Direct migration and penetration in 15 or more day-old biofilms 2–6-day-old biofilms are destroyed by phagocytosis, NETosis, and degranulation
Monocytes	Phagocytosis of foreign bacterial cells
Antibody IgG	Causes opsonization of *S. aureus* biofilms releasing ROS causing destruction of biofilm
Cytokine IL-17	Clears biofilm infections by recognition of biofilm cells

are restricted because of resistant mechanisms of bacterial cells (Peschel 2002); however, cathelicidin LL-37 is known to show an inhibitory effect in biofilm clearance (Table 6.1) (Gertrud Maria Hansch 2012).

White blood cell count and/or elevated concentrations of standard systemic markers like C-reactive protein due to infections are not much reliable in infections caused by implants. Besides, in patients suffering from implant infections, it has been found that T cells get activated (upregulation of CD11 and downregulation of CD28) in peripheral blood, but T cells are not considered to have been majorly involved in countering bacterial infections. Also, invasion of T cells and upregulated gene expression of CD3 have been found in samples of tissues taken from the contaminated site (Dapunt et al. 2014).

Essentially, neutrophil count at the infected site directly correlates with number of bone-resorbing osteoclasts, thereby indicating that a persistent pro-inflammatory response produced by the immune system is destined to end up with osteoclast production and subsequently bone degradation.

The local reaction was noteworthy; as in every patient, leukocytes like PMN (predominant, 50–70%), T lymphocytes (in a lesser quantity, 5–20%), NK cells (5%), and few monocytes have been found at the infected sites. The cells were profoundly activated. Adhesion proteins (CD18, CD11b), chemokine receptors CXCR6 or Fc receptors, and MHC class II molecules were upregulated by neutrophils. In addition to this, the PMN was modified because of the increased generation of oxygen radicals leading to a decreased chemotactic activity (Hansch 2012).

Typically, it was found that most implants were subjected to bacterial colonization without showing any symptoms of infection (Inacio et al. 2015). However, neutrophils are always challenged by growing biofilms because biofilms seem to be the favored lifestyle of bacteria. Probably, infections caused due to biofilms have become more common than it was previously thought and effectively get cleared away without any detection. Biomaterials are per se prone for bacterial cells to adhere and favor formation of biofilms. Adherence of bacteria and their formation of biofilms are affected by the physical and chemical characteristics of the biomaterial. Hansch (2012) observed how polymorphonuclear neutrophils behave toward staphylococcal biofilms formed on orthopedic implant materials and found that neutrophils are able to penetrate biofilms, thereby destroying them as well as damaging the host tissues at the infected site (Wagner et al. 2005). Abundant biofilms growing on other biomaterials are also attacked by neutrophils (Günther et al. 2009).

6.2.2 Systemic Immune Response

6.2.2.1 Immune Response from Host Tissues

Biofilm disruption can lead to the release of detaching bacteria which may find their way into the blood stream subsequently causing systemic infections (Gristina 1987). When biofilms formed by *Staphylococcus* are dispersed, peptide toxins such as extracellular enzymes and phenol-soluble modulins (PSMs) having properties similar to surfactants are released. A robust response walls the infection off to prevent systemic spread.

Traditional criteria for an infection include increased C-reactive protein concentration, enhanced number of leukocytes in peripheral blood, and fever. However, in case of infections caused by implants, it has been observed over the last few years that no such systemic reactions take place routinely; for example, out of 100 patients, only 20% displayed increased leukocyte count and 40% showed increased concentrations of C-reactive proteins (Hansch 2012).

Biofilm activates migration of macrophages to the blood circulation. Circulating leukocytes are called as monocytes which are drawn toward the infected site by systemic production of recruiting chemokines CCL7 and CCL2 and microbe-specific molecules like pro-inflammatory cytokines or N-formyl-methionyl peptides in case of any bacterial infections or inflammation.

Polymorphonuclear leukocytes (PMNs) show the primary response against inflammation and account for the majority of circulating leukocytes. PMNs are considered as the primary defense mechanism employed by the host immune system to interact with biofilm-forming or planktonic bacterial cells. Most of the experiments dealing with biofilms and PMNs have been done in vitro with common biofilm producers, *P. aeruginosa* and *Staphylococcus.* A variety of microscopic techniques showed that the biofilms formed by *S. aureus* after 2, 6, and 15 days were prone to damage caused by phagocytosis via PMN (Günther et al. 2009). Researchers studied that the PMNs directly migrated and penetrated the biofilms of *S. aureus* and also clearance efficiency by PMN subsequently decreases with increasing age of biofilms. This same group of researchers also found that biofilms which are 2 and 6 days old were also damaged by PMNs through phagocytosis, NETosis, and degranulation. These authors also suggested that attachment of PMN, degranulation, and phagocytosis of biofilms formed by *S. aureus* was not affected by complement opsonization carried out by C3bi. The disadvantage of these previous studies is that biofilms were developed with continuous shaking which could lead to improper attachment giving rise to growth of more planktonic cells. However, the capability of PMNs to disrupt biofilms formed by *S. aureus* has been observed to be dependent on IL-β as mice that showed a deficiency of this cytokine exhibited more formation of biofilms with 50% less recruitment of neutrophils. Nguyen et al. (2013)

observed that in the presence of *S. aureus* biofilms, oxidative activity decreased inside a model of diabetic wound infection.

Macrophages in the tissues cause phagocytosis of microbes and any other foreign materials, thereby triggering an immune response followed by the employment of PMNs. After phagocytosis of pathogens by PMNs, and apoptosis, naïve macrophages called as monocytes are involved from the bloodstream. In order to be effective at the infected site, macrophages must get activated either by classical pathway (M1 via IFN-γ) or by alternative pathway (M2 via IL-4) (Watters et al. 2016). Opsonization of bacteria carried out by antibodies is critical to clearance by phagocytosis, and it was found that biofilms formed by *S. epidermidis* when incubated with IgG showed a lesser deposition of IgG causing delay in killing by PMNs (Stroh et al. 2011). On the other hand, deposition of IgG and opsonization were not hampered in biofilms of *S. aureus* and therefore was found to be not much significant in attachment of PMN and phagocytosis of cells in the biofilm; but opsonization by IgG was identified as critical to production of ROS and subsequent destruction of biofilms. The response of antibodies against biofilms of *S aureus* was portrayed in a porcine osteomyelitis infection model showing enhanced levels of antibodies and systemic IL-6 (Jensen et al. 2015). Researchers acclaimed IL-6 as a pro-inflammatory cytokine connected to Th1- as well as Th17-induced responses, thereby suggesting that biofilms of *S. aureus* are able to avoid destruction by Th2 since they are intracellular pathogens (Watters et al. 2016).

In case of infections caused by biofilms, generation of cytokine IL-17 gets increased due to growing infection, and in turn the infected mice produce a strong Th17 response. Bacterial cells inside biofilms are shielded from phagocytosis by macrophages and neutrophils since these bacterial cells are extracellularly embedded within a matrix. Production of inflammatory cytokines by means of Th17 cells initiates the maturation and engagement of neutrophils in order to weaken the surface of biofilm, thereby aiding in clearance of bacterial cells. Additionally, in BME extract immunization, a humoral response is initiated alongside the enhanced IL-17 production helping in clearing bacterial cells in biofilm. To describe the role of induction of IL-17 through administration of BME in the efficacy of this multi-component extract, a prior experiment had been performed where cytokine IL-17 was being neutralized by introduction of an antibody against IL-17. Mice those who are BME-immunized were administered with the antibody against IL-17, and this displayed an insignificant bacterial count retrieved from biofilm-contaminated meshes in comparison with BME-immunized mice used as control. The results obtained from these experiments revealed that cytokine IL-17 plays a well-recognized role in immunity against infections caused by biofilms of *S. aureus.* Again, administration of a monoclonal antibody against cytokine IL-10 to mice showed inhibition of clearance of *S. aureus* biofilms indicating that IL-10 plays a significant role in host immunity against systemic infections caused by *S. aureus* biofilms. Moreover, further experiments are needed to know more about the function of induction of IL-10 by administration of BME for clearing infections due to *S. aureus* biofilms. It has also been noted that some extracts consisting of exoproteins from biofilm matrix initiate a shielding immune response against infections by *S. aureus* biofilms, thereby preventing them from colonizing and persisting. The reason behind this immune response produced by this multicomponent vaccine is presumed to be the eradication of biofilm-mediated infections by connecting cell-mediated immunity with humoral response producing opsonic antibodies helping in neutralizing infections. In future studies, it will be the goal to know about the roles played by every antigen of the BME extract leading to its immunogenicity and also to determine a specific combination of antigen providing a better protection against infections caused by *S. aureus* biofilms.

6.2.2.2 Immune Response due to Implant

Humoral and cellular immunity gets activated by biofilm formation on cardiac tissues in infective endocarditis (IE), which in turn triggers macrophage migration into the blood stream. This macrophage migration in the blood stream initiates the synthesis of hypergammaglobulinemia, rheumatoid factors, and splenomegaly. A variety of flowing antibodies (opsonic and agglutinating fragments) and immune complexes are found during IE contributing to the systemic expositions (Mahr et al. 2014). Examples of complement depositions and immune complexes are glomerulonephritis and Osler nodes.

After primary hip and knee arthroplasties, it has been found that many revision surgeries have been performed to nullify some complications associated with arthroplasties. The interplay between osteoclasts (bone cells) and immune cells has been immensely studied in osteoimmunology field. A chronic infection may develop due to formation of biofilms on surfaces of implants, leading to a wrongly presented adaptive immunity that fails to eradicate the biofilm grown on the implant surfaces. The T-cell-mediated immunity was negatively regulated by regulatory T cells (T_{reg}s), anti-inflammatory macrophages, osteoclasts, and myeloid-derived suppressor cells (MDSCs), and all immune defense mechanisms were suppressed (Seebach and Kubatzky 2019).

As a significantly lesser number of IgG and C3b are found to be deposited on biofilm-embedded bacteria, they are not opsonized and killed by PMNs. This is further aggravated due to decreased production of ROS. Matured biofilm was difficult to be removed by PMNs due to increased tolerance of these cells against phagocytosis and immune cell killing. Data obtained from a mouse undergone post-arthroplasty infection showed that neutrophils are recruited to the infected site by IL-1β. Therefore, it can be concluded that neutrophils can decrease biofilm concentration to some extent at least because the knockout mouse exhibited a smaller number of neutrophils with increased formation of biofilms (Alvarez et al. 2006). In a mouse catheter biofilm model, it was found that expression of IL-1β was suppressed during infections caused by biofilms. Myeloid-derived suppressor cells (MDSCs) are defined as diversified population of cells containing granulocytes (G-MDSCs) and immature monocytes (M-MDSCs) which at first, suppress activation of T cells and aggregation, and their levels keep rising spontaneously after the biofilms start to form, and then these levels get steady when the chronic infection already progressed. As a result, quite a lesser number of T cells are found to be present at the portions of orthopedic biofilm infections both in corresponding mouse model and humans (Heim et al. 2018).

Various in vitro studies revealed that M-MDSCs further differentiate into anti-inflammatory M2-like macrophages due to biofilm formation. Thus, MDSCs not only lead to chronicity of diseases but also maintain the balance between tissue injury and long-lasting inflammation.

6.3 Immune Response for Commensal Microorganisms

Vertebrates have parallelly evolved with a myriad of microbes forming a symbiotic relationship. The skin, gut, mouth, mucosal surfaces, and vagina are filled with a wide variety of commensal microbiota. The intestinal mucosa is filled with the largest number of microorganisms in humans; many of these microbes prove to be beneficial, while some other microbes called as pathobionts that are otherwise harmless but can turn pathogenic due to an imbalance in the microbial community also exist (Spasova and Surh 2014). Immunity in mammals comprises a combination of innate and adaptive constituents in every tissue, thereby playing an important role in providing defense against extremely pathogenic external agents or endogenous disruptions in homeostasis (Zhang et al. 2020). The interactions between development and functioning of mammalian immunity and commensal microbiota involve multiple interactions of disease and homeostasis.

As recognition by innate immunity is microorganism-specific and independent of antigen, thus contact with harmful pathogens and beneficial commensal microbes requires the capability of differentiating between the two (Naren Srinivasan 2010), which relies on differences in their invasiveness (Hooper et al. 2012).

Commensal microbiota is generally not pathogenic, but it is obviously important to keep barriers to stop any opportunistic invasions, and so an immune response is needed to keep these pathobionts in control. Although the host requires essential nutrients and metabolites from microorganisms, it must also protect itself from any infection, thereby maintaining a balance between pro- and anti-inflammatory T-cell populations in the gut. In its steady state, the intestine contains a huge amount of T cells producing IL-22, IL-17, IL-10, and IFN-γ. In germ-free mice, a number of CD4 T cells were reduced which affect T helper 17 (Th17) and T helper 1 (Th1) cells, but the frequency of regulatory T-cells (Tregs) remained constant. On being infected with a mucosal pathogen, intestinal CD4 T cells respond to commensal microbiota as well as pathogens that can pass through the epithelial layer of the intestines revealing the capability of the adaptive immune system to tolerate the commensals and also its capability to contain opportunistic pathogens and pathobionts.

Recent reports show that in addition to the existence of Treg cells and Th17 cells, there are also LTi cells and NKp46$^+$ cells inside the gut (Maloy and Powrie 2011). However, the procedure through which mucosal lining in the intestine produces large number of lymphocytes is not fully known. So, the most prevalent idea is that such diverse types of lymphocytes basically occur due to the interplay between commensal bacteria and host cells, each with a distinctive characteristic (Tanoue et al. 2010).

The generation and function of the immune system in the host body is influenced by commensal microbes. In the absence of commensal microbes, the structure and functions of the lymphoid organs get disrupted. Therefore, responses produced by T cells against systemic antigens also get altered if commensal microbes are not present. Recently, some experiments demonstrated that the immune system is fine-tuned and is made ready for action in steady state only due to the presence of commensal microbes, and so conventional mice give more robust response than mice that are germ-free.

The gut contains a huge amount of Th17 cells producing IL-17 which protect against extracellular pathogens (Ivanov et al. 2008). Since in germ-free experimental animal, population of Th17 cells is largely reduced confirming that Th17 differentiation is primarily dependent on commensal microbiota. Experimental observations show that only the presence of a single commensal microbial species is adequate to initiate differentiation of T cells to Th17. In the skin, residing commensal microbes are known to induce differentiation of Th1 and Th17 to shield the host against opportunistic microorganisms and pathogens (Naik et al. 2012).

Keeping tolerance against residing microbiota is critical to restrict origination of inflammatory diseases within the mucosal tissues. Before the concept of tolerance came into knowledge, it was believed by many that existence of a mysterious pathogen triggered the inflammatory bowel disease (IBD). Afterward, successfully treating diseases with immunosuppressive drugs gave rise to the hypothesis that IBD may get triggered due to faults in tolerance toward normally harmless gut microbiota.

The concept that T cells must become tolerant toward commensal microbes was found decades back when naïve CD4 T cells led to colitis (Powrie and Mason 1990) in adoptive transfer model, and these CD4 T cells were countered by another set of CD4 T cells (Maloy and Powrie 2011) which are now called as regulatory T cells (Tregs).

It is known that Tregs are also T cells expressing a key regulatory transcription factor Foxp3 and also exhibit anti-inflammatory functions like TGF-β and/or IL-10 secretion. These Tregs are needed to restrict responses produced by aberrant T cells against residing microbes. Researchers found that Tregs of the gut (small intestine) are functionally and developmentally independent of commensal microbiota, but inside the colon, commensal microbiota accounts for the major induction of Tregs (Atarashi et al. 2011; Geuking et al. 2011). Experimental reports showed that in normalized hosts, differentiation of Tregs takes place in the gut against foreign antigens and not inside thymus.

Keeping balance between capability of host to produce an immune response to pathogenic and microbes and tolerating commensal microbes are important for the host to survive. In spite of the inclination to induce tolerance responses inside the gut, immune responses also occur rapidly against pathogenic microbes to give protection to the host from any infections. The procedure through which the host is capable of distinguishing between commensal and pathogenic microbes is still not fully understood, and researches on this are still going on.

It was found that ligands for toll-like receptors (TLRs) are not produced only due to pathogens but are also produced due to the presence of commensal microbes during healthy colony formation. TLRs are responsible for maintaining plenty of commensal microbes and tissue integrity and providing defense against pathogens in host (Zhang et al. 2020).

Commensal microbes play an important role in initiating tolerance toward beneficial microbes so that the immune system is capable of defending the host against pathogenic microbes (Spasova and Surh 2014).

Since the connection between mammalian immunity and bacteria present in the intestines is complicated and bidirectional, it is not easy to totally comprehend the molecular mechanisms involved in mounting of an immune response, tolerance, or ignorance in intestinal mucosa (Tanoue et al. 2010).

6.4 Immune Evasion by Biofilm

The biofilm-producing microbes remain implanted in a self-synthesized matrix and adhere to each other and/or to various interfaces and surfaces by developing some intricately designed tolerance procedures against the immune responses. Nonetheless, completely capable eukaryotes attempt to produce a sophisticated response against bacterial biofilms (Gonzalez et al. 2018). To survive the immune response, attacking bacteria must permeate through the epithelial lining, host microbiota, diverse leukocytes, and complements. If pathogenic bacteria can permeate through these initial barriers and cell populations, then they may have an enhanced chance of survival that leads to chronic infections in hosts.

Although planktonic cells get rapidly cleared away, efficiency of macrophages and polymorphonuclear neutrophils get reduced due to biofilms. Additionally, in the presence of PMNs and macrophages, formation of biofilms is upregulated, and the constituents of host immune system get aggregated in the EPS matrix. Although infectious biofilms cause chronic inflammations, biofilms by probiotic *Lactobacillus* initiate an insignificant amount of immune response while mount anti-inflammatory responses during inflammation. Thus, these probiotics are used for biofilm colonization protecting the vagina and gut, and also these probiotic biofilms contribute to rapid wound healing in skin damages. Moreover, biofilm can induce an immune response that is unique in nature and not fully understood till now (Watters et al. 2016).

Once established inside a biofilm, the bacteria gradually start adapting to the host environment, thereby becoming less troublesome lacking virulence and persistence and also downregulating virulence factors like toxins, type III secretion systems, motility, and other parameters necessary to establish severe infection. Biofilm-embedded bacteria also detach themselves and therefore form a mixture of planktonic, detached cells and sessile biofilm bacteria that are found to coexist in biofilm infections. The detached planktonic cells can implant an infection in another site causing **chronic or secondary acute infections.**

Biofilms are able to overstimulate as well as suppress immune responses depending on several factors like location of the biofilm in the host, status of host immunity, particular antigens encountered by immune system, and composition of bacterial species in biofilm. Various EPS components show immune modulation of immune responses mounted by host against biofilm contaminations.

6.4.1 Evasion of Innate Immune Reaction

The innate immune response imparts the capability to respond against pathogens by recognizing a wide range of microbial determinants like components of cell wall, formylated short peptides, and nucleic acids. These determinants get recognized in humans by means of receptors present on surfaces of cells in innate immunity, and these cells include PMNs, dendritic cells (DCs), natural killer cells (NK cells), and gamma-delta subset of T lymphocytes (Bryers 2008; Darisipudi et al. 2018).

PMNs cannot recognize pathogenic bacteria by means of particular antigens and possess no immunological memory (Hansch 2012); but they recognize surface determinants that are common in most of the bacteria, and these surface determinants are called pathogen-associated molecular patterns (PAMPs). PMNs can recognize PAMPs through pattern recognition receptors (PRRs). Other professional phagocytes are also known to express PRRs and the most well-known family of PRR is toll-like receptors (TLRs) that are expressed on the surfaces of immunocompetent cells as well as intracellularly in humans (O'Neill et al. 2013).

The biofilm-forming bacteria have the ability to defend themselves against innate immune responses. Secreted cyclic di-GMP is a bacterial derived potent immune stimulatory molecule (Karaolis et al. 2007). Cui et al. (2018) later discovered that it also possesses an anti-innate immune activity inhibiting lipocalin 2 from sequestering siderophores scavenging iron. Another free radical NO is also a component of innate immune system having a wide concentration supported activity directed against microorganisms starting from a signaling molecule such as anti-biofilm to a bactericidal capacity (Barraud et al. 2015; Howli et al. 2017). NO is continually sensed by two sets of conserved sensors that are non-covalently or covalently linked downstream with quorum sensing and/or secondary messenger cyclic di-GMP signaling output (Römling et al. 2007). Intracellular signaling of cyclic di-GMP as well as extracellular quorum sensing is very tightly linked to regulation of biofilm. Quorum sensing molecules like acyl-homoserine lactones in bacteria not only regulates formation of biofilms and their virulence but also regulates innate immune components of host (Turkina and Vikström 2018).

It was also found that cells like macrophages of the innate immune system respond somewhat differently in severe infection initiating formation of single planktonic microbial cell and also multicellular biofilms that can modify the immune cells to an anti-inflammatory state (Yamada and Kielian 2019).

It is found that after crossing the intestinal barrier, the gastrointestinal pathogen *S. typhimurium* somehow manages to use host macrophages as its shelter for intracellular survival and proliferation and as a mode of dissemination to distal systemic sites (Gogoi et al. 2019). Moreover, *Salmonella* through some novel mechanisms can resist free radical-mediated macrophage antimicrobial activity (Linehan and Holden 2003). Abundant small molecules have numerous functions including acting as intermediate metabolites, nutrients, internal and external signaling molecules, defense and repair molecules, lifestyle regulators, and electron acceptors. Raffatellu et al. (2005) reported that exopolysaccharides in the capsule from *Salmonella typhi* can diminish expression of IL-8 produced by cells of the human intestines, thereby reducing the chance of being phagocytosed.

Multiple studies (Watters et al. 2016) have revealed that some exopolysaccharides of bacteria prevent the onset of normal immune response against biofilms. Exopolysaccharides that are secreted by biofilms of *Lactobacillus plantarum* diminished generation of pro-inflammatory cytokines such as MCP (monocyte chemotactic protein I called CCL2), IL-8, and IL-6 through enhancement of negative regulators of toll-like receptor 4 (TLR4) comprising essential pattern recognition receptors (PRRs) to detect pathogen-associated molecular pattern (PAMPs).

Rhamnolipids were considered as an important parameter being upregulated, and researchers suggested a "rhamnolipid shield model" where rhamnolipid surrounds biofilms and removes incoming PMNs. In *P. aeruginosa* bacterial biofilms, rhamnolipids can antagonize PMNs (Jensen et al. 2007). Major constituents that provide protection to bacteria from innate immunity of host are rhamnolipids that can lead to hemolysis causing lysis of many immune cells like PMNs and macrophages in humans (Alhede et al. 2014).

A way of resisting the host immunity by biofilms of *S. aureus* is that its biofilm can remain embedded and resist NETosis by means of amalgamated activity of toxins such as C-hemolysin AB and leukocidin Panton Valentine. Thus, the activity of either of these toxins can result in NET-mediated death of neutrophils in *S. aureus* biofilms, and this evading mechanism employed by biofilms is unique and cannot be employed by other planktonic cells (Bhattacharya et al. 2018).

One more mechanism employed by *S. aureus* biofilms against neutrophil extraction traps (NETs) is the secretion of an extracellular adherence protein called Eap, which contains some immunomodulatory as well as adhesive properties. Since Eap is a DNA-binding protein, it can inhibit NET formation and the NET trapping process. Thus, the *S. aureus* biofilms are protected against host immune defenses by means of DNA-binding and DNA-aggregating properties of Eap (Eisenbeis et al. 2018).

For *S. aureus* biofilms, numerous mechanisms have been known that confer immunity against macrophages, thereby modifying leukocyte activation level and encouraging persistence of pathogenic bacteria. For example, in a recent study, it has been shown that the synthesis of lipoic acid by *S. aureus* can promote persistence of *S. aureus* by preventing antimicrobial activities of RNS and ROS in macrophages in case of an infection. LipoylE2-PDH is a compound that depends on lipoic acid synthesis by enzyme lipoic acid synthetase (Lip A) and is known to carry out the suppression of phagocytes and therefore in any way disturbing the functionality of TLR2 receptors, which get activated by ROS production (Grayczyk and Alonzo 2019). Other molecules of *S. aureus* helping in modulating TLR2 signals are staphylococcal superantigen-like 3 (SSL3) and PSMs that can cause disruption of ROS production, chemokine receptor expression and adhesion of leukocytes, and generation of IL-8 (Bhattacharya et al. 2018). Additionally, numerous other virulence factors are also produced by *S. aureus*, and these factors are responsible for receptor disruption in chemotactic signaling, antimicrobial protease and transmigration blocking, and lysis of phagocytes by pore-forming leukocidins (Buchan et al. 2019). Recent studies also showed that biofilms also employ other mechanisms like surrounding TLR2 and inhibit recognition by TLR9 to protect themselves from host immunity (Bernthal et al. 2011; Thurlow et al. 2011). This proves the observation that patients having mutations for TLR2 inactivation have no expanded danger of being infected with *S. aureus* infections after arthroplasty (El-Helou et al. 2011). Again, the mechanism by which *S. aureus* protects itself against recognition by TLR9 is different from that employed by *P. aeruginosa* because eDNA is the main pro-inflammatory stimulus in case of *P. aeruginosa* biofilms (Fuxman Bass et al. 2010). This shows that immune response mechanism to efficiently clear infection is dependent on species of bacteria and the growth state of the bacterial biofilms. Biofilms of *S. aureus* may also get recognized by alternate PRRs apart from TLR9 and TLR2. For instance, eDNA is also recognized by other intracellular PRRs like DNA-dependent activator of IFN regulatory factors (DAI) or AIM2 (Vilaysane and Muruve 2009; Hornung and Latz 2010). Additionally, staphylococcal PGN, which is a product of degradation and muramyl peptide can be recognized by PRRs of cytoplasm comprising a nucleotide binding oligomerized domain having protein (NOD2) and initiate a pro-inflammatory release of mediator (Girardin et al. 2003; Volz et al. 2010). The mechanisms that are used by *S. aureus* biofilms to evade TLR9/TLR2 recognition are not yet understood but may be described as an inaccessibility of ligand. Biofilms are encapsulated in a complicated three-dimensional framework with some free bacteria at exposed superficial surface, thereby resisting recognition by PRRs that are expressed on the phagocyte surfaces (Thurlow et al. 2011). Besides all these mechanisms, some constituents of matrix of biofilms such as complex polysaccharide polymers (Flemming and Wingender 2010) are known to interfere with maximum interaction of TLRs with potential ligands.

In case of *Mycobacterium tuberculosis* (Mtb), macrophages are employed for engulfing the bacteria followed by the development of compartment of phagosome containing the bacteria. While maturing, they release enzymes and antimicrobial peptides that can turn the environment acidic in order to destroy the pathogen. However, Mtb make use of an entire toolbox of procedures to efficiently prevent this process, among which one of them is production of the tyrosine phosphatase (PtpA). PtpA can inactivate the vacuolar ATPase in the host, and this ATPase is essential for acidifying phagosomes through interaction with one particular region of protein. Through inhibition of acidification by Mtb comprising a phagosomal compartment, Mtb fabricates a suitable niche for itself for persistence inside macrophages of the host.

For example, innate immune cellular constituents like neutrophils respond to biofilms. Their capacity to recognize invading microbes are supported by the presence of pattern recognition receptors (PRRs) which can detect conserved pathogen-associated molecular patterns (PAMPs) in microbes, thereby activating host immune response. Numerous types of PRRs, secreted as well as membrane bound along with corresponding ligands, are detected in contrast to microbe-specific PRRs in biofilms which are not yet identified. The flexibility of biofilms may also continuously trigger PRRs of the innate immunity. Soluble receptors of complement mainly belong to well-known secreted PRRs but the function of the complement in case of infections due to biofilms has not yet been studied because of the absence of biofilm infections in patients having deficiency of complement. In spite of the complement being activated, development of biofilm infections still occurs in patients having fully activated complement system. *P. aeruginosa* can deactivate the complement mechanism by producing elastase, and alkaline protease and alginate with O-acetylation provide protection from opsonization of complement in mucoid of biofilms formed by *P. aeruginosa* (Pier et al. 2001). In sputum of patients suffering from chronic lung infections caused by *P. aeruginosa*, activation of complement (C3c) was more rapid in case of CF. Nevertheless, if activation of complement is only because of formation of biofilms is still not clear, it has been observed that planktonic cells can activate complement more strongly and also biofilm-forming and planktonic *P. aeruginosa* cells coexist in patients suffering from CF. Moreover, resistance against active complement has been found in *P. aeruginosa* present in sputum samples of CF patients (Malhotra Jr. et al. 2019). Biofilms formed by *Mycoplasma pulmonis* (Simmons and Dybvig 2011) and *Mycobacterium abscessus* (Rhoades et al. 2009) are also known to inhibit complement activation (Table 6.2).

In order to be effective at a particular site inside the tissues, macrophages need to be activated through either the classical pathway (M1 via IFN-γ) or by the alternative pathway (M2 via IL-4). After phagocytosis of the bacteria, macrophages such as neutrophils generate nitric oxide (NO) and ROS in order to engulf the bacteria and kill them. Previously, interactions between macrophages and biofilms of bacteria were not much studied, but recently, there have been some studies highlighting the aspects of such interactions. In an experiment on biofilm and macrophage interaction in *S. epidermidis* model, "bacterial slime" was separated out in three fractions through the method of column separation (Myrvik et al. 1989). The researchers found that after 20 h of incubation with macrophages from rabbit alveoli, two-third of EPS portions lowered the oxidative stress in the macrophages.

A study was carried out where the main constituents like intercellular adhesions (Embp and Aap) and polysaccharide intercellular adhesion (PIA) of biofilms of *S. epidermidis* have been tested to find the role played by these constituents in defense against the immune system. Different isogenic strains with components like Aap, Embp, and PIA containing biofilms were utilized, and when these strains were being exposed to murine macrophages, it was found that all the strains had almost

Table 6.2 Evasion of host innate immunity by microbial biofilm components and their mechanisms

Bacteria	Biofilm components of bacteria	Mechanisms to evade host innate immune response
Lactobacillus plantarum	EPS components	Reduced production of pro-inflammatory cytokine like CCL2, IL-8, IL-6, and negative regulation of TLR4
P. aeruginosa	Rhamnolipids	Hemolysis of immune cells like PMN and macrophages
	eDNA	Prevent recognition by TLR9
	Elastase and alkaline protease	Deactivation of complement system
	Alginate	Protection from opsonization of biofilm mucoid
S. aureus	C-hemolysin AB and leukocidin Panton Valentine	NET-mediated death of neutrophils
	Eap	Binds DNA inhibiting NET process and having immunomodulatory and adhesive properties
	Lipoic acid	Prevents antimicrobial activity of RNS and ROS in macrophages
	LipoylE2-PDH	Phagocyte suppression and disturbed functionality of TLR2
	SSL3 and PSM	Reduced ROS production and leukocyte deactivation by adhesion
	Polysaccharide polymers of cells walls	Interferes with interactions of TLR and ligands
Mycobacterium tuberculosis	Tyrosine phosphatase (ptpA)	Host vacuolar ATPase inactivation followed by phagosome acidification

similar level of protection from phagocytosis. It was also found in this study that biofilms of *S. epidermidis* prevented activation of NF-κB and IL-1β production in comparison to strains that do not form biofilms. Again, when biofilms of *S. epidermidis* were made to produce some dormant cells that cannot be cultured, then after incubation of these dormant cells with macrophages, it was found that the pro-inflammatory cytokine (IL-1, IL-6, and TNF-α) levels released by these macrophages decreased, in contrast to those incubated with biofilms containing less dormant cells. Thus, it was concluded from these studies that biofilms of *S. epidermidis* employ various mechanisms to protect themselves against detection by macrophages in host immune system.

6.4.2 Evasion of Adaptive Immune Response

The acquired immune system also gets activated parallelly with innate immune system, moreover with a considerable inertia. Initiation of acquired immune response from infections due to biofilms follows similar pathways as though the infection with an identical microbe is because of a severe infection not caused by biofilms. Acquired immune responses get activated through the activities of macrophages and particularly dendritic cells (DCs) that help in rapid activation (Moser et al. 2017).

The adaptive immunity comprises both humoral and cell-mediated immunity. It was observed that as IL-6 is a pro-inflammatory cytokine connected to Th17 as well as Th1 responses so biofilms of *S. aureus* have the ability to resist killing by Th2 since they are intercellular pathogens. Moreover, they have shown experimentally that biofilms of *S. aureus* can modify the immune system to resist destruction. Undoubtedly, biofilms provide *S. aureus* with the resistivity against cell-mediated and humoral immunity so that they can continue to persist inside the host. Thus, it can be said that biofilms protect *S. aureus* from any macrophage responses by several mechanisms such as preventing activation of macrophages, phagocytosis impairment, modifying phenotype of macrophages, and generation of leukocidin effects.

S. aureus is an ordinary commensal microbe but often causes severe infections. Weak adaptive immunity like reduced production of antibodies can cause frequent infections from *S. aureus.* This is because of the presence of effector proteins like staphylococcal protein A (SpA) in *S. aureus* that prevent efficient immune responses. SpA employs two different mechanisms to suppress adaptive immune response against *S. aureus.* One of the mechanisms is the direct binding of SpA with antibodies, thereby preventing recognition and killing of *S. aureus.* Another mechanism is the binding of SpA with B-cell receptors, therefore inactivating them from generating a protective immunity.

The interaction of biofilm with host immunity particularly with regard to adaptive immunity has been well observed in CF patients developing chronic lung infections from *P. aeruginosa.* Till preliminary antibiotic system of treatment, it is important to maintain that antibiotic treatment in planned or exacerbations and elective courses of antibiotic.

Bacteria adapt to persist in the host tissues by shifting from planktonic to biofilm cells (Yamada and Kielian 2019).

Recently, it has been found that staphylococcal biofilms do not stop humoral constituent diffusion, but their huge biomass in comparison to planktonic cells dilutes the particular antibodies, thus interrupting opsonophagocytic (Cerca et al. 2006). Other components such as antimicrobial peptides and complement of humoral immune system are also evaded by biofilms. As, for example, biofilms of *P. aeruginosa* can gear up elastases and alkaline proteases, which cause direct inactivation of proteins in the complement. In addition, ECM can directly bind with antimicrobial peptides that are positively charged and form a shield on the surface of bacteria or deactivate the alternate pathway of complement activation in case of *Pseudomonas aeruginosa.* Some studies also showed that macrophages can phagocytose *S. aureus* cells from destroyed biofilms but not from the intact one (Yamada and Kielian 2019). This implicates that the size of biofilms serves as a physical barrier leading to a phenomenon called "frustrated phagocytosis" (Thurlow et al. 2011; Leid et al. 2005) and the production of intracellular molecules that cause

Table 6.3 Bacterial biofilm components and their mechanisms to evade adaptive host immunity

Bacteria	Biofilm components	Mechanism for adaptive immune evasion
S. aureus	Biofilm constituents	Immune system modification and resistance against killing by Th2
	SpA	Directly binds with antibodies preventing recognition and killing of bacterial cells
	Leukocidin AB	Phagocytosis prevention leading to macrophage killing
	Hemolysin-α toxin	Evade immune cells in phagosomes
Pseudomonas	Elastases and alkaline proteases	Direct inactivation of complement proteins
	ECM	Deactivation of alternative pathway in complement system by directly binding with positively charged AMPs
	Rhamnolipids	Killing neutrophils
	Alginate	Initiation of frustrated phagocytosis

bystander toxicity of neighboring stromal and immune cells. Apart from these, biofilms of *S. aureus* can resist recognition by toll-like receptor-9 (TLR9) and TLR2, and *Pseudomonas* can downregulate expression of PAMPs through biofilm growth in order to counter immune response (Rada 2017; Bernthal et al. 2011).

All toxins produced by biofilms are controlled by quorum sensing which gets enhanced due to biofilm development directly killing macrophages, neutrophils, and leukocytes to prevent recognition by immune cells and microbicidal activity. As, for example, biofilms of *S. aureus* produce leukocidin AB and α-hemolysin (α-toxin), which can prevent phagocytosis by macrophages, leading to death of macrophages (Scherr et al. 2015). Moreover, α-toxin helps in evading immune cells in phagosomes and hence restricting killing of phagocytosed bacteria intracellularly (Koziel et al. 2015). Mutants of *S. aureus hla/lukAB* showed a huge reduction of bacterial count and enhanced macrophage invasion in case of infections caused due to biofilms on orthopedic implants, proving a complementary action of two toxins in vivo. The toxins are regulated by mechanisms of quorum sensing that are interrupted by disruption of biofilm framework. This indicates significance of intercellular reactions inside the biofilm network showing communal virulence determinant (Scherr et al. 2015). Biofilms produce detergent-like molecules showing cytotoxic activity against reacting immune effector cells. For example, although rhamnolipids help in biofilm network formation and integration in *Pseudomonas*, these are poisonous for neutrophils and neighboring host tissues around the site of infections caused by biofilms (Table 6.3). The cell debris from host serves as substrates for ECM development. Moreover, heavy matrix constituents like alginate serves as a virulence factor in biofilms of *P. aeruginosa* by initiating frustrated phagocytosis just like biofilms from *Staphylococcus*, already discussed above.

The concept of immune polarization initiated from in vitro studies of activation of macrophages led to the recognition of classical (M1) and alternative (M2) pathways explaining the difference between anti-inflammatory and pro-inflammatory states of macrophages. In mice and human models, recent studies show that infections caused due to biofilms of *S. aureus* result from MDSC expansion, anti-inflammatory macrophages, and scarcity of T cells (Hanke et al. 2013; Heim et al. 2014).

6.5 Skewing of Immune Response

Th1 and Th2 cells are responsible for initiating a proper immune response against different pathogens. The main feature of Th1-Th2 model is cross regulation: Th2 cell differentiation is inhibited by IFN-γ, whereas Th1 cell differentiation is inhibited by IL-4. Such balance is disturbed resulting in a skewed immune response (Fig. 6.2) in the presence of an infection only (Philip and Calder 2007). In some instances, biofilms can skew T-cell reactivity with regard to the balance promoting a standstill

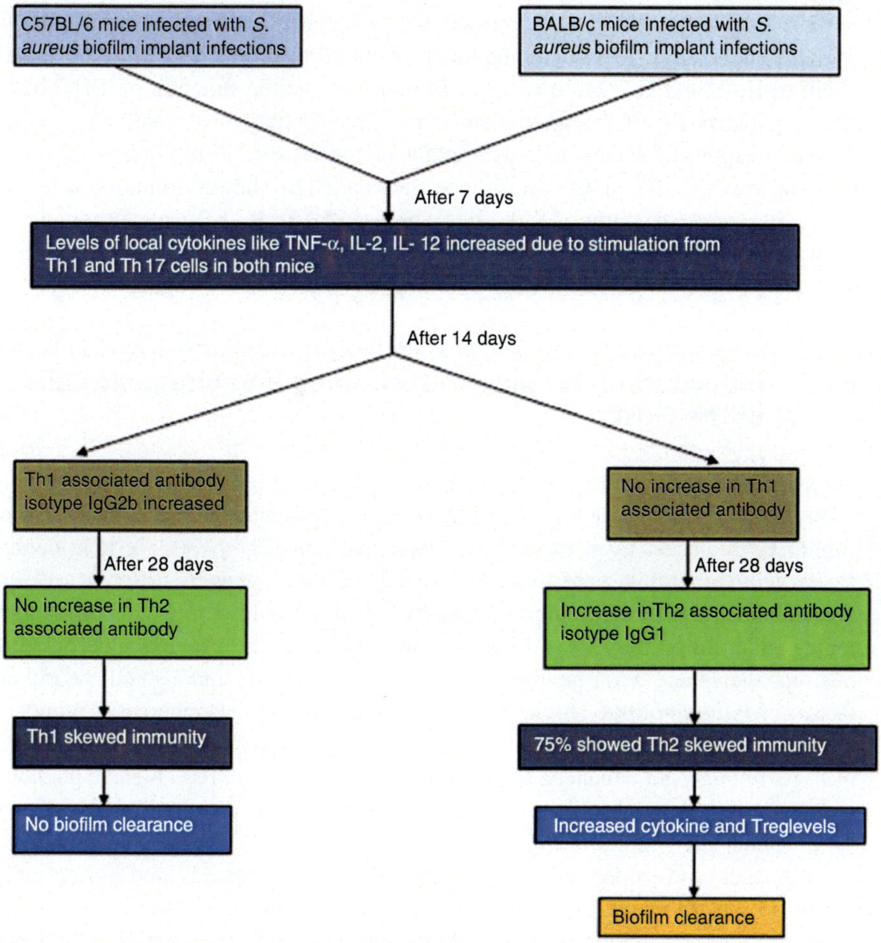

Fig. 6.2 Skewed immune response for biofilm clearance

state between pathogens and host, and thus infections can persist (Gonzalez et al. 2018).

Macrophages are activated either by the classical pathway resulting in an enhanced pro-inflammatory subset (M1) or by the alternative pathway resulting in a regulatory/anti-inflammatory and pro-fibrotic subset (M2). It was found in the mouse catheter model that biofilms cause invading macrophages to skew from M1 to M2, proved by the reduction of inducible nitric oxide synthases (iNOS) and increased production of arginase-1 (Arg-1). Finally, an enhanced anti-inflammatory and pro-fibrotic responses prevent phagocytosis and killing of pathogens. This accumulation of pro-fibrotic matrix along with alternative response of macrophage surrounding the biofilms prevents cells of the immune system from invading the infected site further increasing pathogenic persistence.

The response of antibodies was further studied in a biofilm implant of *S. aureus* model which showed increased amounts of Th1-associated antibody isotype called IgG2b during previous stages of infection till the 14th day. When the infection remained up to the 28th day, increased Th2-associated antibody isotype IgG1 was found. After 7 days, local cytokine level increased to produce Th17 and Th1 cells (TNF-α, IL-2, and IL-12). In order to further observe the function of Th1/Th17 in stopping clearance of biofilms, researchers investigated mice displaying skewed immune response for clearance of implant infections due to biofilms of *S. aureus* (Prabhakara et al. 2011). C57BL/6 mice showed a Th1 skewed immunity and were unable in clearing biofilms of *S. aureus*, whereas 75% of BALB/c mice showing Th2 skewed immunity could clear such infections. This clearance by Th2 cells is mainly due to the increased corresponding cytokine levels and also T-regulatory cells.

6.6 Induction of Biofilm Production by Immunogenic Cells of the Host

On the other hand, biofilms of *P. aeruginosa* producing alginate could not be killed even after incubating for 4 h with macrophages of humans (Leid et al. 2005) and cannot be removed by a mechanism dependent on IFN-γ. As six non-identical *P. aeruginosa* strains were treated with murine macrophages, different cytokine levels almost got doubled due to formation of biofilms in comparison with planktonic cells. Surprisingly, as biofilms were cultured in supernatant containing macrophages, there was doubling of production of alginate and also other virulence factors of *P. aeruginosa.* This supernatant of macrophages also increased number of biofilm cells that are viable and planktonic cells. Thus, it was concluded that biofilms of *P. aeruginosa* are enhanced by macrophages as well as PMNs. A same phenomenon occurred when biofilms of *Candida albicans* were also incubated with supernatants of monocyte/macrophages. The researchers further found that macrophages and supernatant from macrophage cultures favored formation of biofilms by *C. albicans*, but the macrophage constituents from the surfaces of cells did not. When supernatant from developed biofilms of *S. aureus* were treated with macrophages, phagocytosis due to macrophages decreased while cytotoxicity was increased. This consequence is due to leukocidin AB (lukAB) and alpha toxin (hla). In a double mutant of *S. aureus* called *lukAB/hla* orthopedic infection model, it was observed that, in the absence of the virulence factors, bacterial count decreased while macrophage count increased. These researchers also studied previously the interplay of macrophages and biofilms of *S. aureus* in a murine catheter infection model. It was observed that biofilm infections from *S. aureus* reduced invading macrophages, levels of CCL2, and production of NO but arginase-1 level was increased. Since arginase-1 and NO compete for arginine, so by enhancing levels of arginase-1, fibrosis is preferred over clearance of bacteria. Then, the significance of MyD88 signaling pathway (TLR and IL-1 receptor) was studied in containing infections due to biofilms of *S. aureus.* It was observed that in a murine catheter model, bacterial count, populations of invading and alternatively active M2 macrophages and fibrosis were increased, while in MyD88 knockout mice, pro-inflammatory cytokine level was decreased. However, in the catheter model, it was noticed that when M1 macrophages were exogenously added, it inhibited formation of biofilms from *S. aureus* and increased levels of pro-inflammatory cytokines. Introduction of MyD88 mutant macrophages (M2) could not reduce formation of biofilms, thereby indicating the significance of classically active M1 macrophages in clearance of biofilms of *S. aureus.*

In a study, workers proposed an in vitro model of biofilms of *P. aeruginosa* and human PBMC co-culture appropriate for assessing immune responses (cytokine release and activation marker expression) in various types of PBMC. By fine adjustment of the experimental specifications like bacterial and PBMC count, time of incubation, medium type for formation of biofilm, and mature biofilms of *P. aeruginosa* can be obtained in PBMC co-culture after 24 h of incubation with very low death rate of cells. Surprisingly, these results indicated a reciprocal interplay between biofilms of *P. aeruginosa* and human PBMC. Biofilms of *P. aeruginosa* supported activation of PBMC and cytokine production, but it has been observed that PBMC constituents/PBMC increased the number of *P. aeruginosa* cells in their biofilms after 24 h of co-culture (Fig. 6.3). The results, thus obtained from these experiments, showed that bacteria can successfully defend themselves and can persist continually against host immunity (Kaya et al. 2020).

6.7 Conclusion

Although both innate and adaptive immune systems are found to be ineffective or under-effective against biofilms, these biofilms are not naturally shielded from host immune responses. When experiments are carried out in similar conditions on *S. epidermidis*, their biofilms were found to be a little less sensitive in the presence of immunological reactions in the host. According to Hansch (2012) host protection from biofilms is not a "myth," but that is possible only by means of an early intervention. Furthermore, during chronic biofilm infections, the host immunity cannot clear and destroy the biofilm within a specific period of time, thereby resulting in a continuous inflammatory condition damaging host tissues. It can be

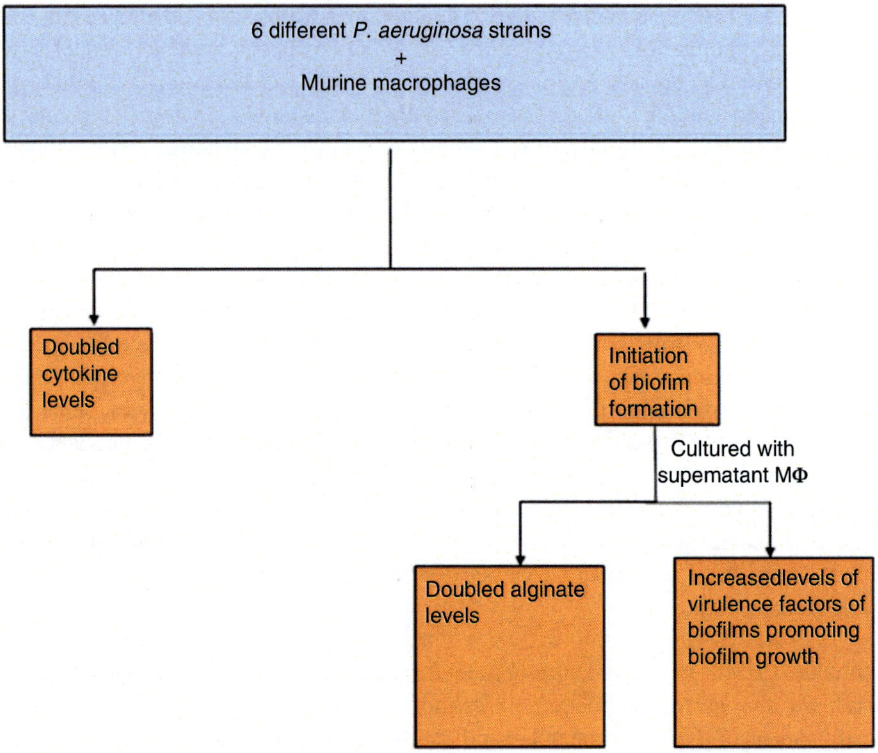

Fig. 6.3 Induction of biofilm growth of *P. aeruginosa* by immunological host cells

concluded that implant biofilm infections are prone to become chronic and the damage to persist (Dapunt et al. 2016).

References

Alhede M, Bjarnsholt T, Givskov M, Alhede M (2014) Pseudomonas aeruginosa biofilms: mechanisms of immune evasion. Adv Appl Microbiol 86:1–40. https://doi.org/10.1016/B978-0-12-800262-9.00001-9

Alvarez ME, Fuxman Bass JI, Geffner JR, Fernandez Calotti PX, Costas M, Coso OA et al (2006) Neutrophil signaling pathways activated by bacterial DNA stimulation. J Immunol 177:4037–4046. https://doi.org/10.4049/jimmunol.177.6.4037

Arciola CR, Campoccia D, Speziale P, Montanaro L, Costerton JW (2012) Biofilm formation in *Staphylococcus* implant infections. A review of molecular mechanisms and implications for biofilm-resistant materials. Biomaterials 33(26):5967–5982. https://doi.org/10.1016/j.biomaterials.2012.05.031

Atarashi K, Tanoue T, Shima T, Imaoka A, Kuwahara T, Momose Y (2011) Induction of colonic regulatory T cells by indigenous Clostridium species. Science 331:337–341. https://doi.org/10.1126/science.1198469

Barraud N, Kelso MJ, Rice SA, Kjelleberg S (2015) Nitric oxide: a key mediator of biofilm dispersal with applications in infectious diseases. Curr Pharm Des 21(1):31–42. https://doi.org/10.2174/13816128206661409051l2822

Bernthal NM, Pribaz JR, Stavrakis AI, Billi F, Cho JS, Ramos RI et al (2011) Protective role of IL-1 β against post-arthroplasty *Staphylococcus aureus* infection. J Orthop Res 29(10):1621–1626. https://doi.org/10.1002/jor.21414

Bhattacharya M, Evelien TM, ETM B, Chan R, Schwab E, Roy S, Sen CK, Torres VJ, Wozniak DJ (2018) *Staphylococcus aureus* biofilms release leukocidins to elicit extracellular trap formation and evade neutrophil-mediated killing. PNAS 115(28):7416–7421. https://doi.org/10.1073/pnas.1721949115

Brinkmann V, Zychlinsky A (2012) Neutrophil extracellular traps: is immunity the second function of chromatin. J Cell Biol 198:773–783. https://doi.org/10.1083/jcb.201203170

Bryers JD (2008) Medical biofilms. Biotechnol Bioeng 100:1–18. https://doi.org/10.1002/bit.21838

Buchan KD, Foster SJ, Renshaw SA (2019) Staphylococcus aureus: setting its sights on the human innate immune system. Microbiology 165:367–385. https://doi.org/10.1099/mic.0.000759

Cerca N, Jefferson KK, Oliveira R, Pier GB, Azeredo J (2006) Comparative antibody-mediated phagocytosis of *Staphylococcus epidermidis* cells grown in a biofilm or in the planktonic state. Infect Immun 74(8):4849–4855. https://doi.org/10.1128/IAI.00230-06

Ciornei CD, Novikov A, Beloin C, Fitting C, Caroff M, Ghigo JM et al (2010) Biofilm-forming *Pseudomonas aeruginosa* bacteria undergo lipopolysaccharide structural modifications and induce enhanced inflammatory cytokine response in human monocytes. Innate Immun 16 (5):288–301. https://doi.org/10.1177/1753425909341807

Cui T, Cang H, Yang B, He ZG (2018) Cyclic dimeric guanosine monophosphate: activation and inhibition of innate immune response. J Innate Immune 11:242–248. https://doi.org/10.1159/000492679

Dapunt U, Giese T, Prior B, Gaida MM, Hänsch GM (2014) Infectious versus non-infectious loosening of implants: activation of T lymphocytes differentiates between the two entities. Int Orthop 38(6):1291–1296. https://doi.org/10.1007/s00264-014-2310-5

Dapunt U, Gaida MM, Meyle E, Prior B, Hänsch GM (2016) Activation of phagocytic cells by *Staphylococcus epidermidis* biofilms: effects of extracellular matrix proteins and the bacterial stress protein GroEL on NETosis and MRP-14 release. Pathog Dis 74:5. https://doi.org/10.1093/femspd/ftw035

Darisipudi MN, Nordengrun M, Broker BM, Peton V (2018) Messing with the sentinels-the interaction of *Staphylococcus aureus* with dendritic cells. Microorganisms 6(87):1–19. https://doi.org/10.3390/microorganisms6030087

de Vos WM (2015) Microbial biofilms and the human intestinal microbiome. NPJ Biofilms Microbiomes 1:15005. https://doi.org/10.1038/npjbiofilms.2015.5

Eisenbeis J, Saffarzadeh M, Peisker H, Jung P, Thewes N, Preissner KT, Herrmann M, Molle V, Geisbrecht BV, Jacobs K, Bischoff M (2018) The *Staphylococcus aureus* extracellular adherence protein Eap is a DNA binding protein capable of blocking neutrophil extracellular trap formation. Front Cell Infect Microbiol 8:235. https://doi.org/10.3389/fcimb.2018.00235

El-Helou O, Berbari EF, Brown RA, Gralewski JH, Osmon DR, Razonable RR (2011) Functional assessment of Toll-like receptor 2 and its relevance in patients with *Staphylococcus aureus* infection of joint prosthesis. Hum Immunol 72:47–53. https://doi.org/10.1016/j.humimm.2010.10.001

Flemming HC, Wingender J (2010) The biofilm matrix. Nat Rev Microbiol 8:623–633. https://doi.org/10.1038/nrmicro2415

Fuxman Bass JI, Gabelloni ML, Alvarez ME, Vermeulen ME, Russo DM, Zorreguieta A et al (2008) Characterization of bacterial DNA binding to human neutrophil surface. Lab Investig 88 (9):926–937. https://doi.org/10.1038/labinvest.2008.59

Fuxman Bass JI, Russo DM, Gabelloni ML et al (2010) Extracellular DNA: a major proinflammatory component of *Pseudomonas aeruginosa* biofilms. J Immunol 184 (11):6386–6395. https://doi.org/10.4049/jimmunol.0901640

Geuking MB, Cahenzli J, Lawson MA, Ng DC, Slack E, Hapfelmeier S et al (2011) Intestinal bacterial colonization induces mutualistic regulatory T cell responses. Immunity 34:794–806. https://doi.org/10.1016/j.immuni.2011.03.02

Girardin SE, Boneca IG, Carneiro LA, Antignac A, Jehanno M, Viala J et al (2003) Nod1 detects a unique muropeptide from gram-negative bacterial peptidoglycan. Science 300:1584–1587. https://doi.org/10.1126/science.1084677

Gogoi MM, Shreenivas MM, Chakravortty D (2019) Hoodwinking the big-eater to prosper: the *Salmonella*-macrophage paradigm. J Innate Immun 11(3):289–299. https://doi.org/10.1159/000490953

Gonzalez JF, Hahn MM, Gunn JS (2018) Chronic biofilm-based infections: skewing of the immune response. Pathog Dis 76:3. https://doi.org/10.1093/femspd/fty023

Grayczyk JP, Alonzo F (2019) *Staphylococcus aureus* lipoic acid synthesis limits macrophage reactive oxygen and nitrogen species production to promote survival during infection. Infect Immun 87:10. https://doi.org/10.1128/IAI.00344-19

Gristina AG (1987) Biomaterial-centered infection: microbial adhesion versus tissue integration. Science 237(4822):1588–1595. https://doi.org/10.1126/science.3629258

Günther F, Wabnitz GH, Stroh P, Prior B, Obst U, Samstag Y, Wagner C, Hänsch GM (2009) Host defence against *Staphylococcus aureus* biofilms infection: phagocytosis of biofilms by polymorphonuclear neutrophils (PMN). Mol Immunol 46:1805–1813. https://doi.org/10.1016/j.molimm.2009.01.020

Hanke ML, Heim CE, Angle A, Sanderson SD, Kielian T (2013) Correction: targeting macrophage activation for the prevention and treatment of staphylococcus aureus biofilm infections. J Immunol 190(12):6709–6710. https://doi.org/10.4049/jimmunol.1390029

Hansch GM (2012) Host defence against bacterial biofilms: "mission impossible"? Int Schol Res Netw ISRN Immunol 2012:853123. https://doi.org/10.5402/2012/853123. 1–17

Heim CE, Vidlak D, Scherr TD, Kozel JA, Holzapfel M, Muirhead DE et al (2014) Myeloid-derived suppressor cells contribute to *Staphylococcus aureus* orthopedic biofilm infection. J Immunol 192(8):3778–3792. https://doi.org/10.4049/jimmunol.1303408

Heim CE, Vidlak D, Odvody J, Hartman CW, Garvin KL, Kielian T (2018) Human prosthetic joint infections are associated with myeloid-derived suppressor cells (MDSCs): implications for infection persistence. J Orthop Res 36:1605–1613. https://doi.org/10.1002/jor.23806

Herman A, Herman AP (2018) Antimicrobial peptides activity in the skin. Skin Res Technol 25 (2):111–117. https://doi.org/10.1111/srt.12626

Hooper LV, Littman DR, Macpherson AJ (2012) Interactions between the microbiota and the immune system. Science 336:1268–1273. https://doi.org/10.1126/science.1223490

Hornung V, Latz E (2010) Intracellular DNA recognition. Nat Rev Immunol 10:123–130. https://doi.org/10.1038/nri2690

Howli RP, Cathie K, Hall-Stoodley L, Cornelius V, Duignan C, Allan RN, Fernandez BO, Barraud N, Bruce KD, Jefferies J, Kelso M et al (2017) Low-dose nitric oxide as targeted anti-biofilm adjunctive therapy to treat chronic pseudomonas aeruginosa infection in cystic fibrosis. Molecul Therap 25(9):2104–2116. https://doi.org/10.1016/j.ymthe.2017.06.021

Inacio RC, Klautau GB, MAS M, Da Silva CB, Nigro S, Antonio Rivetti L, Lopes Pereira W, Costa Salles MJ (2015) Microbial diagnosis of infection and colonization of cardiac implantable electronic devices by use of sonication. Int J Infect Dis 38:54–59. https://doi.org/10.1016/j.ijid.2015.07.018

Ivanov II, Frutos Rde L, Manel N, Yoshinaga K, Rifkin DB, Sartor RB et al (2008) Specific microbiota direct the differentiation of IL-17-producing T-helper cells in the mucosa of the small intestine. Cell Host Microbe 4:337–349. https://doi.org/10.1016/j.chom.2008.09.009

Jensen PO, Bjarnsholt T, Phipps R, Rasmussen TB, Calum H, Christoffersen L et al (2007) Rapid necrotic killing of polymorphonuclear leukocytes is caused by quorum-sensing-controlled production of rhamnolipid by Pseudomonas aeruginosa. Microbiology 153:1329–1338. https://doi.org/10.1099/mic.0.2006/003863-0

Jensen LK, Jensen HE, Koch J, Bjarnsholt T, Eickhardt S, Shirtliff M (2015) Specific antibodies to *Staphylococcus aureus* biofilm are present in serum from pigs with osteomyelitis. In Vivo 29 (5):555–560

Karaolis DK, Means TK, Yang D, Takahashi M, Yoshimura T, Muraille E et al (2007) Bacterial c-di-GMP is an immunostimulatory molecule. J Immunol 178(4):2171–2181. https://doi.org/10.4049/jimmunol.178.4.2.2171

Kaya E, Grassi L, Benedetti A, Maisetta G, Pileggi C, Di Luca M, Batoni G, Esin S (2020) *In vitro* interaction of *Pseudomonas aeruginosa* biofilms with human peripheral blood mononuclear cells. Front Cell Infect Microbiol 10:187. https://doi.org/10.3389/fcimb.2020.00187

Killick J, Morisse G, Sieger D, Astier AL (2017) Complement as a regulator of adaptive immunity. Semin Immunopathol 40(1):37–38. https://doi.org/10.1007/s00281-017-0644-y

Kotsougiani D, Pioch M, Prior B, Heppert V, Hänsch GM, Wagner C (2010) Activation of T lymphocytes in response to persistent bacterial infection: induction of CD11b and of toll-like receptors on T cells. Int J Inf Secur 2010:526740. https://doi.org/10.4061/2010/526740. 1–10.

Koziel J, Chmiest D, Bryzek D, Kmiecik K, Mizgalska D, Maciag-Gudowska A et al (2015) The Janus face of α-toxin: a potent mediator of cytoprotection in staphylococci-infected macrophages. J Innate Immun 7(2):187–198. https://doi.org/10.1159/000368048

Leid JG, Willson CJ, Shirtliff ME, Hassett DJ, Parsek MR, Jeffers AK (2005) The exopolysaccharide alginate protects Pseudomonas aeruginosa biofilm bacteria from IFN-gamma-mediated macrophage killing. J Immunol 175(11):7512–7518. https://doi.org/10.4049/jimmunol.175.11.7512

Linehan SA, Holden DW (2003) The interplay between salmonella typhimurium and its macrophage host-what can it teach us about innate immunity? Immunol Lett 85(2):183–192. https://doi.org/10.1016/s0165-2478(02)00227-4

Mahr A, Batteux F, Tubiana S, Goulvestre C, Wolff M, Papo T et al (2014) Brief report: prevalence of antineutrophil cytoplasmic antibodies in infective endocarditis. Arthritis Rheum 66:1672–1677. https://doi.org/10.1002/art.38389

Malhotra S Jr, Hayes D, Wozniak DJ (2019) Cystic fibrosis and *Pseudomonas aeruginosa*: the host-microbe interface. Clin Microbiol Rev 32:3. https://doi.org/10.1128/CMR.00138-18

Maloy KJ, Powrie F (2011) Intestinal homeostasis and its breakdown in inflammatory bowel disease. Nature 474:298–306. https://doi.org/10.1038/nature10208

Maurice NM, Bedi B, Sadikot RT (2018) *Pseudomonas aeruginosa* biofilms: host response and clinical implications in lung infections. Am J Respir Cell Mol Biol 58(4):428–439. https://doi.org/10.1165/rcmb.2017-0321TR

Meyle E, Brenner-Weiss G, Obst U, Prior B, Hänsch GM (2012) Immune defense against S. epidermidis biofilms: components of the extracellular polymeric substance activate distinct bactericidal mechanisms of phagocytic cells. Int J Artif Organs 35:700–712. https://doi.org/10.5301/ijao.5000151

Moser C, Lerche CJ, Kolpen L, Line L, Thomsen K, Høiby N, Jensen PO (2017) Biofilms and host response—helpful or harmful. Acta Pathol Microbiol Immunol Scand 125:320–338. https://doi.org/10.1111/apm.12674

Myrvik QN, Wagner W, Barth E, Wood P, Gristina AG (1989) Effects of extracellular slime produced by *Staphylococcus epidermidis* on oxidative responses of rabbit alveolar macrophages. J Investig Surg 2(4):381–389. https://doi.org/10.3109/08941938909018263

Naik S, Bouladoux N, Wilhelm C, Molloy MJ, Salcedo R, Kastenmuller W et al (2012) Compartmentalized control of skin immunity by resident commensals. Science 337:1115–1119. https://doi.org/10.1126/science.1225152

Nguyen KT, Seth AK, Hong SJ, Geringer MR, Xie P, Leung KP, Mustoe TA, Galiano RD (2013) Deficient cytokine expression and neutrophil oxidative burst contribute to impaired cutaneous wound healing in diabetic, biofilm-containing chronic wounds. Wound Repair Regen 21:833–841. https://doi.org/10.1111/wrr.12109

O'Neill LA, Golenbock D, Bowie AG (2013) The history of Toll-like receptors-redefining innate immunity. Nat Rev Immunol 13:453–460. https://doi.org/10.1038/nri3446

Peschel A (2002) How do bacteria resist human antimicrobial peptides? Trends Microbiol 10 (4):179–186. https://doi.org/10.1016/s0966-842x(02)02333-8

Philip C, Calder PC (2007) Immunological parameters: what do they mean? J Nutr 137(3):773S–780S. https://doi.org/10.1093/jn/137.3.773S

Pier GB, Coleman F, Grout M, Franklin M, Ohman DE (2001) Role of alginate O-acetylation in resistance of mucoid Pseudomonas aeruginosa to opsonic phagocytosis. Infect Immun 69:1895–1901. https://doi.org/10.1128/IAI.69.3.1895-1901.2001

Powrie F, Mason D (1990) OX-22high CD4+ T cells induce wasting disease with multiple organ pathology: prevention by the OX-22 low subset. J Exp Med 172:1701–1810. https://doi.org/10.1084/jem.172.6.1701

Prabhakara R, Harro JM, Leid JG, Keegan AD, Prior ML, Shirtliff ME (2011) Suppression of the inflammatory immune response prevents the development of chronic biofilm infection due to methicillin-resistant *Staphylococcus aureus*. Infect Immun 79(12):5010–5018. https://doi.org/10.3389/fimmu.2019.01724

Rada B (2017) Interactions between neutrophils and *Pseudomonas aeruginosa* in cystic fibrosis. Pathogens 6(1):6. https://doi.org/10.3390/pathogens6010010

Raffatellu M, Chessa D, Wilson RP, Dusold R, Rubino S, Baumler AJ (2005) The Vi capsular antigen of *Salmonella enterica* serotype Typhi reduces Toll-like receptor-dependent interleukin-8 expression in the intestinal mucosa. Infect Immun 73(6):3367–3374. https://doi.org/10.1128/IAI.73.6.3367-3374.2005

Rhoades ER, Archambault AS, Greendyke R, Hsu FF, Streeter C, Byrd TF (2009) Mycobacterium abscessus Glycopeptidolipids mask underlying cell wall phosphatidyl-myo-inositol mannosides blocking induction of human macrophage TNF-alpha by preventing interaction with TLR2. J Immunol 183:1997–2007. https://doi.org/10.4049/jimmunol.0802181

Römling U, Galperin MY, Gomelsky M (2007) Cyclic di-GMP: the first 25 years of a universal bacterial second messenger. Microbiol Mol Biol Rev 77(1):1–52. https://doi.org/10.1128/MMBR.00043-12

Ryder C, Byrd M, Wozniak DJ (2007) Role of polysaccharides in *Pseudomonas aeruginosa* biofilm development. Curr Opin Microbiol 10:644–648. https://doi.org/10.1016/j.mib.2007.09.010

Scherr TD, Hanke ML, Huang O, James DB, Horswill AR, Bayles KW et al (2015) *Staphylococcus aureus* biofilms induce macrophage dysfunction through leukocidin AB and alpha-toxin. MBio 6(4):6. https://doi.org/10.1128/mBio.01021-15

Schierholz JM, Beuth J (2001) Implant infections: a haven for opportunistic bacteria. J Hosp Infect 49:87–93. https://doi.org/10.1053/jhin.2001.1052

Seebach E, Kubatzky KF (2019) Chronic implant-related bone infections—can immune modulation be a therapeutic strategy? Front Immunol 10:1724. https://doi.org/10.3389/fimmu.2019.01724

Simmons WL, Dybvig K (2011) Biofilms protect Mycoplasma pulmonis cells from lytic effects of complement and gramicidin. Infect Immun 75:3696–3699. https://doi.org/10.1128/IAI.00440-07

Spasova DS, Surh CD (2014) Blowing on embers: commensal microbiota and our immune system. Front Immunol 5:318. https://doi.org/10.3389/fimmu.2014.00318

Srinivasan N (2010) Telling apart friend from foe: discriminating between commensals and pathogens at mucosal sites. Innate Immun 16(6):391–404. https://doi.org/10.1177/1753425909357577

Stroh P, Gunther F, Meyle E, Prior B, Wagner C, Hansch GM (2011) Host defence against *Staphylococcus aureus* biofilms by polymorphonuclear neutrophils: oxygen radical production but not phagocytosis depends on opsonisation with immunoglobulin G. Immunobiology 216 (3):351–357. https://doi.org/10.1016/j.imbio.2010.07.009

Sun F, Qu F, Ling Y, Mao P, Xia P, Chen H, Zhou D (2013) Biofilm-associated infections: antibiotic resistance and novel therapeutic strategies. Future Microbiol 8:877–886. https://doi.org/10.2217/fmb.13.58

Tanoue T, Umesaki Y, Kenya Honda K (2010) Immune responses to gut microbiota-commensals and pathogens. Gut Microbes 1(4):224–233. https://doi.org/10.4161/gmic.1.4.12613

Thurlow LR, Hanke ML, Fritz T, Angle A, Aldrich A, Williams SH et al (2011) *Staphylococcus aureus* biofilms prevent macrophage phagocytosis and attenuate inflammation in vivo. J Immunol 186(11):6585–6596. https://doi.org/10.4049/jimmunol.1002794

Trøstrup H, Thomsen K, Christophersen LJ, Hougen HP, Bjarnsholt T, Jensen PO et al (2013) *Pseudomonas aeruginosa* biofilm aggravates skin inflammatory response in BALB/c mice in a novel chronic wound model. Wound Repair Regen 21:292–299. https://doi.org/10.1111/wrr.12016

Turkina MV, Vikström E (2018) Bacteriahost crosstalk: sensing of the quorum in the context of *Pseudomonas aeruginosa* infections. J Innate Immun 11(3):263–279. https://doi.org/10.1159/000494069

Vilain S, Cosette P, Hubert M, Lange C, Junter GA, Jouenne T (2004) Proteomic analysis of agar gel-entrapped *Pseudomonas aeruginosa*. Proteomics 4:1996–2004. https://doi.org/10.1002/pmic.200300707

Vilaysane A, Muruve DA (2009) The innate immune response to DNA. Semin Immunol 21:208–214. https://doi.org/10.1016/j.smim.2009.05.006

Volz T, Nega M, Buschmann J et al (2010) Natural *Staphylococcus aureus*-derived peptidoglycan fragments activate NOD2 and act as potent costimulators of the innate immune system exclusively in the presence of TLR signals. FASEB J 241(10):4089–4102. https://doi.org/10.1096/fj.09-151001

Wagner C, Kaksa A, Müller W, Denefleh B, Heppert V, Wentzensen A, Hänsch GM (2004) Polymorphonuclear neutrophils in posttraumatic osteomyelitis: cells recovered from the inflamed site lack chemotactic activity but generate superoxides. Shock 22(2):108–115. https://doi.org/10.1097/01.shk.0000132488.71875.15

Wagner C, Obst U, Hänsch GM (2005) Implant-associated post traumatic osteomyelitis: collateral damage by local host defense? Int J Artif Organs 28(11):1172–1180. https://doi.org/10.1177/039139880502801115

Watters C, Fleming D, Bishop D, Rumbaugh KP (2016) Host responses to biofilm. Prog Mol Biol Transl Sci 142:193–239. https://doi.org/10.1016/bs.pmbts.2016.05.007

Yamada KJ, Kielian T (2019) Biofilm-leukocyte cross-talk: impact on immune polarization and immunometabolism. J Innate Immun 11:280–288. https://doi.org/10.1159/000492680

Zasloff M (2002) Antimicrobial peptides of multicellular organisms. Nature 415:389–395. https://doi.org/10.1038/415389a

Zhang K, Li X, Yu C, Wang Y (2020) Promising therapeutic strategies against microbial biofilm challenges. Front Cell Infect Microbiol 10:359. https://doi.org/10.3389/fcimb.2020.00359

Biofilm on Medical Appliances

7

Dibyajit Lahiri, Moupriya Nag, Dipro Mukherjee, Sayantani Garai, Ritwik Banerjee, Ankita Dey, and Rina Rani Ray

Abstract

Biofilm-forming microbes are the root cause of various almost incurable chronic, nosocomial, and medical device-associated infections that are of serious concern in present-day condition. With the advent of science, a number of diseases, disorders, and abnormalities can be effectively managed by the use of various medical devices including pacemakers, vascular catheters, chronic hemodialysis catheters, prosthetic heart valves, and prosthetic joints. But the effectiveness of these medical devices is seriously hampered by the biofilm grown aggressively on these devices. The interactions existing between the microbial cells, host, and the biomaterials result in the development, persistence, and failure in treating these device-associated infections. The present chapter would focus on various medical devices-associated biofilm infections that are affecting the host immune system leading to chronic infections and failure of the objective of this implant operation.

7.1 Introduction

The mechanism of the formation of biofilm upon various biotic and abiotic surfaces is mediated by the presence of glycocalyxes and a polymeric matrix being produced by the sessile colonies (Costerton et al. 1981; Dudman 1977). The microcolonies that

D. Lahiri · M. Nag · D. Mukherjee · S. Garai · R. Banerjee
Department of Biotechnology, University of Engineering & Management, Kolkata, West Bengal, India

A. Dey · R. R. Ray (✉)
Department of Biotechnology, Maulana Abul Kalam Azad University of Technology, West Bengal, Haringhata, India

R. R. Ray et al. (eds.), *Biofilm-Mediated Diseases: Causes and Controls*,
https://doi.org/10.1007/978-981-16-0745-5_7

Table 7.1 Various biofilm-producing organisms

Common biofilm-forming bacterial species	Nature of species	Infections caused by them	In vitro biofilm formation	References
Escherichia coli	Rod-shaped Gram-negative bacteria	Community infections such as urinary tract infections (UTIs), prostatitis, and other nosocomial infections	Form biofilm in vitro	De Kievit (2001)
Pseudomonas aeruginosa	Gram-negative notorious opportunistic pathogen	Responsible for chronic infections	Form biofilm in vitro	Grkovic (2002)
Staphylococcus epidermidis	Opportunistic pathogen	Causes chronic infections	Doesn't form biofilm in vitro	Nickel (1987)
Staphylococcus aureus	Gram-positive bacteria	Nosocomial diseases caused by multidrug-resistant organisms	Form biofilm in vitro	Hellstrtom (1938)
Staphylococcus epidermidis	Opportunistic pathogen	Leads to the pathogenesis	Form in vitro biofilm	Falanga (2000)
Enterobacter cloacae	Gram-positive bacteria	Bacterial infections at lower respiratory tract and other nosocomial diseases	Does not form in vitro biofilm	Koch and Hoiby (1983)
Klebsiella pneumoniae	Gram-negative bacterium	Causing nosocomial infections	Form in vitro biofilm	Neut (2007)
Actinomyces israelii	Gram-positive, rod-shaped bacteria	Causes actinomycosis	Doesn't form in vitro biofilm	Mendoza (2004)
Haemophilus influenzae	*Bacillus* bacteria	Some of these illnesses, like ear infections, are mild, while others, like bloodstream infections, are very serious	Doesn't form biofilm in vitro	Hellstrtom (1938)
Burkholderia cepacia	Aerobic Gram-negative bacillus	Cause various types of infections, including catheter-associated infections and respiratory tract infections	Doesn't form biofilm in vitro	Lewis (2008)

are associated with the biofilm are responsible for the development of various chronic bacterial infections like that of osteomyelitis, cystic fibrosis (Marrie et al. 1979), as well as various device-associated chronic infections (Marrie and Costerton 1983) (Table 7.1). The outer surface being exposed to the microbes plays an important role in the development of pathogenesis and the survival of the organisms. The attachment of the bacterial cell is dependent upon the type and characteristics of its outer coat. Other factors that also play important roles in the process of attachment of microbial cells upon the device surface include Brownian motion, surface charge, or chemical bonding. The development of biofilm upon medical devices results in the development of financial burden on healthcare sectors by the enhancement of mortality and morbidity of the patients (Donlan 2008). The development of biofilm upon the healthcare devices has been problematic since it involves near about 65% of the total nosocomial infections, and they are usually associated with various prosthetic and medical devices (Otter et al. 2015).

Studies showed that biofilm formed upon the medical devices is also composed of both Gram-positive and Gram-negative bacterial cells. Some of the notable Gram-positive bacteria are *Staphylococcus aureus, Enterococcus faecalis, Staphylococcus epidermidis*, and *Streptococcus viridans* (Stickler 1996). The group of coagulase-negative staphylococci comprising of *Staphylococcus epidermidis* is responsible for the nosocomial diseases and also for the development of implanted biomaterial-related chronic infections (US Department of Health 1996). Various medical devices comprising of central venous catheters, cardiac pacemakers, prosthetic joints, artificial lenses, heart valves, and CSF shunts are the common sites for the development of biofilm (Table 7.2) and have been a worldwide concern (Mack et al. 2006). Gram-negative bacteria like *Pseudomonas aeruginosa*, *Escherichia coli*, and *Klebsiella pneumoniae* are found responsible for the development of biofilm upon the surfaces of the medical devices (Stickler 1996). Among all the various bacterial cells, the common species which are responsible for 40–50% of various prosthetic heart valve infections are comprised of *S. epidermidis* and *S. aureus*. These organisms are also associated with 50–70% biofilm-associated catheter infections and about 87% of bloodstream-associated infections (Chen et al. 2013). The accumulation of the sessile communities on the surface of cardiac pacemakers, catheters, prosthetic heart valves, chronic ambulatory peritoneal dialysis catheters, and prosthetic joints with the transformation to highly infective conditions has resulted in the development of various illnesses and morbidity (Gristina et al. 1993). The prosthetic devices are usually comprised of synthetic materials and are recognized as foreign materials to the body. The interaction of the host body with that of the foreign material occurs by encompassing the substances with the help of glycoproteinaceous components comprising of fibronectin, fibrin, albumin, vitronectin, and laminin that act as receptors for the purpose of microbial attachment. It has been observed that organisms like *Staphylococcus aureus* and *Candida* sp. show high amount of attraction and observed to adhere tightly with fibrin and fibronectin (Herrmann et al. 1988; Vandaux et al. 1989). The formation of biofilm upon the various medical devices is an immense concern due to the development of the antimicrobial resistance among the sessile colonies of the attached bacterial cells. The condition in which an organism acquires resistances against various types of antibiotics results in the development of the superbugs. This chapter would focus on the various medical device-associated biofilm conditions and the genetics associated with their attachment with special reference to *Staphylococcus epidermidis*.

Table 7.2 Biofilm-producing microorganism on medical implants

Medical implants	Microorganisms	References
Breast implants	*S. aureus, Enterococcus*, and *S. epidermidis*	Bryers (2008)
Cardiac pacemakers	*S. aureus*	Darouiche (2001)
Central venous catheters	*S. epidermidis, S. aureus, E. faecalis, K. pneumoniae, P. aeruginosa, C. albicans*	Bryers (2008) Darouiche (2001) Rodrigues et al. (2007)
Cerebrospinal fluid shunts	*S. aureus, S. epidermidis, Enterococcus*	Darouiche (2001)
Contact lenses	*P. aeruginosa* and Gram +ve cocci	Bryers (2008) Darouiche (2001) Rodrigues et al. (2007)
Dental implants	Acidogenic Gram +ve cocci, Gram -ve anaerobic oral bacteria	Bryers (2008) Darouiche (2001) Rodrigues et al. (2007)
Endotracheal tubes	*S. aureus, S. epidermidis, C. albicans, P. aeruginosa*	Darouiche (2001)
Hip prosthesis	*Staphylococcus* sp., *Proteus mirabilis, Bacteroides* species, *Streptococcus, E. coli, P. aeruginosa*	Darouiche (2001)
Intrauterine contraceptive devices	*Micrococcus* sp., *Enterococcus* sp., *C. albicans,* Group B *Streptococcus*	Rodrigues et al. (2007)
Orthopedic implants	Hemolytic *Streptococcus, Enterococcus, P. mirabilis, Bacteroides* sp., *P. aeruginosa, E. coli*	Rodrigues et al. (2007)
Peritoneal dialysis catheters	*Streptococcus* sp., *Staphylococcus* sp.	Bryers (2008) Darouiche (2001) Rodrigues et al. (2007)
Prosthetic heart valves	*Streptococcus viridans,* coagulase-ve *Staphylococcus, Enterococcus, S. aureus*	Rodrigues et al. (2007)
Prosthetic implants for erectile dysfunction	*S. aureus* and *S. epidermidis*	Bryers (2008) Darouiche (2001) Rodrigues et al. (2007)

(continued)

Table 7.2 (continued)

Medical implants	Microorganisms	References
Replacement joints	*S. aureus* and *S. epidermidis*	Bryers (2008)
Urinary catheters	*S. epidermidis, Klebsiella pneumoniae, Enterococcus faecalis, P. mirabilis*	Bryers (2008) Darouiche (2001) Rodrigues et al. (2007)
Voice prostheses	*C. albicans, C. tropicalis, Streptococcus mitis, Streptococcus salivarius, Rothia dentocariosa, Streptococcus sobrinus, Staphylococcus epidermidis, Stomatococcus mucilaginous*	Bryers (2008) Rodrigues et al. (2007)

7.2 Infections Associated with Medical Devices

7.2.1 Biofilm of Gram-Positive Bacteria and Infections

The infections caused by Gram-positive group of bacterial cells are more predominant than the Gram-negative organisms which are enunciated by various components of the human blood. The extracellular matrix that is formed is dependent on the various carbohydrate and protein components along with the presence of extracellular DNA (eDNA). Different types of biofilm-associated diseases caused by Gram-positive bacteria like gastrointestinal ulcers, chronic otitis, endocarditis, urinary tract infections, osteomyelitis, and periodontitis are commonly caused by *Staphylococcus*, *Streptococcus*, and *Enterococcus* species (Heilmann and Götz 2009). Staphylococci thrive upon the mucous membrane and epithelium in both animals and humans. The most potent organism to cause nosocomial infections in human is *Staphylococcus aureus*. Among the various types of staphylococci, *S. epidermidis* is responsible for various foreign body-associated infections, whereas *S. aureus* develops colonies upon the surface of the epithelial tissues. It has been observed that eDNA is responsible for the development of biofilm by *S. aureus* by interacting with the positively charged molecules of the polymeric substances of the surfaces thus increasing the formation of biofilm matrix for better attachment (Heilmann and Götz 2009). The mechanism of detachment of the sessile communities from the surface results in the metastasis of the infection and colonization of the cells at some other site resulting in the spread of infections to a great extent. Some types of enzymes can bring about disintegration in the architecture of the biofilm which includes glycosyl hydrolase that causes disintegration of polysaccharide intercellular adhesin (PIA) proteoses that have the ability to interact with the proteinaceous components of the matrix and nucleases that bring about the denaturation of the eDNA (Heilmann and Götz 2009). Studies showed that *S. aureus* possesses the ability to prevent protease-associated degradation of biofilm by the mechanism of quorum sensing (QS). The development of biofilm upon the surface of the medical device is mediated by the use of pili being present at the outer surface of the cell wall (Murphy et al. 2014). Usually the pili are the elongated proteinaceous structures found in Gram-negative bacteria but except *Corynebacterium renale*, a Gram-positive one. The development of biofilm is mediated by the proteins present upon the surface that predominantly act as adhesives and help in the mechanism of biofilm development (Danne and Dramsi 2012).

7.2.2 Biofilm of Gram-Negative Bacteria and Infections

At present-day scenario, the development of resistance within bacteria against various antimicrobial drugs and therapies is of serious concern. The persister cells existing within the biofilm are responsible for the tolerance against the antibiotics (Lewis 2012). Hence, the formed biofilm shows large amount of resistance against host immune system and the antimicrobial drugs (Barlow 2009). Various types of Gram-negative bacteria like *P. aeruginosa* possess the ability to get converted to mucoid variants that are usually responsible for various types of chronic infections (Mathee et al. 1999). The greater amount of production of alginate by the groups of Gram-negative bacterial cells results in the development of greater amount of resistances (Cabral et al. 1987). *P. aeruginosa* possesses the ability to exhibit modified phenotypic character at the time of formation of biofilm. Initially the cells get attached reversibly with the surface followed by irreversible attachment with various tissue fragments being available. The sessile microcolonies encompass themselves with the help of a protective coating made up of lipopolysaccharides. This coating mediates the mechanism of adhesion with the surface and surrounding tissues (Sauer et al. 2002). The adhesion by biofilm of Gram-negative bacteria upon the surface is mediated by extracellular polymeric substances (EPS) comprised of homo- and heteropolysaccharides (Sutherland 2001). *P. aeruginosa* possesses type IV pili which help in the mechanism of surface adhesion, comprising of subunits of proteins that results in the development of fimbrial polymers (Bohn et al. 2009). Depending upon the nature of surface, they can alter the phenotype of the lipopolysaccharides enhancing the hydrophobicity resulting in the better attachment of the cells with the surface (Makin and Beveridge 1996). The contraction and relaxation of the type IV pili helps in the movement of the bacterial cells upon the surface (Barken et al. 2008). Planktonic cells also produce a chemical substance known as lectins that enhance the adhesion of cells upon any surfaces resulting in the better biofilm formation (Laverty et al. 2014).

7.3 Device-Associated Biofilm Formation

The European Commission has grouped non-invasive medical devices into four main categories; firstly are the devices which are not in contact with the skin or those which come in total contact with the skin. Such devices include waste collection devices from body parts, devices providing comforts to the people, or the devices which are immobilized with the body parts. Non-invasive devices are also comprised of eye occlusion plasters, incision drapes, non-invasive electrodes, and conductive gels. Another type of classification includes devices that are indirectly acting as invasive structures. The classification also includes the devices which possess the ability to bring about chemical or biological modifications for being infused within the body, example of which are hemodialyzers.

7.3.1 Non-invasive Medical Devices

The medical devices like bandages and mechanical barriers which usually come in direct contact with the injured skin are grouped under the category of the non-invasive devices. The attachment of the sessile communities with the surfaces of the devices marks the first stage of biofilm formation. This further helps in the proliferation of the microcolonies existing within the biofilm. The sessile communities existing within the biofilm encompass themselves with the help of EPS that facilitates better attachment of the sessile colonies. The EPS not only provides nourishment to the existing colonies but prevents the penetration of drugs resulting in antibiotic and antimicrobial resistances (Table 7.3) (Davey and O'Toole 2000). Various researchers have shown that inanimate surfaces attracted most of the microbial cells to get attached and develop various chronic infections (Kramer et al. 2006).

7.3.2 Invasive Medical Devices

The adherence of the pathogenic organisms on the surfaces of medical devices causes significant problem to the hosts. The invasive devices are either reusable or can be sterilized. The knowledge of biofilm development on the surfaces of such transient devices is important (Garrett et al. 2008). A study was performed to identify the various degrees of contaminations of the medical devices by quantifying specific proteins being present upon the surfaces of such devices. It was further observed that detergents were unable to reduce the load of organisms by 99.9% (Murdoch et al. 2006). A very common phenomenon is the contamination by catheters as the immobilized microcolonies dwelling upon its surface result in the development of serious infections among the users.

7.3.3 Short-Term Use of Medical Devices

The most common type of medical device for such type of infection is found on intravascular catheters. These provide a surface for easy congregation of bacterial cells leading to severe health aspects (Akbari and Kjellerup 2015). The most common nosocomial diseases associated with such medical devices are

Table 7.3 Antimicrobial resistance observed in microorganisms associated with medical device biofilms

Species	Observed resistance	Medical device	References
Mycobacterium avium	Resistant to clarithromycin compared to planktonic bacterial	Catheter	Falkinham (2007)
Candida albicans	Alterations in the expression of several drug resistance genes		Samaranayake et al. (2015)
Escherichia coli, Klebsiella pneumoniae, Pseudomonas aeruginosa, Acinetobacter lwoffii, coagulase-negative *Staphylococcus*	Biofilm producing bacteria had higher resistance than planktonic bacteria to ampicillin (83.3% vs. 60%), cefotaxime (73.3% vs. 35%), norfloxacin (80% vs. 60%), and nalidixic acid (93.3% vs. 70%)	Urinary catheter	Subramanian et al. (2012)
Enterococcus	Resistance to amoxicillin, co-trimoxazole, ciprofloxacin, gentamicin, cefotaxime, and cefuroxime		Akhter et al. (2014)
Staphylococcus lugdunensis	Strains were antimicrobial susceptible, but they carry virulence factors such as fbl, ica, atlL, vwbl, and slush	Intravenous and peritoneal catheters	Giormezis et al. (2015)
P. aeruginosa, S. aureus, P. mirabilis, K. pneumoniae, Enterobacter, and *E. faecalis*	Less susceptible to tobramycin, cefotaxime, and cefuroxime than tracheal isolates	Endotracheal tubes	Adair et al. (1999)
E. faecalis and *E. faecium*	Mature biofilms had higher tolerance to antibiotics such as rifampicin-containing combinations than new biofilms	Prosthetic joint	Holmberg and Rasmussen (2016)
S. aureus	Resistance to cefazolin		Dastgheyb et al. (2015)

predominantly caused by *Staphylococcus* and *Candida* spp. resulting approximately in 71% of morbidity (Moran et al. 2009).

Contact lenses, chemically made up of hydrogel etafilcon A and silicone hydrogel senofilcon, serve as another common site for bacterial attachment. The attachment of *P. aeruginosa* occurs on such surface due to its attraction for etafilcon A, and they get bound to the surface more effectively than that of *S. aureus* (Dutta and Willcox 2013). Studies showed that there is an enhancement of surface adhesion of the bacterial cells from 2 to 18 h of incubation. Another study showed that *S. aureus* and *P. aeruginosa* are attracted towards the silicone hydrogel contact lenses to develop the biofilm (Subbaraman et al. 2011).

7.3.4 Long-Term Use of Medical Devices

Another major problem of biofilm-associated infections is related to medical devices such as artificial joints, mechanical heart valves or stents, dental implants, and catheters where the sessile cells easily colonize and develop various chronic infections. The most common type of bacterial cells associated with such infections comprises Gram-positive organisms like *S. aureus*, *E. faecalis*, *S. epidermidis*, and *Streptococcus viridans* (Shigeru Fujimura 2015) and Gram-negative organisms like *Klebsiella pneumoniae, P. mirabilis, E. coli*, and *P. aeruginosa* (Rohit Ruhal 2015). The medical devices that are used for a longer period of time are referred to as invasive as they remain inside the body for a longer period of time. The development of biofilm on such devices causes adverse effect upon the health of the host (Hyun et al. 2015). Table 7.2 lists the major biofilm-forming species on medical implants.

7.4 Mechanisms of Bacterial Attachment on the Surface of Appliances

A number of factors like bacterial motility, geometrical conformation of the surface, nature of surface material, and other environmental factors regulate the bacterial attachment on the surface. Since the exact process of biofilm formation and propagation on medical appliances is not yet fully known, extensive research is warranted on the understanding of the factors regulating the growth of biofilm on these devices.

7.4.1 Bacterial Motility and Surface Attachment

7.4.1.1 Brownian Motion

The movement of bacterial species devoid of locomotory organelles, like *Klebsiella* sp. and *Streptococcus* sp. is controlled by the Brownian motion, a random unregulated movement of particles within the fluid with constant collision (Mitchell and Kogure 2006). It helps in regulating the movement among these groups of bacterial cells on the surface.

7.4.1.2 Motility of the Flagella

A flagellum of bacterial cell comprises three important parts: the basal part or body, the hook, and the flexible region known as filament (Berg and Anderson 1973). The filament is comprised of 20,000 subunits of proteins known as flagellin which protrudes out approximately thrice the size of the bacterial cell (Macnab 2003). It has been observed that some of the bacterial species like *Helicobacter pyroli* and *Vibrio cholerae* possess an external sheath in addition to the flagella. This persisting structure is found to protect *H. pyroli* from the acidic environment of the stomach (Geis et al. 1993). The filament of the flagella remains connected with the hook providing stability to the filament and in contrary connects to the basal body which provides energy being required for the movement of the flagella (Silverman and

Simon 1974). The basal region of the flagella comprises various rings that extend from the cytoplasmic region up to the membrane (Osterman et al. 2015; Chevance and Hughes 2008). The flagella is comprised of three major types of proteins, FliM, FliN, and FliG, along with MotA and MotB for the mechanism of motor rotation found predominantly within the *E. coli* (Kojima and Blair 2004). The bacterial cells employ various cationic components for the generation of energy required for the flagellar movement (Paul et al. 2008). Some studies have also shown that most of the bacterial species use proton motive force for the generation of the driving power for flagellar movement (Berg 2003), but the exception is *Vibrio cholerae* which utilizes sodium ions for the purpose of flagellar rotation (Atsumi et al. 1992).

7.4.1.3 Non-flagellar Motility

The bacterial cells that are devoid of flagella perform various types of movements that include swarming, swimming, gliding, and twitching or sliding motility. The extension and retraction of the type IV pili are responsible for the twitching motility (Merz et al. 2000; Skerker and Berg 2001). A large number of bacterial species like *Pseudomonas*, *Myxococcus*, and *Neisseria* possess the type IV pili (Skerker and Berg 2001). Thus the mechanism of twitching involves the movement in a group via cell-to-cell contact (Mattick 2002). The mechanism of twitching motility helps the bacterial cells to form microcolonies and disperse out from the state of biofilm (Harshey 2003). The mechanism of smooth movement of bacterial cells without the use of pili or flagella upon a surface is referred to as gliding that shows similarity to swarming and twitching motilities. *Mycobacteria* and *Cyanobacteria* perform the gliding movement (Henrichsen 1972). These various types of movements performed by the bacterial cells are dependent on various types of external stimuli.

7.4.2 Environmental Factors Regulating Bacterial Movements

7.4.2.1 Chemotaxis

The cell poles are usually comprised of chemosensory receptors comprising the transmembrane chemoreceptor systems that possess the ability to form bonds with the ligands and undergo the mechanism of differential methylations (Briegel et al. 2009). The mechanism of chemotaxis in *E. coli* is regulated by chemoattractants that are being present within the media that results in CheA autophosphorylation and causes phosphotransfer to CheY cytoplasmic receptor. The flagellar protein FliM binds to the phosphorylated CheY and allows the clockwise movement of the flagella (Porter et al. 2011). The mechanism of clockwise and counterclockwise movements of the flagella allows the movement of the bacterial cells toward the higher concentrations of the chemoattractants (Wadhams and Armitage 2004). This movement allows the bacteria to precisely select their favorable position on devices to form the biofilm.

7.4.2.2 Quorum Sensing

The mechanism of quorum sensing (QS) is a density-dependent cell-to-cell communication between the bacterial cells with the help of chemical signals known as autoinducers (AI). The amount of secretion of the AI increases with the increase in bacterial population up to a threshold. The interaction of AI with that of the "two-component system" comprised of transcriptional regulator or signal transduction system results in variation of gene expression and density-dependent change of the behavior in bacterial cells (Marles-Wright and Lewis 2007). The mechanism of QS observed within the Gram-negative group of bacterial cells is regulated by LuxI/LuxR systems (Henke and Bassler 2004). Acyl-homoserine lactones (HSL) are being produced by the LuxI. *Streptococcus pneumoniae* and *Staphylococcus aureus* are the group of Gram-positive bacterial cells that produce peptides instead of HSLs which are being identified via the two-component system and transported with the help of ABC transporters (Miller and Bassler 2001).

7.4.2.3 Bis-(3′-5′)-Cyclic Dimeric Guanosine Monophosphate (c-di-GMP)

This is a type of secondary messenger that determines the transition from the planktonic to sessile phases by the process of flagellar downregulation. This promotes the mechanism of the formation of biofilm (Povolotsky and Hengge 2015). The conversion of the planktonic forms to the sessile structures is controlled by the high levels of c-di-GMP. In *E. coli*, degradation of c-di-GMP is done by YhjH, a phosphodiesterase. Hence by regulating its concentration the formation of the biofilm can be regulated (Pesavento et al. 2008).

7.4.3 Mechanism of Development of Biofilm upon Medical Implants

The major biofilm-associated infection related to medical devices affecting human health is usually comprised of microbial consortia resulting in the formation of polymicrobial structure. These types of interactions occurring among various groups of organisms play an important role in determining the type of biofilm to be formed on different medical devices available. The persistence of polymicrobial species results in the mechanism of horizontal gene transfer which causes the development of antimicrobial resistance among the various existing organisms (Table 7.3) (Marsh et al. 2000). This is confirmed by the presence of a large number of vancomycin-resistant *S. aureus* from the nephrostomy tube biofilm (Weigel et al. 2007). Again, the horizontal gene transfer between *E. coli* and *Streptococcus gordonii* present at the dental root canal results in the development of antibiotic resistances (Sedgley et al. 2008). This exchange of genetic components among the microbial consortia results in the development of complete resistance against various antibiotics and antimicrobials.

7.4.3.1 Initiation of the Biofilm Formation

The initial stage of the development of biofilm upon the surfaces of the medical devices is by the mechanism of surface adhesion. The adhesion on implants by staphylococci is a complex process and is dependent on various types of specific and non-specific factors. The mechanism of attachment can occur on various types of native surfaces of implanted biomaterials or the surfaces which have been already modified by the host plasma protein derivatives, coagulation proteins like platelets, and the proteins being present within the extracellular matrix (Vaudaux et al. 1994). The development of the biofilm and its adhesion is governed by various factors like the London-van der Waals forces, polarity, and hydrophobic interactions (Dickinson and Bisno 1989). Studies have also shown that the hydrophobic interactions between the staphylococcal cells with the biomaterial surface are greatly responsible for its attachment (Espersen et al. 1990). The mechanism of adhesion is dependent on various matrix proteins like vitronectin, fibronectin, and thrombospondin that are greatly activated by the platelets and help in the adhesion of bacterial cells like *S. epidermidis* (Allignet et al. 1999). The biofilm is comprised of various types of electrolytes, proteins, and unidentified molecules (Habash and Reid 1999). The mechanism of initial attachment is also regulated by a type of capsular polysaccharides/adhesins and slime (Muller et al. 1993). *S. epidermidis* produces adhesins or autolysins, which apart from having a major role in the metabolism of cell wall also help in the adhesion of the bacterial cells to the surface of polystyrene. Although the high amount of autolysin (AtlE) shows difficulty of adhesion upon the surface of polystyrene materials, it still shows effective biofilm formation upon the glass surface (Heilmann et al. 1997). It has been observed that AtlE helps in the attachment of *S. epidermidis* upon the surface of the medical devices by binding with vitronectin (Li et al. 2001).

7.4.3.2 Adhesion by Proteins

The family of the bacterial cell surface adhesins is comprised of microbial surface components recognizing adhesive matrix (MSCRAMMs) that facilitates the adhesion of the bacterial cell to the host cell surface. In Gram-positive bacterial cells, MSCRAMMs recognize the extracellular matrix of the host cell by the presence of lectin, fibronectin, and collagen (Patti et al. 1994a). The binding of MSCRAMMs with the peptidoglycan, present on the bacterial cells, occurs by covalent interactions (Chagnot et al. 2012). A single bacterium may possess multiple MSCRAMM sites which are specific for a particular ligand. For instance, the fibronectin-binding proteins like FnBPA and FnBPB, collagen binding proteins (Can), and fibrinogen binding proteins (ClfA) are present within *S. aureus* (Clarke and Foster 2006). The infections associated with *S. aureus* are dependent on the expression of the adhesion proteins as it is found that expression of adhesion expression gene *cna* leads to septic arthritis (Patti et al. 1994b). Gram-negative bacteria are usually comprised of outer membrane proteins, which facilitate the elaboration of the pili (Soto and Hultgren 1999). The Gram-negative cells also contain adhesins which are produced by the type V secretion pathway. These are usually the groups of autotransporters comprising N-terminal signal sequence, C-terminal transmembrane domain, and the internal passage domain that help in the performance of the specific functions by the autotransporters (Chagnot et al. 2013).

7.4.3.3 Amyloid Fibers

A large number of bacterial species possess a class of beta sheet fibers known as amyloid fibers that show homology to the amyloid which is actually associated with human diseases like the Alzheimer's and Parkinson's diseases (Cohen and Kelly 2003). The amyloid fibers being present within the bacterial cells show a wide differentiation from the eukaryotic ones on the basis of their functional aspect and important role played in the mechanism of surface adhesion (Barnhart and Chapman 2006). The amyloids being present within *Gordonia, Corynebacterium*, and *Mycobacterium* spp. are characterized functionally, yet their roles in the mechanism of pathogenesis have not been fully studied (Ramsugit and Pillay 2015). The amyloid type of protein subunits being observed in *Salmonella* sp. and *E. coli* is transported to the periplasm via sec translocon in association with the membrane protein CsgG. The mechanism of nucleation accruing by the CsgA results in the assemblage of the curli fibers upon the minor subunit SsgB which is located upon the cell surface (Evans et al. 2015). These types of amyloid fibers being present in *E. coli* and *Salmonella* sp. allow better adhesion with the abiotic surface with the help of the matrix (Evans et al. 2015).

7.4.3.4 Maturation of Biofilm

The maturation of the biofilm takes place by the mechanism of interactions taking place among each other in the presence of the AI and resulting in the expression of the biofilm-forming specific genes (Gupta et al. 2016). The enhancement in the production of AI results in the development of virulence and regulation of the genes (Gu et al. 2013). This phase is marked by the production of EPS that helps in encompassing the cells and protects them from drugs and other antimicrobial agents (Veerachamy et al. 2014).

7.4.3.5 Dispersal of Biofilm

This is an important phase as it results in the metastasis of infection within the host body. The mechanism of dispersal is responsible for the development of chronic infection within the host (Veerachamy et al. 2014). The maturation of the biofilm results in the accumulation of various toxic products, and in order to get rid of these toxic waste materials, the cells try to disperse to any other place of host body or to other section of the implant. The availability of nutrients from the host system allows the sessile microcolonies to grow and get dispersed to various parts of the host system (Oppenheimer-Shaanan et al. 2013).

7.5 Characteristics of Medical Device-Associated Infections

7.5.1 Host Factors

The device may create a pressure to the adjacent host tissues,and the chemicals coming out of the implant may trigger some immune response. The implant-associated chronic infections may occur by the damage of tissue or loss in the local immunity and improper integration of tissues at the interface of biomaterials and tissue resulting in inflammation. The defense mechanism of the body is altered due to the presence of these biomaterials. The surface properties of the biomaterial regulate the deposition and orientation of protein (Busscher et al. 1991a, b), enzymes, and other biomolecules and may cause the change in local pH and temperature.

7.5.2 Nature of Biomaterials

In response to the environmental changes, the chemical nature of the implant biomaterial may change leading to its premature degradation. As the adhesion of the bacterial cells upon the surface of the biomaterials is dependent upon the physicochemical properties of the appliance or implant, change in the nature of implant will lead to the alteration in the cell surface dynamics and metabolic pattern of the bacteria (Engel 1986; Fletcher 1980). The surface of the biomaterials gets modified with respect to the topography, manufacturing processes, texture, trace chemicals, and its hydrophobic nature that in turn allow better binding of the sessile cells causing formation of huge amount of biofilm (Pulverer et al. 1987).

7.5.3 Microbial Factors

The bacterial cells possess the ability of getting adhered upon an exposed surface with the help of EPS and glycocalyx (Anwar et al. 1989). Various forces like chemical bonding or Brownian motion are responsible for the better adherence of the organism upon the abiotic surface (Costerton and Lappin-Scott 1989). The various chemical and physical materials help in attracting the microbial cells and develop a biofilm upon the surfaces of the medical devices.

7.6 Common Appliances and Implants with Biofilm

The common medical devices which are often reported to be infected by biofilm-forming bacteria include various catheters, artificial heart valves, lenses, etc. (Fig. 7.1). Infection of these devices poses a serious threat mainly to the surgical treatments. Some of them are as follows:

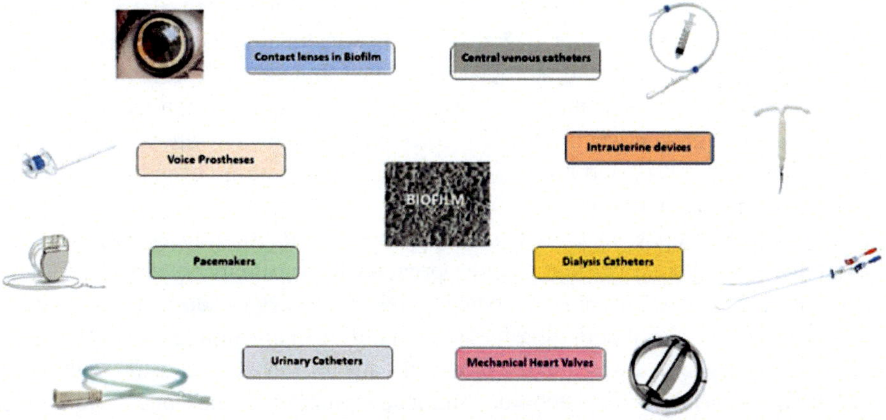

Fig. 7.1 Medical devices susceptible to biofilm attack

7.6.1 Contact Lenses

Biofilm formation by different types of microorganisms on general contact lenses and lens storage cases can be a dangerous phenomenon in respect to contact lens-associated corneal infections. Commercially there are various types of care solutions available for different types of contact lens, which are used to kill or decrease microbial population and their contamination (Pulcini 2001). There have been many comparisons with the growth of biofilm on the different contact lenses and lens cases, in different experimental studies conducted across the world. Determination of the effects of lens care solutions and bacteriophage on these biofilms is also a very popular experimental study field for researchers researching with different types of biofilms. Lenses are of two types, hard lens and soft lens (Flemming et al. 2000). The organisms which are responsible for such biofilm formations are generally *Staphylococcus aureus, Pseudomonas aeruginosa, Candida albicans*, *Escherichia coli,* etc. Biofilm production process can be performed by any of the following methods such as modified O'Toole and Kolter method. The obtained results can be visualized and viewed using scanning electron microscopy. The obtained biofilms can also be quantified if desired, by colony counting method or by spectrophotometric measurement (Sutherland 2001). Also, it should be noted that the soft lens solution, used generally for cleaning, has a significant inhibitory effect on biofilm formation on soft lenses and on lens cases too. There is no significant inhibitory effect by bacteriophages on contact lenses as possessing biofilms (Hussain et al. 1993).

7.6.2 Central Venous Catheters (CVC)

The use of CVC or which is otherwise known as central venous access catheters is often associated with the danger of getting infected by microbes. The infecting microorganisms gradually cover the surface of the CVC with the extracellular matrix to form the biofilm (Leriche et al. 2000). As the invading microorganisms of biofilm are well equipped with antimicrobial resistance and are able to evade the host immune reactions, these catheters become a source of detrimental infections. Therefore, the best preventive method is to avoid any catheterization which is not immediately necessary or trying to decrease the duration of use of indwelling device when CVC is required (Donlan 2000a).

In spite of aseptic measurement and use of antibiotic-infused catheters, for example, minocycline or rifampin, the preventive measurements may be inserted in some cases, but in case of non-biocidal cases, the strategies adopted are still under observation, like use of anti-adhesive or competitive interaction approaches. Generally, in the common cases, the diagnosis of catheter-related bloodstream infection (CRBSI) is usually detected on clinical symptoms and needs a special microbiological confirmation by corresponding blood cultures in order to prevent the frequent removal of catheter unnecessarily (Tolker-Nielsen and Molin 2000). The treatment of catheter-related bloodstream infection relies on catheter removal and a systematic removal of antimicrobials. But anyhow, the ALT or the antibiotic lock technique can be used as a method or a way to remove biofilm from the interior surface of the catheter lumen in case of an uncomplicated long-term catheter-related BSI caused by coagulate-negative staphylococci or CoNS or *Enterobacteriaceae* (Lewandowski 2000). But in the recent times, newer assays and strategies have been developed which are very promising and have been specifically made to advance the process of biofilm removal; these methods depend on matrix degradation or destabilization or the development of anti-persister compounds. These methods target the most stable bacterial cells which are located inside the biofilm (Stoodley et al. 1997). The biofilm formation has to be understood at the molecular level so that it may guide toward the development of a new approach to prevent or treat these frequent infections.

7.6.3 Endotracheal Tubes (ETT)

Biofilm in ETT is related to ventilated patients, mainly for the patients of v**entilator-associated pneumonia (VAP)**. The purpose of this is to observe the formation of ETT biofilm and its application in relapse and response of VAP (James et al. 1995). The general results in these studies are obtained through surveillance of ETA or endotracheal aspirates at exudation. ETT are generally looked upon for any kinds of microbial assessment and are examined by scanning electron microscopy. The process of airway bacterial colonization and the process of biofilm formation on ETT are events which are too early and in most cases cannot be predicted in ventilated patients (Tolker-Nielsen and Molin 2000). There is a presence of several microbes and their continuity in between the airway colonization and the biofilm formation and also the VAP development. Biofilm performs a pathogenic mechanism in the occurrence of microbial community and it is impaired through the response of the treatment in VAP (Durack 1975).

7.6.4 Intrauterine Devices (IUD)

Many microorganisms which are pathogenic may form biofilms on inert surfaces of implanted devices, like intrauterine devices or IUD. Many studies are carried out to observe the formation and location of biofilm on the surface of intrauterine devices in patients who have often vulvovaginal candidiasis. These studies are usually done to observe the susceptibility profile of any isolated species. For example, the susceptibility of an isolated yeast strain to amphotericin B and fluconazole is often tested. *Candida albicans* can be recovered from the IUD, and it is often found to be susceptible to the antifungal agents when they are tested upon some planktonic growing conditions (Donlan 2000b). These results can clearly indicate that the presence of the biofilm on surface of the IUD is a sign of danger and can result into recurrent vulvovaginal candidiasis (Ehlers and Bouwer 1999), a condition that affects about 75% of women on at least once in a lifetime.

7.6.5 Mechanical Heart Valves

Microorganisms in a consortium are considered as the biggest enemies of human, as they become stronger and hence more difficult to eradicate from the surface they are attached to. To lower the morbidity level out of cardiac problems, mechanical heart valves are often implanted.

Development of biofilms on components of mechanical heart valves and surrounding tissues of the heart leads to a condition which is known as prosthetic valve endocarditis (Roberts et al. 1999). The primary organisms which are responsible for this disease is *Staphylococcus aureus.*

7.6.6 Pacemakers

Cardiac pacing is a common operational phenomenon which takes place by the mechanism of cardiac pacemaker. It is one of the most efficient and advantageous methods in comparison to others in the treatment of arrhythmias. However, it can also cause medical reactions, whether it is natural or complex reactions (Hausner and Wuertz 1999). It occurs either early or late. The scientific evidence for the chances to have the infection is studied in research laboratories, and biofilm formation which is associated with cardiac pacemaker is also studied nowadays in many research laboratories all over the world. The study mainly takes place in different laboratories in three different categories. The first one is the diagonal complications and therapies of infection; the second one being the microbiological complications and therapies of infection, and the third one is the clinical complications and therapies of infection. *Staphylococcus epidermidis* and *Staphylococcus aureus* are the microorganisms most frequently used in this research study by isolating them. It is not an easy process to find out the incidence of infection, which is generally associated with pacemakers (Ghigo 2001). By various processes of therapies, the process of complete removal of pacemakers is done in special cases where there is any biofilm suspected. But the use of systemic antibiotic prophylaxis in decreasing the incidence of infection related with implantation of a pacemaker remains controversial.

7.6.7 Dialysis Catheters

Biofilm on the surface of tunneled hemodialysis (HD) catheters is hard to remove and moreover it is associated with recurrent bacteremia (Xie et al. 2000). The outer lining surface of the extravascular segment of tunneled dialysis catheters in patients with both bacteremia and non-bacteremia hemodialysis catheter patients has the thickest biofilm and the highest microbiological yield in them. Especially the biofilm is thicker in patients with bacteremia (Davies et al. 1998). This idea is important for the designing of preventive strategies and it is also important for the management of patients having catheter infection.

Dialysis membrane is often found to be permeable for the bacterial components, produced by biofilms. These bacteria can trigger an inflammatory response in the patient. The ultrafiltration membrane of hemodialysis may also get infected by biofilm-forming microbes, due to improper and insufficient sterilization (Suman et al. 2013).

7.6.8 Urinary Catheters

The predominant life forms for the majority of microbes, which are hydrated in a biologic system, are usually a community which is cooperative and it is termed as "biofilm." On an indwelling urinary catheter, a biofilm may consist of a number of microorganisms (Stickler et al. 1998). These can be adherent in nature (Tunney et al. 1999), and their extracellular products may also be present within the biofilm, and the host components which are generally deposited on the catheter are also present within the biofilm. The mode of life which is conducted by the biofilms usually follows a survival strategy which is advantageous to the microorganisms related to it. The biofilm on urinary catheters causes the infections to persist which are also resistant to antimicrobial therapy. This occurs because of the chronic characters and it leads to bacteriuria which is almost inevitable (Yung-Hua et al. 2001). The routine treatment of asymptomatic bacteriuria in people who possesses these characters is not suggested. Person developing urinary tract infection, urine collection is performed before the changing of catheter to enhance the accuracy of the result. Now, when we are changing the catheter characteristics, it can cause to improve the response to antibiotic therapy by the removal of the biofilm (Murga et al. 2001). This biofilm may sometimes contain the infectious organisms, and it also can stay and can be used as a nidus for reinfection. Till date, there are no effective strategies which are proven and can exist for the prevention of catheter-associated urinary tract infection in people which are specially characterized and mentioned chronically important (McLaughlin-Borlace et al. 1998).

7.6.9 Voice Prostheses

The predominant life forms that comprise majority of microbes, which are hydrated in a biologic system, are usually a community which is cooperative and it is termed as "biofilm." There are many complications which can be effective to reduce the voice prostheses, but one of its specific types that decrease voice prosthesis device life is the formation of microbial biofilm plaque and invasive growth in silicone on the prosthesis valve that can cause ultimately the aspiration by the help of this device. Again, this process may also take place by the process of increased airflow resistance (Stewart et al. 1997).

7.6.10 Intracardiac Prostheses

The prosthetic valve endocarditis is mostly used because of the implantation of it in the microbes while undergoing the process of surgery or the process can also be undergone by the perioperative period (Raad et al. 1992). The reason for this kind of contamination consists of damages in the aseptic surgical procedures, contaminated cardiac bypass device, and infection because of the indwelling of intravascular catheters. Also in few cases, it is seen in the contamination of the valves too. In case of porcine valves, in many studies it has been seen that this will cause its contamination with the microorganism *Mycobacterium chelonae* (Wirtanen et al. 1996). The frequently used organisms which are related to the prosthetic valve endocarditis prove to be one of the reasons for coagulase-negative staphylococci which also cause fungi, for example, *Candida* or *Aspergillus*, and various other organisms are also applied over there. Patients with prosthetic valves remain at definite though decreased risk of endocarditis in the years following implantation (Buswell et al. 1996). The recent development of the bacteriogenic strategies of the prosthetic valve endocarditis reminds us of the raw valve endocarditis and *Streptococcus viridans* strains which is in used often as the etiological agents (Camper et al. 1998). These endocardial infections are caused as a result of the different bacterial strains that can be spontaneous or can take place during a specific process, for example dental cleaning (Hood and Zottola 1997). The different microorganisms other than the abovementioned ones are the result of the beginning of prosthetic valve endocarditis which possesses enterococci, coagulase-negative staphylococci, *Staphylococcus aureus,* and fastidious Gram-negative coccobacilli, which are *Haemophilus* species such as *Actinobacillus actinomycetemcomitans, Cardiobacterium hominis, Eikenella corrodens,* and *Kingella kingae* (Watnick and Kolter 1999). The process of eradication of infection from a prosthetic valve is an extremely hard method, and this occurs when a long method of bacterial antimicrobial therapy is performed (Stark et al. 1999).

7.7 Conclusion and Future Prospects

The polymicrobial biofilms are the group of sessile communities which are ubiquitous in nature and are frequently found to be formed on the abiotic surface of medical implants placed in a host body. In this entire chapter we tried to categorically describe the various mechanisms that are associated with the formation of biofilm upon the surfaces of the medical devices. The biofilm grown on the surface of implants invites more opportunistic pathogens to invade, and this becomes the major obstruction for achievement of success after many surgical operations. Such biofilm attack is the root cause of several nosocomial diseases also. Such trend in ever-increasing biofilm-associated diseases demands adoption of serious measures to combat such microbes, growing on lifesaving medical devices.

References

Adair CG, Gorman SP, Feron BM, Byers LM, Jones DS, Goldsmith CE, Moore JE, Kerr JR, Curran MD, Hogg G, Webb CH, Mccarthy GJ, Milligan KR (1999) Implications of endotracheal tube biofilm for ventilator-associated pneumonia. Intensive Care Med 25:1072–1076

Akbari F, Kjellerup BV (2015) Elimination of bloodstream infections associated with Candida albicans biofilm in intravascular catheters. Pathogens 4:457–469

Akhter J, Ahmed S, Saleh AA, Anwar S (2014) Antimicrobial resistance and in vitro biofilm-forming ability of Enterococci spp. isolated from urinary tract infection in a tertiary care hospital in Dhaka, Bangladesh. Med Res Counc Bull 40:6–9

Allignet J, Galdbart JO, Morvan A, Dyke KGH, Vaudaux P, Aubert S, Desplaces N, Solh NE (1999) Tracking adhesion factors in *Staphylococcus caprae* strains responsible for human bone infections following implantation of orthopaedic material. Microbiology 145:2033–2042

Anwar H, Van Biesen T, Dasgupta M, Lain K, Costerton JW (1989) Interaction of biofilm bacteria with antibiotics in a novel in vitro chemostat system. Antimicrob Agents Chemother 33:1824–1826

Atsumi T, Mccarter L, Imae Y (1992) Polar and lateral flagellar motors of marine vibrio are driven by different ion-motive forces. Nature 355:182–184

Barken KB, Pamp SJ, Yang L, Gjermansen M, Bertrand JJ, Klausen M, Givskov M, Whitchurch CB, Engel JN, Tolker-Nielsen T (2008) Roles of type IV pili, Flagellum mediated motility and extracellular DNA in the formation of mature multicellular structures in *Pseudomonas aeruginosa* biofilms. Environ Microbiol 10:2331–2343

Barlow M (2009) What antimicrobial resistance has taught us about horizontal gene transfer. Methods Mol Biol 532:397–411

Barnhart MM, Chapman MR (2006) Curli biogenesis and function. Annu Rev Microbiol 60:131–147

Berg HC (2003) The rotary motor of bacterial flagella. Annu Rev Biochem 72:19–54

Berg HC, Anderson RA (1973) Bacteria swim by rotating their flagellar filaments. Nature 245:380–382

Bohn YS, Brandes G, Rakhimova E, Horatzek S, Salunkhe P, Munder A, Van Barneveld A, Jordan D, Bredenbruch F, Haussler S (2009) Multiple roles of *Pseudomonas aeruginosa* TBCF10839 PilY1 in motility, transport and infection. Mol Microbiol 71:730–747

Briegel A, Ortega DR, Tocheva EI, Wuichet K, Li Z, Chen S, Müller A, Iancu CV, Murphy GE, Dobro MJ, Zhulin IB, Jensen GJ (2009) Universal architecture of bacterial chemoreceptor arrays. Proc Natl Acad Sci U S A 106:17181–17186

Bryers JD (2008) Medical biofilms. Biotechnol Bioeng 100(1):1–18

Busscher HJ, Van Der Mei HC, Schakenraad JM (1991a) Analogies in the two-dimensional spatial arrangements of adsorbed proteins and adhering bacteria: bovine serum albumin and *Streptococcus sanguis* 12. J Biomat Sci Polymer EDN 3:85094

Busscher HJ, Stokroos I, Schakenraad JM (1991b) Two-dimensional, spatial arrangement of fibronectin adsorbed to biomaterials with different wettabilities. Cell Mater 1:49–57

Buswell CM, Herlihy YM, Lawrence LM, McGuiggan JTM, Marsh PD, Keevil CW (1996) Extended survival and persistence of Campylobacter spp. in water and aquatic biofilms and their detection by immunofluorescent-antibody and -rRNA staining. Appl Environ Microbiol 5:319–326. https://doi.org/10.1016/0927-7765(95)01226-5

Cabral DA, Loh BA, Speert DP (1987) Mucoid *Pseudomonas aeruginosa* resists nonopsonic phagocytosis by human neutrophils and macrophages. Pediatr Res 22:429–431

Camper AK, Warnecke M, Jones WL, McFeters GA (1998) Pathogens in model distribution system biofilms. American Water Works Association Research Foundation, Denver

Chagnot C, Listrat A, Astruc T, Desvaux M (2012) Bacterial adhesion to animal tissues: protein determinants for recognition of extracellular matrix components. Cell Microbiol 14:1687–1696

Chagnot C, Zorgani MA, Astruc T, Desvaux M (2013) Proteinaceous determinants of surface colonization in bacteria: bacterial adhesion and biofilm formation from a protein secretion perspective. Front Microbiol 4:303

Chen M, Yu Q, Sun H (2013) Novel strategies for the prevention and treatment of biofilm related infections. Int J Mol Sci 14(9):18488–18501

Clarke SR, Foster SJ (2006) Surface adhesins of Staphylococcus aureus. Adv Microb Physiol 51:187–224

Cohen FE, Kelly JW (2003) Therapeutic approaches to protein-misfolding diseases. Nature 426:905–909

Costerton JW, Lappin-Scott HM (1989) Behavior of bacteria in biofilms. ASM News 1989:650

Costerton JW, Irvin RT, Cheng KJ (1981) The bacterial glycocalyx in nature and disease. Annu Rev Microbiol 35:299–324

Danne C, Dramsi S (2012) Pili of Gram-positive bacteria: roles in host colonization. Res. Microbiology 163:645–658

Darouiche RO (2001) Device-associated infections: a macroproblem that starts with microadherence. Clin Infect Dis 33:1567–1572

Dastgheyb SS, Hammoud S, Ketonis C, Liu AY, Fitzgerald K, Parvizi J, Purtill J, Ciccotti M, Shapiro IM, Otto M, Hickok NJ (2015) Staphylococcal persistence due to biofilm formation in synovial fluid containing prophylactic cefazolin. Antimicrob Agents Chemother 59:2122–2128

Davey ME, O'Toole A (2000) Microbial biofilms: from ecology to molecular genetics. Microbiol Mol Biol Rev 64:847–867

Davies DG, Parsek MR, Pearson JP, Iglewski BH, Costerton JW, Greenberg EP (1998) The involvement of cell-to-cell signals in the development of a bacterial biofilm. Science 280:295–298. https://doi.org/10.1126/science.280.5361.295

De Kievit TR (2001) Multidrug efflux pumps: expression patterns and contribution to antibiotic resistance in Pseudomonas aeruginosa biofilms. Antimicrob Agents Chemother 45:1761–1770

Dickinson GM, Bisno AL (1989) Infections associated with indwelling devices: infections related to extravascular devices. Antimicrob Agents Chemother 33(5):602–607

Donlan RM (2000a) Role of biofilms in antimicrobial resistance. ASAIO J 46:S47–S52. https://doi.org/10.1097/00002480-200011000-00037

Donlan RM (2000b) Biofilm control in industrial water systems: approaching an old problem in new ways. In: Evans LV (ed) Biofilms: recent advances in their study and control. Harwood Academic Publishers, Amsterdam, pp 333–360

Donlan RM (2008) Biofilms on central venous catheters: is eradication possible? Curr Top Microbiol Immunol 322:133–161

Dudman WF (1977) The role of surface polysaccharides in natural environments. In: Sutherland IW (ed) Surface carbohydrates of the prokaryotic cell. Academic Press, New York, pp 357–414

Durack DT (1975) Experimental bacterial endocarditis. IV Structure and evolution of very early lesions. J Pathol 115:81–89. https://doi.org/10.1002/path.1711150204

Dutta D, Willcox MD (2013) A laboratory assessment of factors that affect bacterial adhesion to contact lenses. Biology 2:1268–1281

Ehlers LJ, Bouwer EJ (1999) RP4 plasmid transfer among species of Pseudomonas in a biofilm reactor. Water Sci Technol 7:163–171. https://doi.org/10.1016/S0273-1223(99)00164-X

Engel T (1986) Low-energy atom scattering from surfaces. Science 234:327–333

Espersen F, Wilkinson BJ, Gahrn-Hansen B, Thamdrup Rosdahl V, Clemmensen I (1990) Attachment of staphylococci to silicone catheters in vitro. APMIS 98:471–478

Evans ML, Chorell E, Taylor JD, Åden J, Götheson A, Li F, Koch M, Sefer L, Matthews SJ, Wittung-Stafshede P, Almqvist F, Chapman MR (2015) The bacterial curli system possesses a potent and selective inhibitor of amyloid formation. Mol Cell 57:445–455

Falanga V (2000) Classifications for wound bed preparation and stimulation of chronic wounds. Wound Repair Regen 8:347–352

Falkinham JO (2007) Growth in catheter biofilms and antibiotic resistance of Mycobacterium avium. J Med Microbiol 56:250–254

Flemming H-C, Griegbe WJ, Mayer C (2000) Physico-chemical properties of biofilms. In: Evans LV (ed) Biofilms: recent advances in their study and control. Harwood Academic Publishers, Amsterdam, pp 19–34

Fletcher M (1980) Adherence of marine microorganisms to smooth surfaces. In: Beachey EH (ed) Bacterial adherence: receptors and recognition. Series B. Chapman and Hall, London, pp 345–374

Garrett TR, Bhakoo M, Zhang Z (2008) Bacterial adhesion and biofilms on surfaces. Prog Nat Sci 18:1049–1056

Geis G, Suerbaum S, Forsthoff B, Leying H, Opferkuch W (1993) Ultrastructure and biochemical studies of the flagellar sheath of Helicobacter pylori. J Med Microbiol 38:371–377

Ghigo J-M (2001) Natural conjugative plasmids induce bacterial biofilm development. Nature 412:442–445. https://doi.org/10.1038/35086581

Giormezis N, Kolonitsiou F, Makri A, Vogiatzi A, Christofidou M, Anastassiou ED, Spiliopoulou I (2015) Virulence factors among Staphylococcus lugdunensis are associated with infection sites and clonal spread. Eur J Clin Microbiol Infect Dis 34:773–778

Gristina AG, Giridhar G, Gabriel BL, Naylor PT, Myrvik QN (1993) Cell biology and molecular mechanisms in artificial device infections. Int J Artif Organs 16:755–764

Grkovic S (2002) Regulation of bacterial drug export systems. Microbiol Mol Biol Rev 66:671–701

Gu HS, Hou C, Yongyat S, De Tore Ren D (2013) Patterned biofilm formation reveals a mechanism for structural heterogeneity in bacterial biofilms. Langmuir 29(35):11145–11153

Gupta PS, Sarkar B, Das S, Bhattacharjee P (2016) Tribedi, Biofilm, pathogenesis and prevention e a journey to break the wall: a review. Arch Microbiol 198(1):1–15

Habash M, Reid G (1999) Microbial biofilms: their development and significance for medical device-related infections. J Clin Pharmacol 39(9):887–898

Harshey RM (2003) Bacterial motility on a surface: many ways to a common goal. Annu Rev Microbiol 57:249–273

Hausner M, Wuertz S (1999) High rates of conjugation in bacterial biofilms as determined by quantitative in situ analysis. Appl Environ Microbiol 65:3710–3713

Heilmann C, Götz F (2009) Cell–cell communication and biofilm formation in gram-positive bacteria. In: Krämer R, Jung K (eds) Bacterial signaling. Wiley-VCH Verlag GmbH & Co. KGaA, Weinheim, Germany. https://doi.org/10.1002/9783527629237.ch1

Heilmann C, Hussain M, Peters G, Götz F (1997) Evidence for autolysin-mediated primary attachment of *Staphylococcus epidermidis* to a polystyrene surface. Mol Microbiol 24:1013–1024

Hellstrtom J (1938) The significance of staphylococci in the development and treatment of renal and urethral stones. Br J Urol 10:348–372

Henke JM, Bassler BL (2004) Bacterial social engagements. Trends Cell Biol 14:648–656

Henrichsen J (1972) Bacterial surface translocation: a survey and a classification. Bacteriol Rev 36:478–503

Herrmann M, Vaudanx PE, Pittet D (1988) Fibronectin, fibrinogen and laminin act as mediators of adherence of clinical staphylococcal isolates to foreign material. J Infect Dis 158:693–701

Holmberg A, Rasmussen M (2016) Mature biofilms of *Enterococcus faecalis* and *Enterococcus faecium* are highly resistant to antibiotics. Diagn Microbiol Infect Dis 84:19–21

Hood SK, Zottola EA (1997) Adherence to stainless steel by foodborne microorganisms during growth in model food systems. Int J Food Microbiol 37:145–153. https://doi.org/10.1016/S0168-1605(97)00071-8

Hughes KT (2008) Coordinating assembly of a bacterial macromolecular machine. Nat Rev Microbiol 6:455–465

Hussain M, Wilcox MH, White PJ (1993) The slime of coagulase-negative staphylococci: biochemistry and relation to adherence. FEMS Microbiol Rev 104:191–208. https://doi.org/10.1111/j.1574-6968.1993.tb05867.x

Hyun H-K, Salehi S, Ferracane JL (2015) Biofilm formation affects surface properties of novel bioactive glass-containing composites. Dent Mater. www.intl.elsevierhealth.com/journals/dema

James GA, Beaudette L, Costerton JW (1995) Interspecies bacterial interactions in biofilms. J Ind Microbiol 15:257–262. https://doi.org/10.1007/BF01569978

Koch C, Hoiby N (1983) Pathogenesis of cystic fibrosis. Lancet 341:1065–1069

Kojima S, Blair DF (2004) The bacterial flagellar motor: structure and function of a complex molecular machine. Int Rev Cytol 233:93–134

Kramer A, Schwebke I, Kampf G (2006) How long do nosocomial pathogens persist on inanimate surfaces? A systematic review. BMC Infect Dis 6:130. https://doi.org/10.1186/1471-2334-6-130. PMID: 16914034; PMCID: PMC1564025.

Laverty G, Gorman SP, Gilmore BF (2014) Biomolecular mechanisms of *Pseudomonas aeruginosa* and *Escherichia coli* biofilm formation. Pathogens 3:596–632

Leriche V, Sibille P, Carpentier B (2000) Use of an enzyme-linked lectin sorbent assay to monitor the shift in polysaccharide composition in bacterial biofilms. Appl Environ Microbiol 66:1851–1856. https://doi.org/10.1128/AEM.66.5.1851-1856.2000

Lewandowski Z (2000) Structure and function of biofilms. In: Evans LV (ed) Biofilms: recent advances in their study and control. Harwood Academic Publishers, Amsterdam, pp 1–17

Lewis K (2008) Multidrug tolerance of biofilms and persister cells. Bacterial biofilms. Springer. Antimicrob Agents Chemother 51:1934–1938

Lewis K (2012) Persister cells: molecular mechanisms related to antibiotic tolerance. In: Antibiotic resistance, vol 211. Springer, Berlin, Germany, pp 121–133

Li DQ, Lundberg F, Ljungh A (2001) Characterization of vitronectin binding proteins of *Staphylococcus epidermidis*. Curr Microbiol 42:361–367

Mack D, Horstkotte MA, Rohde H, Knobloch JKM (2006) Coagulase-negative staphylococci. In: Pace JL, Rupp ME, Finch RG (eds) Biofilms, infection, and antimicrobial therapy. CRC Press, Boca Raton, pp 109–153

Macnab RM (2003) How bacteria assemble flagella. Annu Rev Microbiol 57:77–100

Makin SA, Beveridge TJ (1996) The influence of A-band and B-band lipopolysaccharide on the surface characteristics and adhesion of Pseudomonas aeruginosa to surfaces. Microbiology 142:299–307

Marles-Wright J, Lewis RJ (2007) Stress responses of bacteria. Curr Opin Struct Biol 17:755–760

Marrie TJ, Costerton JW (1983) A scanning and transmission electron microscopic study of the surface of intrauterine contraceptive devices. Am J Obstet Gynecot 146:384–394

Marrie TJ, Harding GKM, Ronald AR, Dikkema J, Lain J, Hoban S, Costerton JW (1979) Influence of antibody coating of *Pseudomonas aeruginosa*. J Infect Dis 19:357–361

Marsh PD, Moter A, Devine DA (2000) Dental plaque biofilms: communities, conflict and control. Periodontology 55:16–35

Mathee K, Ciofu O, Sternberg C, Lindum PW, Campbell JI, Jensen P, Johnsen AH, Givskov M, Ohman DE, Molin S et al (1999) Mucoid conversion of *Pseudomonas aeruginosa* by hydrogen peroxide: a mechanism for virulence activation in the cystic fibrosis lung. Microbiology 145:1349–1357

Mattick JS (2002) Type IV pili and twitching motility. Annu Rev Microbiol 56:289–314

McLaughlin-Borlace L, Stapleton F, Matheson M, Dart JKG (1998) Bacterial biofilm on contact lenses and lens storage cases in wearers with microbial keratitis. J Appl Microbiol 84:827–823

Mendoza MT (2004) El papel del biofilm en el proceso infeccioso y la resistencia. Nova 2:71–80

Merz AJ, So M, Sheetz MP (2000) Pilus retraction powers bacterial twitching motility. Nature 407:98–102

Miller MB, Bassler BL (2001) Quorum sensing in bacteria. Annu Rev Microbiol 55:165–199

Mitchell JG, Kogure K (2006) Bacterial motility: links to the environment and a driving force for microbial physics. FEMS Microbiol Ecol 55:3–16

Moran C, Grussemeyer CA, Spalding JR, Benjamin DK Jr, Reed SD (2009) *Candida albicans* and non-albicans bloodstream infections in adult and pediatric patients: comparison of mortality and costs. Pediatr Infect Dis J 28:433–435

Muller E, Hubner J, Gutierrez N (1993) Isolation and characterization of transposon mutants of Staphylococcus epidermidis deficient in capsular polysaccharide/adhesion and slime. Infect Immun 61(2):551–558

Murdoch H, Taylor D, Dickinson J, Walker JT, Perrett D, Raven NDH, Sutton JM (2006) Surface decontamination of surgical instruments: an ongoing dilemma. J Hosp Infect 63:432–438

Murga R, Forster TS, Brown E, Pruckler JM, Fields BS, Donlan RM (2001) The role of biofilms in the survival of Legionella pneumophila in a model potable water system. Microbiology 147:3121–3126

Murphy EC, Janulczyk R, Karlsson C, Mörgelin M, Frick IM (2014) Identification of pili on the surface of *Finegoldia magna*—a Gram positive anaerobic cocci. Anaerobe 27:40–49

Neut DC (2007) The role of small-colony variants in failure to diagnose and treat biofilm infections in orthopedics. Acta Orthop 78:299–308

Nickel JC (1987) An ecological study of infected urinary stone genesis in an animal model. Br J Urol 59:21–30

Oppenheimer-Shaanan Y, Steinberg N, Kolodkin-Gal I (2013) Small molecules are natural triggers for the disassembly of biofilms. Trends Microbiol 21(11):594–601

Osterman IA, Dikhtyar YY, Bogdanov AA, Dontsova OA, Sergiev PV (2015) Regulation of flagellar gene expression in bacteria. Biochemistry (Mosc) 80:1447–1456

Otter JA, Vickery K, Walker JT, Delancey Pulcini E, Stoodley P, Goldenberg SD, Salkeld JAG, Chewins J, Yezli S, Edgeworth JD (2015) Surface-attached cells, biofilms and biocide susceptibility: implications for hospital cleaning and disinfection. J Hosp Infect 89:16–27

Patti JM, Allen BL, Mcgavin MJ, Höök M (1994a) MSCRAMM-mediated adherence of microorganisms to host tissues. Annu Rev Microbiol 48:585–617

Patti JM, Bremell T, Krajewska-Pietrasik D, Abdelnour A, Tarkowski A, Rydén C, Höök M (1994b) The *Staphylococcus aureus* collagen adhesin is a virulence determinant in experimental septic arthritis. Infect Immun 62:152–161

Paul K, Erhardt M, Hirano T, Blair DF, Hughes KT (2008) Energy source of flagellar type III secretion. Nature 451:489–492

Pesavento C, Becker G, Sommerfeldt N, Possling A, Tschowri N, Mehlis A, Hengge R (2008) Inverse regulatory coordination of motility and curli-mediated adhesion in Escherichia coli. Genes Dev 22:2434–2446

Porter SL, Wadhams GH, Armitage JP (2011) Signal processing in complex chemotaxis pathways. Nat Rev Microbiol 9:153–165

Povolotsky TL, Hengge R (2015) Genome-based comparison of c-di-GMP signaling in pathogenic and commensal Escherichia coli strains. J Bacteriol 198(1):111–126

Pulcini E (2001) The effects of initial adhesion events on the physiology of *Pseudomonas aeruginosa* [Ph.D. dissertation]. Montana State University, Bozeman, MT

Pulverer G, Quie P, Peters G (eds) (1987) Pathogenesis and clinical significance of coagulase-negative staphylococci. Zentralbl Bakteriol. Mikobiol. Hyg. (A) Suppl 16. Gustav Fischer, Stuttgart

Raad II, Sabbagh MF, Rand KH, Sherertz RJ (1992) Quantitative tip culture methods and the diagnosis of central venous catheter-related infections. Diagn Microbiol Infect Dis 15:13–20. https://doi.org/10.1016/0732-8893(92)90052-U

Ramsugit S, Pillay M (2015) Pili of *Mycobacterium tuberculosis*: current knowledge and future prospects. Arch Microbiol 197:737–744

Roberts AP, Pratten J, Wilson M, Mullany P (1999) Transfer of a conjugative transposon, Tn5397 in a model oral biofilm. FEMS Microbiol Lett 177:636. https://doi.org/10.1111/j.1574-6968.1999.tb13714.x

Rodrigues L, Banat IM, Teixeira J (2007) Strategies for the prevention of microbial biofilm formation on silicone rubber voice prostheses. J Biomed Mater Res B Appl Biomater 81 (2):358–370

Rohit Ruhal HA (2015) A multivariate approach to correlate bacterial surface properties to biofilm formation by lipopolysaccharide mutants of *Pseudomonas aeruginosa*. Colloids Surf B: Biointerfaces 127:182–191

Samaranayake LP, Anil S, Hashem M, Vellappally S, Cheung BP (2015) Human serum potentiates the expression of genes associated with antifungal drug resistance in *C. albicans* biofilms on central venous catheters. Mycopathologia 179:195–204

Sauer K, Camper AK, Ehrlich GD, Costerton JW, Davies DG (2002) *Pseudomonas aeruginosa* displays multiple phenotypes during development as a biofilm. J Bacteriol 184:1140–1154

Sedgley CM, Lee EH, Martin MJ, Flannagan SE (2008) Antibiotic resistance gene transfer between *Streptococcus gordonii* and *Enterococcus faecalis* in root canals of teeth ex vivo. J Endod 34:570–574

Shigeru Fujimura TS (2015) Antimicrobial efficacy of combined clarithromycin plus daptomycin against biofilms-formed methicillin-resistant *Staphylococcus aureus* on titanium medical devices. J Infect Chemother 21:756–759

Silverman M, Simon M (1974) Flagellar rotation and the mechanism of bacterial motility. Nature 249:73–74

Skerker JM, Berg HC (2001) Direct observation of extension and retraction of type IV pili. Proc Natl Acad Sci U S A 98:6901–6904

Soto GE, Hultgren SJ (1999) Bacterial adhesins: common themes and variations in architecture and assembly. J Bacteriol 181:1059–1071

Stark RM, Gerwig GJ, Pitman RS, Potts LF, Williams NA, Greenman J (1999) Biofilm formation by Helicobacter pylori. Lett Appl Microbiol 28:121–126. https://doi.org/10.1046/j.1365-2672.1999.00481.x

Stewart PS, Camper AK, Handran SD, Huang C-T, Warnecke M (1997) Spatial distribution and coexistence of *Klebsiella pneumoniae* and *Pseudomonas aeruginosa* in biofilms. Microb Ecol 33:2–10. https://doi.org/10.1007/s002489900002

Stickler DJ (1996) Bacterial biofilms and the encrustation of urethral catheters. Biofouling 94:293–305

Stickler DJ, Morris NS, McLean RJC, Fuqua C (1998) Biofilms on indwelling urethral catheters produce quorum-sensing signal molecules in situ and in vitro. Appl Environ Microbiol 64:3486–3490

Stoodley P, Boyle JD, Dodds I, Lappin-Scott HM (1997) Consensus model of biofilm structure. In: Wimpenny JWT, Gilbert PS, Lappin-Scott HM, Jones M (eds) Biofilms: community interactions and control. Bioline, Cardiff, UK, pp 1–9

Subbaraman LN, Borazjani R, Zhu H, Zhao Z, Jones L, Willcox MD (2011) Influence of protein deposition on bacterial adhesion to contact lenses. Optom Vis Sci 88:959–966

Subramanian P, Shanmugam N, Sivaraman U, Kumar S, Selvaraj S (2012) Antibiotic resistance pattern of biofilm-forming uropathogens isolated from catheterised patients in Pondicherry, India. Aust Med J 5:344–348

Suman E, Varghese B, Joseph N, Nisha K, Kotian MS (2013) The bacterial biofilms in dialysis water systems and the effect of the sub inhibitory concentrations of chlorine on them. J Clin Diagn Res 7(5):849–852. https://doi.org/10.7860/JCDR/2013/5118.2956

Sutherland IW (2001) Biofilm exopolysaccharides: a strong and sticky framework. Microbiology 147:3–9

Tolker-Nielsen T, Molin S (2000) Spatial organization of microbial biofilm communities. Microb Ecol 40:75–84

Tunney MM, Jones DS, Gorman SP (1999) Biofilm and biofilm-related encrustations of urinary tract devices. In: Doyle RJ (ed) Methods in enzymology, Biofilms, vol 310. Academic Press, San Diego, pp 558–566

U.S. Department of Health and Human Services (1996) Public Health Service: National nosocomial infections surveillance (NNIS) report. Data summary from October 1986–April 1996, issued May. Am J Infect Control 24:380–388

Vandaux P, Pittet D, Haeberli A (1989) Host factors selectively increase staphylococcal adherence on inserted catheters: a role for fibronectin and fibrinogen or fibrin. J Infect Dis 160:865–875

Vaudaux PE, Lew DP, Waldvogel FA (1994) Host factors predisposing to and influencing therapy of foreign body infections. In: Bisno AL, Waldvogel FA (eds) Infections associated with indwelling medical devices. American Society for Microbiology, Washington DC, pp 1–29

Veerachamy S, Yarlagadda T, Manivasagam G, Yarlagadda PK (2014) Bacterial adherence and biofilm formation on medical implants: a review. Proc Inst Mech Eng H 228(10):1083–1099

Wadhams GH, Armitage JP (2004) Making sense of it all: bacterial chemotaxis. Nat Rev Mol Cell Biol 5:1024–1037

Watnick PI, Kolter R (1999) Steps in the development of a vibrio cholerae El Tor biofilm. Mol Microbiol 34:586–595. https://doi.org/10.1046/j.1365-2958.1999.016234

Weigel LM, Donlan RM, Shin DH, Jensen B, Clark NC, Mcdougal LK, Zhu W, Musser KA, Thompson J, Kohlerschmidt D, Dumas N, Limberger RJ, Patel JB (2007) High-level vancomycin-resistant Staphylococcus aureus isolates associated with a polymicrobial biofilm. Antimicrob. Agents Chemother 51:231–238

Wirtanen G, Alanko T, Mattila-Sandholm T (1996) Evaluation of epifluorescence image analysis of biofilm growth on stainless steel surfaces. Colloids Surf B: Biointerfaces 5:319–326. https://doi.org/10.1016/0927-7765(95)01226-5

Xie H, Cook GS, Costerton JW, Bruce G, Rose TM, Lamont RJ (2000) Intergeneric communication in dental plaque biofilms. J Bacteriol 182:7067–7069. https://doi.org/10.1128/JB.182.24.7067-7069

Yung-Hua L, Lau PCY, Lee JH, Ellen RP, Cvitkovitch DG (2001) Natural genetic transformation of Streptococcus mutans growing in biofilms. J Bacteriol 183:897–908. https://doi.org/10.1128/JB.183.3.897-908.2001

Biofilm and Antimicrobial Resistance

8

Dibyajit Lahiri, Moupriya Nag, Anushka Ghosh, Deboleena Das, Ankita Dey, Dipro Mukherjee, Sayantani Garai, and Rina Rani Ray

Abstract

An antimicrobial is an agent that destroys microorganisms or hinders their growth. They can be grouped according to the microorganisms they act primarily against. Microbes, mainly bacteria, develop a number of ways to check the action of these antimicrobials, which is often genetically controlled. One of such way is achievement of biofilm-mediated antimicrobial resistance. Biofilms are the sessile groups of organism that possess the property of adhering to the biotic and abiotic surfaces by synthesizing extracellular polymeric substances (EPS) that not only provide nourishment to the developing microcolonies but also render resistance against antibiotics and antimicrobial agents. Varieties of molecular mechanisms are responsible for the development of high-degree recalcitrance that determines the characteristics of biofilm. The mechanisms that are involved in these processes are various types of interactions between the antimicrobials and the biofilm matrix that reduces the rate of growth, and various gene expressions result in the formation of antibiotic resistance.

D. Lahiri · M. Nag · A. Ghosh · D. Das · D. Mukherjee · S. Garai
Department of Biotechnology, University of Engineering & Management, Kolkata, West Bengal, India

A. Dey · R. R. Ray (✉)
Department of Biotechnology, Maualana Abul Kalam Azad University of Technology, West Bengal, Kolkata, West Bengal, India

R. R. Ray et al. (eds.), *Biofilm-Mediated Diseases: Causes and Controls*,
https://doi.org/10.1007/978-981-16-0745-5_8

8.1 Introduction

The bacterial biofilm has resulted in a serious health concern due to its inherent ability to tolerate against antibiotics, host defense systems, and various external stresses; as a result it causes various chronic infections (Costerton et al. 1999; de la Fuente-Núnez et al. 2013). Biofilms are the consortia of sessile communities that remain attached to biotic or abiotic surfaces and with other cells with the help of extracellular polymeric substances (EPS). They are predominantly found growing upon the surfaces of catheters, medical devices, prosthetics, and dental implants with the help of EPS which acts as a glue for the adherence of the sessile communities on a surface, and these result in the development of unaffordable treatment as well as deterioration of the mental conditions among the afflicted person (Costerton et al. 2005; Hoiby et al. 2011). The biofilm contains sessile colonies of bacterial cells containing EPS that possess various types of polymeric substances produced by the bacterial cells, extracellular DNA (e-DNA), proteins, and other types of amyloid proteins (Whitchurch et al. 2002; Wingender et al. 2001). This has been recognized as the key factor for the persistence of the varied infections (Wingender et al. 2001). The development of the biofilm is a multistep process that involves adsorption of macromolecules and micromolecules upon the surfaces, attachment of the bacteria upon the surface with the help of EPS, development of colony by sessile communities, and maturation of the biofilm. This also results in the change in the genetic and metabolic expression of the planktonic to sessile cells with the enhancement of the production of EPS (Flemming et al. 2007). The biofilm has the property of protecting the bacterial cells from adverse environmental conditions like alteration in pH, scarcity of nutrients, pH, osmolarity, and shear stresses (Costerton and Lewandowski 1995; Fux et al. 2005; McCarty et al. 2012) and also prevents the penetration of antibiotics up to the indwelling cells existing within it as well as resists the immune reaction of the host (Costerton et al. 1999; Stewart and William Costerton 2001). Thus the EPS provides the protective shield to the sessile communities leading to the development of resistance and also development of multidrug resistance, total drug resistance, and extensive drug resistance.

The increase in the nosocomial infections by the persistence of biofilm is increasing with time; much of the research has been directed toward the effects that the antimicrobial agents have on attached communities surfaces. Thus this chapter involves with the question: what are the mechanisms that the biofilm adopt for the development of antimicrobial compounds?

8.2 Ultrastructure of Biofilm

The microbial biofilm is the group of sessile communities which possess the ability of being attached with the substratum and remain embedded with a self-producing noncrystalline extracellular polymeric substances forming a matrix (Hoiby et al. 2011). The sessile communities show a wide difference from the planktonic communities of cells in context to the gene expression, growth rate, transcription,

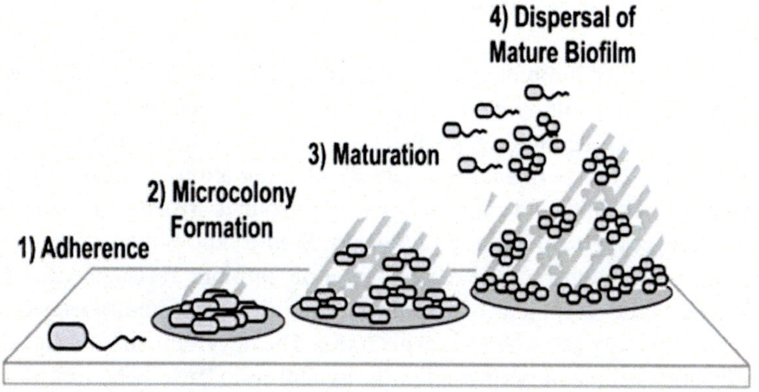

Fig. 8.1 Development of biofilm

and translation as the former thrive within varied microenvironments that may possess higher osmolarity, scarcity of nutrients, and higher density of cells derived from heterogenous bacterial communities. The development of three-dimensional structures of the biofilm helps in providing the bacterial cells protection from environmental stresses such as antimicrobial attacks and desiccation and from the ingestion by protozoa (Wilkins et al. 2014). The cellular interaction existing between the sessile communities within the biofilm is termed as quorum sensing (QS) and this occurs due to the regulation of specific genes that are responsible for the development of the biofilm (Solano et al. 2014). The formation of biofilm is a multistep process that starts with attachment of the bacterial cells upon a solid surface followed by colonization and modification in genes or proteins resulting in the exponential growth. The EPS facilitates the supply of nutrients to the indwelling cells resulting in the pathogenesis and multidrug resistance (Fig. 8.1).

8.3 Structure of Biofilm and Associated Places

Various infections on the outer and the inner surface of the body are associated with the bacterial cells forming the biofilm. For example, the development of biofilm can occur upon the surface of the teeth (Kolenbrander and Palmer Jr 2004), development of cystic fibrosis of lungs leading to chronic bronchopneumonia (Bjarnsholt et al. 2009), development of middle ear infections (Hall-Stoodley et al. 2006), development of prosthetic joint infections and chronic osteomyelitis (Gristina et al. 1985; Trampuz and Zimmerli 2005; Del Pozo and Patel 2009), and development of infections associated with catheters and stents (Taconelli et al. 2009).

8.4 Antibiotic Resistance by Biofilm

The development of antibiotic resistances among the sessile communities dwelling within the biofilm result in the development of chronic infections. The development of resistance in the sessile organisms is not similar to the planktonic forms which include lower cell permeability, target site mutations, efflux pumps, drug-neutralizing proteins, and drug-modifying enzymes (Beauclerk and Cundliffe 1987; Kumar et al. 2013; Lata et al. 2015). Studies show that the concept on the conventional mechanism of the development of antibiotic resistance cannot be used to describe the development of resistance within the biofilm-associated infections (Alekshun and Levy 1999; Anderl et al. 2000). The development of resistance within the microorganisms can be predicted by its ability to grow when bactericidal and bacteriostatic antimicrobial are present at a concentration which generally brings about inhibition to the growth of the microbial cells. The measurement of resistance is determined within the planktonic cultures by determining the minimum inhibitory concentration (MIC). It is the least concentration of an antimicrobial possessing the ability to inhibit the growth of the microorganisms. The development of resistance is often considered to be the attribute to mutation or the mechanism of exchange of the antibiotic resistance genes, dependent on various intrinsic factors or the wild-type and the innate genetic composition of the organisms (Cox and Wright 2013; Blair et al. 2015). The Gram-negative bacteria shows more intense intrinsic resistance against vancomycin (antibiotic) than the Gram-positive bacteria due to the persistent impermeability of the outer membrane being present within Gram-negative bacterial cells. Lewis (2008) described resistance as the mechanism by means of which antimicrobial agents are prevented from interacting with its target (Lewis 2008). Some researches showed that biofilm developed by *Pseudomonas aeruginosa* exhibit resistance against ceftazidime by the conventional type of development of resistance (Bagge et al. 2000). Another study showed that the continuous exposure of ceftazidime to the biofilm growing *Pseudomonas aeruginosa* resulted in the development of conventional type of intrinsic biofilm-associated infections (Taconelli et al. 2009). The antibiotic resistance of the sessile communities within the biofilm occurs due to various strategies which includes improper penetration of antibiotics (Conibear et al. 2009), variation in the microchemical environment within the biofilm (Driffield et al. 2008), and the various subpopulation of microbial cells existing within the biofilm by the development of spores (Bjarnsholt et al. 2007). These conditions result in the multi-cellularity of the biofilm that leads to development of antibiotic-resistant sessile communities (Keren et al. 2004). This ultimately results in the failure of the use of antibiotics in treating the cells.

The increase in concern on the development of antibacterial resistance is the mechanism of spread of the antibiotic resistance gene (ARGs) from one bacteria to another by a process known as horizontal gene transfer which enhances the number of antibiotic-resistant organisms (Domingues et al. 2012; Musovic et al. 2006). The process of the horizontal gene transfer (HGT) is a mechanism of nonvertical acquisition of genetic material by different processes which include conjugation (the mechanism of the transmission of DNA by direct physical contact), and

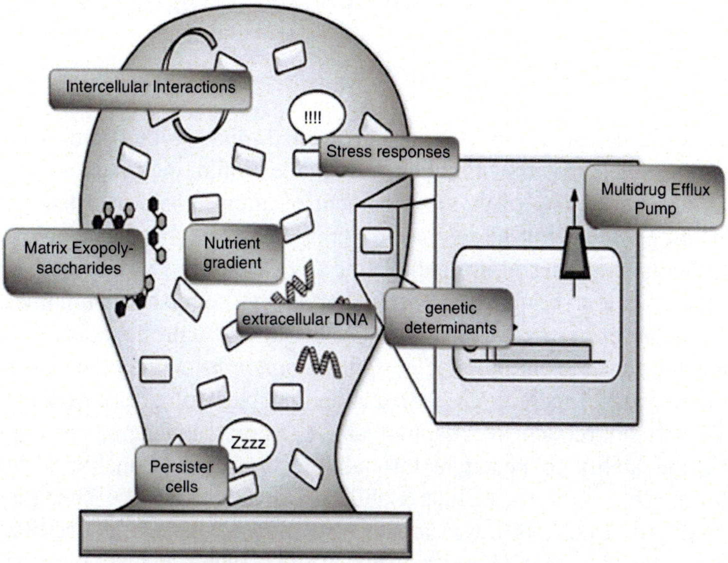

Fig. 8.2 Various mechanisms of development of antimicrobial resistance

transduction (by the mechanism of viral insertion), transformation (by the mechanism of the extracellular DNA uptake) or by agents which help in gene transfer (by phage-like particles) (Von Wintersdorff et al. 2016). The ability of the ARGs to be passed horizontally may be dependent upon the location of the bacterial genome or its ability of being associated with the mobile genetic elements (MGEs) which comprise of transposons, integrin and insertion sequences, conjugative plasmids, and integrative conjugative elements (ICEs) (Partridge et al. 2018). The MGEs have the ability of enabling the mechanism of HGT of the ARGs among the bacterial cells which are diverse phylogenetically, and as a result it helps them in acquiring resistance by enhancing their adaptability against antibacterial therapy (Klumper et al. 2015; Bertram et al. 1991; Yamamoto et al. 2016). It has been observed that microbes are the originators of the clinical antibiotic resistance and the environmental resistome helps in acting as the reservoir for the mechanism of horizontal gene transfer among the pathogens (Wright 2010; Perry and Wright 2013; Hu et al. 2016). Some studies showed there are strong resemblance between environmental resistome and clinical resistance in various cases of extended β-lactamases, vancomycin, and quinolone resistances (Messi et al. 2006; Hawkey and Jones 2009) (Fig. 8.2).

8.5 Mechanisms of Development of Antibiotic Resistance

8.5.1 Antimicrobial Penetration

Antimicrobial recalcitration is very rarely cited, and various studies showed that decrease in penetration of antibiotics through the biofilm matrix does not completely explain the mechanism of the development of antimicrobial resistance among the sessile communities. For instance, antibiotics like tetracyclines are able to reach all the sessile communities of uro-pathogenic *E. coli* biofilms within an exposure time of 10 min without bringing any effect upon the cellular communities, whereas ampicillin (absence of the enzyme β-lactamase) and ciprofloxacin can effectively penetrate through the biofilm of *Klebsiella pneumoniae* at higher MIC values for the planktonic cells (Anderl et al. 2000; Zahller and Stewart 2002). Some studies also showed that antibiotics like daptomycin, rifampin, amikacin, and ciprofloxacin possess the ability to penetrate through the biofilm but unable to bring about significant effect upon the cellular viability (Dunne et al. 1993; Zheng and Stewart 2002; Stewart et al. 2009). Researches also showed that antibiotics like oxacillin, vancomycin, and cefotaxime are prevented from penetrating up to the cells due to the presence of the biofilms formed by *S. epidermidis* and *S. aureus* since the biofilm acts as a barrier for preventing the movement of the antibiotics (Jefferson et al. 2005; Singh et al. 2010). Thus the inability of diffusion of antibiotics depends on variable aspects like the type of bacterial strain, the experimental setup, and the growth conditions of the biofilm. A remarkable instance is the reduced ability of antibiotic penetration like tobramycin through the biofilm formed by *P. aeruginosa.* It was observed that the biofilm formed by *P. aeruginosa* reduces the penetration of antibiotic tobramycin and the penetration barrier could be altered by the supplementation of growth medium with cations (Walters et al. 2003). The studies also showed that the cations were able to interact with e-DNA and phage particles, further decreasing its ability to penetrate due to increased tolerance of *P. aeruginosa* toward aminoglycoside antibiotics (Tseng et al. 2013) (Table 8.1).

8.5.2 Polysaccharide-Associated Development of Resistance Within the Biofilm

Biofilms are composed of extracellular polymeric substances comprising of repeated units of D-mannose, D-glucose, and L-rhamnose (Byrd et al. 2009). *P. aeruginosa* produces Psl, an exopolysaccharide that plays an important role in the development of resistance against polymyxin B, tobramycin, colistin, and ciprofloxacin at initial stages of the biofilm formation (Billings et al. 2013), whereas enhanced expression of Psl did not enhance the development of antibiotic resistance like tobramycin (Colvin et al. 2011). Psl is predominantly responsible for the development of resistance against the non-ionic surfactant that hinders the formation of biofilm (Zegans et al. 2012). Psl acts as the sequesters of antibiotics within the biofilm by electrostatic interactions (Billings et al. 2013). Studies has shown that Psl also results

Table 8.1 Various groups of antibiotics which are commonly used

Class of antibiotic	Examples	References
β-Lactams	Penicillins (e.g., penicillin G, ampicillin), cephalosporins (e.g., cefotaxime, ceftazidime, cefazolin), carbapenems (e.g., meropenem, imipenem)	Singh et al. (2010)
Glycopeptides	Vancomycin	Taber et al. (1987)
Lipopeptides	Daptomycin	Schuster and Bertram (2016)
Polymyxins	Polymyxin B, colistin	Taber et al. (1987)
Aminoglycosides	Tobramycin, gentamicin, kanamycin, amikacin	Singh et al. (2010)
Tetracyclines	Tetracycline, doxycycline	Schuster and Bertram (2016)
Oxazolidinones	Linezolid	Suci et al. (1994)
Chloramphenicol	Chloramphenicol	Schuster and Bertram (2016)
Macrolides	Azithromycin, clarithromycin	Suci et al. (1994)
Steroid-like antibiotics	Fusidic acid	Schuster and Bertram (2016)
Quinolones	Fluoroquinolones (e.g., ciprofloxacin, ofloxacin, norfloxacin)	Suci et al. (1994)
Rifamycins	Rifampin	Suci et al. (1994)
Sulfonamides	Sulfamethoxazole	Suci et al. (1994)
Dihydrofolate reductase (DHFR) inhibitors	Trimethoprim	Singh et al. (2010)

in the development of resistance against antibiotics which are cationic like colistin and tobramycin in the presence of high NaCl concentration (Billings et al. 2013). Another structural component being present within the glycocalyx of the biofilm formed by *P. aeruginosa* is Pel. It is expressed by the gene product being located at *pelABCDEFG* locus (Franklin et al. 2011). Pel not only provides additional structural support to the strains lacking Psl; it provides resistance against antibiotics like gentamicin and tobramycin (Colvin et al. 2011). Studies showed the presence of Pel enhances the resistance against aminoglycoside by fourfold (Khan et al. 2010).

8.5.3 Development of Resistance by the Antibiotic-Modifying Enzymes Being Present in Matrix

The matrix of the biofilm contains enzyme, namely, β-lactamases, which possess the ability to degrade the antimicrobials and thus prevent the antibiotics to act upon the target cells. It has been observed that *K. pneumoniae* biofilms produce the β-lactamases possessing the ability to degrade the ampicillin and prevent the antibiotics to reach up to the sessile communities of the biofilm (Anderl et al. 2000). Studies showed that excess production of chromosomally encoded AmpC cephalosporinase is reasoned to be the main cause of the development of resistance within cystic fibrosis isolates of *P. aeruginosa* against the β-lactam antibiotics (Giwercman et al. 1990). Researches showed that fusion of unstable green-fluorescent protein with ampC in the absence of ceftazidime and imipenem induced the expression of ampC in *P. aeruginosa* (Bagge et al. 2004).

8.5.4 Intrinsic Resistances: Acquired and Adaptive Resistance

The development of intrinsic resistances represents the mechanisms of limiting the susceptibility of antibiotics upon wild type of organisms. The emergence of intrinsic resistances in the Gram-negative bacteria is dependent upon the outer membrane's semi-permeability which limits the uptake of the antibiotics. The bacterial cell's outer membrane is asymmetric as it is composed of an outer layer of polyanionic polysaccharide (LPS) and a layer of phospholipids. It has been observed that the outer membrane permeability of *P. aeruginosa* increases by 12–100-fold than other Gram-negative group of bacteria like *E. coli* (Hancock 1981). The depletion in the permeability of *P. aeruginosa* serves as a barrier to the penetrance of antibiotic, but small antibiotic molecules like lactams and quinolones have the ability to pass through water-filled porin channels leading to the development of intrinsic resistance within the bacteria.

Pseudomonas aeruginosa is mostly resistant to various antibiotics and is considered to be a "superbug" due to its huge capacity of generating resistance. It can readily achieve resistance due to its reduced outer membrane permeability along with adaptive mechanisms and is therefore less susceptible to antibiotics. Using new technologies such as mutant library screens, mutation frequency analysis, and microarray, it has been found out that very large collections of genes (the resistome) lead to resistance when they are mutated and new forms of adaptive resistance can be triggered by antibiotics themselves when there is in vivo growth conditions; adaptations such as swarming motility and growth of biofilm take place (Breidenstein Elena et al. 2011).

The psrA gene encodes a transcriptional regulator which, in response to subinhibitory concentrations of various antimicrobial peptides that are cationic in nature, is upregulated. The comparison between the complemented mutant and the wild type shows that *P. aeruginosa* PAO1 *psrA*Tn*5* mutant has intrinsic supersusceptibility to polymyxin B and an antimicrobial used against infections which are multidrug resistant and the bovine neutrophil antimicrobial peptide indolicidin; this supersusceptibility phenotype is interrelated with high-degree outer membrane permeabilization by such agents. In simple biofilm formation, the cloned *psrA* gene complements all the processes like rapid attachment and swarming motility. *psrA* is considered to be a key regulator for antimicrobial peptide resistance and virulence as suggested in various studies. The PhoQ gene regulates biofilm formation and motility along with antimicrobial peptide resistance. In order to test the ability of the psrA mutant to form simple biofilms, static microtiter biofilm assays were used to demonstrate that the psrA mutant showed significant (fourfold; P 0.05 by Student's t test) impairment in biofilm formation at 18 h. This impairment of biofilm could be overcome by WT psrA allele by introducing it into the mutant.

8.5.5 Mutation

The development of mutation within the sessile communities of bacterial cells is increased as compared to the isogenic cells which are growing planktonically (Driffield et al. 2008), and extent of horizontal gene transfer has enhanced within the biofilm (Molin and Tolker-Nielsen 2003). These changes explain the development of the multidrug resistance from the traditional resistance mechanisms against various groups of antibiotics like aminoglycosides, lactam group of antibiotics, and fluoroquinolones. These can be determined by the regular mode of susceptibility testing in the microbiology laboratory. The newly formed bacterial cells possess the ability of the production of enzymes that simultaneously degrade the antibiotics, change the targets for the action of the antibiotics, and also cause the efflux of the drugs out of the cells. It has been observed that the mutations in bacterial population lead to the development of cystic fibrosis (Høiby et al. 2001). The cystic fibrosis also shows hypermutable phenotypes of *P. aeruginosa* that cause alteration in the DNA repair system genes which are a part of the mismatch repair system (MMR). This includes the genes such as *uvrD*, *mutS*, and *mutL* or the oxidative repair system of DNA with genes like *mutT*, *mutT*, and *mutM* (Oliver et al. 2002; Mandsberg et al. 2009). The mutations that occur in either of the two systems result in the development of antibiotic resistance due to the expression of the multidrug efflux pump (Macia et al. 2005). The enhancement in endogenous reactive oxygen species (ROS) production and deficient antioxidant system results in the development of imbalance between antioxidant defense and oxidative burden leading to the development of mutability within the biofilm (Driffield et al. 2008; Conibear et al. 2009; Hassett et al. 1999; Mai-Prochnow et al. 2008). Studies have shown that the endogenous oxidative stresses in the microcolonies result in the enhancement of genetic adaptations and evolutionary changes (Conibear et al. 2009). Researchers also showed that the antibiotic resistance is promoted by virtue of the endogenous oxidative stress within the biofilm (Boles and Singh 2008). For instance, the sessile bacterial cells causing cystic fibrosis develop endogenous oxidative stress as well as remain exposed to ROS from activated polymorphonuclear leukocytes (PMNs) (Ciofu et al. 2005). These hypermutability features exhibited by the bacterial cells result in the development of antibiotic resistance (Ciofu et al. 1994). β-Lactam antibiotic resistance development mainly occurs by the changes in the regulator genes controlling the production of β-lactamases leading to the decreased production of AmpC-lactamases (Ciofu 2003).

8.5.6 Extracellular DNA-Dependent Development of Resistance

Extracellular DNA or eDNA is one of the most important components of bacterial biofilm matrix. The endogenous synthesis of eDNA occurs by the process of quorum sensing and is produced exogenously from the polymorphonuclear leukocytes at the site of infection (Allesen-Holm et al. 2006). It has been observed that eDNA contributes an important role in the development of antimicrobial resistance (Chiang et al. 2013). In *S. aureus* biofilm, the release of eDNA was enhanced when it was treated with the subinhibitory concentrations of methicillin which is dependent on AtlA autolysin (Kaplan et al. 2012). It has also been observed that subinhibitory concentrations of vancomycin enhanced the release of eDNA by twofold, thus decreasing the probability of the development of antimicrobial resistance (Doroshenko et al. 2014). These evidences reveal that the antibiotic treatment at a sublethal dose causes the release of eDNA from the matrix of the biofilm. The extracellular environment is being altered by eDNA to form resistance. eDNA is an anionic molecule possessing the ability of being chelated with cations like magnesium that results in the depletion of the effective concentration of Mg^{2+} ions within the environment (Mulcahy et al. 2008). It has been further observed that with organisms like *P. aeruginosa* and *Salmonella enterica*, the reduction in magnesium triggers the activation of environmental signals that counter the activation of PhoPQ and PmrAB systems (McPhee et al. 2003, 2006). Some studies revealed that accumulation of eDNA within the matrix of *P. aeruginosa* results in the development of acidic microdomain which results in limiting the concentration of Mg^{2+} ions, thus activating PhoPQ and PmrAB signaling pathway which helps in antimicrobial resistance development (Wilton et al. 2015). The transcriptional effect of PmrAB and PhoPQ triggers the mechanism of upregulation of PA3552–3559 (*arn* BCADTEF *ugd* or *pmr*HFIJKLM-*ugd*) operon (McPhee et al. 2006). The protein-encoded enzymatic action of PA3552–3559 operon results in the development of aminoglycoside and cationic antimicrobial peptide resistance by the mechanism of addition of amino-arabinose to lipid A moiety of lipopolysaccharide (LPS) (Lewenza 2013). Studies have also shown that addition of the eDNA to the wild-type *P. aeruginosa* resulted in the development of antimicrobial peptides, namely, polymyxin B and colistin by four times and eight times, respectively (Mulcahy et al. 2008).

8.5.7 Stress Responses Resulting in the Development of Antibiotic Resistance and Hypoxia-Associated Antimicrobial Resistance

Hypoxia imparts tolerance in biofilm toward the antimicrobials which was supported by the fact that the biofilms exposed to antimicrobials at anaerobic condition showed more resistance than those exposed at aerobic condition (Borriello et al. 2004). It has been further confirmed by the increase in resistance of *P. aeruginosa* biofilm toward colistin under the anaerobic conditions (Kolpen et al. 2016). The accumulation of colistin-resistant subpopulation within the biofilm determines the toxic region within the biofilm that in turn indicates the decreased growth of the biofilm and enhanced tolerance toward the antimicrobials (Haagensen et al. 2006; Pamp et al. 2008). The hypoxia results in the reduction in the potential of the outer membrane of *P. aeruginosa* resulting in the development of antibiotic resistance against aminoglycosides (Taber et al. 1987; Stewart et al. 2015). Researches show that hypoxia results in multidrug efflux gene expression within *P. aeruginosa* by the upregulation of *mexEF-oprN* efflux pump genes that actually contribute to the development of the antibiotic resistance within the organism (Schaible et al. 2012). On the other hand, upregulation of *mexCDoprJ* efflux locus was found to occur at low oxygen concentration (Tata et al. 2016). The reactive oxygen species (ROS) plays a crucial part in the antibiotic-induced cell death; thus the development of hypoxia leads to the development of antibiotic resistance.

8.5.8 Oxidative Stress Responses Resulting in the Development of Resistance

The antibiotic-induced ROS is created with the help of the development of the antibiotic resistance within the biofilm. It has been observed that the use of subinhibitory concentrations of antibiotics results in the enhancement of ROS levels on the treatment with ciprofloxacin, and enhancement in the ROS results in the reduction of the viability of the cells (Jensen et al. 2014). The development of antibiotic-induced ROS is being detoxified (partially) by KatA catalase. The katA mutant present in the wild type of biofilm produces more amount of ROS when it is being acted upon by ciprofloxacin (Jensen et al. 2014). The addition of thiourea to the culture results in the scavenging up of ROS and also results in the improvement in the viability of *P. aeruginosa* (Jensen et al. 2014). Another mechanism by which *P. aeruginosa* develops antibiotic resistance is by the reduction of the pro-oxidant molecules and upregulating the antioxidant mechanisms such as KatA (Nguyen et al. 2011; Khakimova et al. 2013). It has been also observed that there is significant enhancement of ROS with the exposure of tobramycin or ciprofloxacin to *Burkholderia cepacia* biofilm (Van Acker et al. 2013). But addition of antioxidants like glutathione significantly enhances the survival of the biofilm cells (Van Acker et al. 2016).

8.5.9 Starvation of Amino Acids Results in the Development of Resistance

The starvation of amino acids results in the development of stringent conditions that help in conserving the signal pathway required for the survival of the bacterial cells at limiting nutrient conditions (Hauryliuk et al. 2015). The mechanism of starvation results in the enhancement of ribosomes which are able to carry deacetylated tRNA molecules that help in recruiting the ribosome-associated pentaphosphate guanosine (pppGpp) synthase RelA. The enzyme pppGp phosphohydrolase is responsible for the conversion of pppGpp to tetraphosphate guanosine (ppGpp). The amount of pppGp and ppGp varies from one bacterial cell to another, and both of them are responsible for the development of the stringent responses by the cells. RNA polymerase interacts with (p)ppGpp and induces the development of transcriptional changes. It causes downregulation of the ribosomal protein genes and upregulation of amino acid biosynthetic genes that permits the cells to survive under starvational stresses. These conditions result in the enhancement of resistance of biofilm against the antibiotics (Zheng and Stewart 2004). The activation of the stringent response within the biofilm results in the development of tolerance toward the antimicrobial agents (Nguyen et al. 2011). Researchers showed that biofilm formed by the relAspoT double knockout strains of *P. aeruginosa* lacking (p)ppGpp is remarkably more susceptible than the wild type of biofilm to various types of antibiotics like colistin, gentamycin, meropenem, and ofloxacin (Nguyen et al. 2011).

8.5.10 Cell Wall-Modifying Enzyme-Associated Development of Resistance

In *Streptococcus mutans* dltABCD operon is the positive hit for the screening of the gentamycin tolerance (Nilsson et al. 2016). It has been observed that dltA has no effect upon the formation of biofilm, but it is found to have eight times less tolerance to gentamycin as compared to the wild-type strains growing within the biofilm (Nilsson et al. 2016). Homologues of dltABCD play a crucial role in D-alanylation of teichoic acid in Gram-positive bacterial cells (Neuhaus and Baddiley 2003).

It has been observed in the biofilm formed by *Pseudomonas aeruginosa* that it contains ndvB locus which results in the formation of antibiotic resistance. Different studies have confirmed the existence of genetic determinants of biofilm-specific antibiotic resistance (Table 8.2).

8.5.11 Genetic Basis of Resistance: Mobile Genetic Elements

The latest concerning health problems in the world are the increasing antibiotic resistance of both strains of Gram-positive and Gram-negative bacteria. The main challenge for us is to understand the collection of genes and various processes involved that help the biofilm-forming bacteria to fight against the antimicrobial

Table 8.2 Effect of wild-type and mutant genes on the formation and antibiotic resistance

PA number	Mutant phenotype resistance	Other mutant phenotypes	Growth phenotype of biofilm	Gene name	References
PA0084	TOBS	Type 6 secretion system deficiency	Normal	tssC1	Allesen-Holm et al. (2006)
PA0355	CIPR	Swarming deficient	Decreased	pfpI	Ahiwale et al. (2011)
PA0401	PXS	—	Decreased	pyrB	Fernández et al. (2011)
PA0402	COLR, LL37R, PXR	—	Decreased	asrA	Falagas and Bliziotis (2007)
PA0756/57	GENS, TOBS	—	Normal	ndvB	Fernández et al. (2011)
PA0779	TOBS	Reduced heat shock response	Decreased	phoQ	Musken et al. (2010)
PA1163	GENS, TOBS	Reduced expression of ethanol oxidation pathways	Increased	sucC	Fernández et al. (2011)
PA1180	AMKR, KANR, PXR	LPS modification deficiencies	Normal	parR	Falagas and Bliziotis (2007)
PA1588	PXS	—	Decreased	clpP	Shah et al. (2006)
PA1799	COLS, PXS	—	Decreased	lon	Shah et al. (2006)
PA1801	CIPS	Swarming deficient	Normal	—	Musken et al. (2010)
PA1803	CIPS	Reduced cytotoxicity, swarming deficient	Decreased	—	Falagas and Bliziotis (2007)
PA1875–PA1877	CIPS, GENS, TOBS	—	Decreased	czcR	Bagge et al. (2004)
PA2070	GENS	—	Normal	clpS	Shah et al. (2006)
PA2523	IPMR	Reduced uptake of Zn2+, Cd2+, Co2+, Cu2+; Catabolite repression	Normal	psrA	Balaban et al. (2004)
PA2621	ATMR, CAZR, PIPR	—	Decreased	—	Mijnendonckx et al. (2013)
PA3006	INDS, PXS	Reduced swarming	Decreased	lptC	Thomas and Nielsen (2005)
PA3920	TOBR	—	Decreased	cbrA	Balaban et al. (2004)

(continued)

Table 8.2 (continued)

PA number	Mutant phenotype resistance	Other mutant phenotypes	Growth phenotype of biofilm	Gene name	References
PA4459	COLS, INDS, LL37S, PXS	Deficient in catabolite repression, swarming, cytotoxicity, etc.	Decreased	—	Musken et al. (2010)
PA4725	CIPR, PXR, TOBR	—	Increased	ureB	Mijnendonckx et al. (2013)
PA4867	TOBS	—	Decreased	aroB	Shah et al. (2006)
PA5033	GENS	—	Normal	—	Thomas and Nielsen (2005)
PA5038	PX	—	Decreased	—	Balaban et al. (2004)

chemotherapy. The genetic basis of resistance is found to be due to certain mobile elements (genetic) that move within DNA molecules, more specifically known as the lateral gene transfer (LGT) (Ragan and Beiko 2009). These DNA molecules constituting transposons, insertion sequences, and integrons and the transfer between bacterial cells are brought about by various plasmids and conjugative elements (Partridge et al. 2018). All these components together play a significant role in accretion and propagation of resistance genes. LGT is defined as the transfer of genetic elements from one cell to another such that no cell division is required in the process and the gene is inherited by the recipient in a stable manner (Stokes and Gillings 2011).

The mobile genetic elements (MGEs)—transposons (Tn) and insertion sequences (IS)—are DNA segments which are able to transfer from one place to another either within a DNA molecule or between two different DNA molecules. This phenomenon occurs almost randomly. On the other hand, the site-specific recombinations are carried out by the integrons. Homologous recombinations are also facilitated by these MGE as they are present at different locations within the genome. The process of genetic exchange occurs in three different ways which are transduction, transformation, and conjugation/mobilization among the cells thriving within the biofilm (Stokes and Gillings 2011).

8.5.12 Insertion Sequence (IS)

IS are generally small mobile elements carrying more than one transposase gene (*tnp*). At first IS were not considered as passenger genes, but later it was found that they can carry resistance genes as an element of a composite transposon. This composite transposon consists of a region that is bounded by two similar copies of the same IS, and this acts as a single unit. A strong promoter is found in many IS which helps in the expression of the capture gene (Kamruzzaman et al. 2015), and the intrinsic chromosomal gene insertion at the upstream can also influence the antibiotic resistance (Table 8.3).

8.5.13 Transposons

Unit transposons are larger than the IS elements and are bounded by a pair of IR. It also consists of a transposase gene and an internal "passenger" gene that may contain the encoding gene having antibiotic resistance site. The Tn3 family of transposons is mostly associated with antibiotic resistance genes and consists of a resolvase gene, *tnpR*, and a resolution (res) site which may contain the passenger genes. Tn1, Tn2, and Tn3 were the very first transposons that were discovered in Gram-negative bacteria (Partridge and Hall 2005). The res site lies between the *tnpA* and *tnpR* genes that are transcribed in the opposite directions. In either side of res region, homologous recombination takes place, and it is then followed by res-mediated recombination. In Tn1, Tn2, and Tn3 variants, segments, or hybrids, various inhibitor resistant (IRT) variants have been found. A streptomycin-resistant gene, *strAB*, is carried by Tn5393 and is found in a position which is similar to the position in bla_{TEM} in Tn*3*. *Tn1546 is an important member of the Tn3 family as it is associated with antibiotic resistance in Gram-positive bacteria. It has a vanA gene cluster, and this helps in encoding genes against vancomycin. The expression of this gene cluster is regulated by vanRS gene* (López et al. 2010). Tn1721 consists of Tn1722 which has a tetracycline-resistant determinant at the end of IR_R. Mercury-resistant (mer) operon is present in Tn21 family and is very crucial in the movement of antibiotic-resistant genes. The β-lactamase genes in staphylococci origin are believed to be Tn552-like elements as penicillin-resistant *S. aureus* strain have emerged (Gregory et al. 2003).

8.5.14 Plasmids

The link between plasmids and resistance genes was found for the first time in *P. aeruginosa*, and since then it has emerged as an important Gram-negative organism (Holloway 1969). Through various studies it was found that R plasmids may be able to translocate pieces of antibiotic-resistant genes containing DNA, and this process was independent of homologous recombination pathways (Bennett and Richmond 1976). Conjugative plasmids consist of systems that are genetically complex for horizontal plasmid transfer, and this significantly increases the size of their conserved backbone. The transfer (tra) region of conjugative plasmids consists of genes that encode for proteins that are used in mating pair formation (MPF), and this in turn helps in Type IV secretion system pore (T4SS). It also helps in DTR (DNA transfer replication) proteins that help in processing the plasmid DNA. The T4SS forms a conjugative pilus, a filamentous surface appendage in Gram-negative bacteria, and therefore helps in establishing interactions between the cells. The earliest plasmid type which was found to be associated with antibiotic resistance

Table 8.3 Mechanism of action of antibiotics and proposed mechanism of antibiotic resistance

Gene(s)	Antibiotics	Gene product(s)	Proposed mechanism of protection	References
brlR	Tobramycin, norfloxacin, trimethoprim, tetracycline, kanamycin, chloramphenicol	Transcriptional regulator	Upregulation of multidrug efflux pumps	Spoering and Lewis (2001)
sagS	Tobramycin, norfloxacin	Two-component hybrid	Activation of BrlR by promoting increased c-di-GMP levels	Webb et al. (2003)
ndvB	Tobramycin, gentamicin, ciprofloxacin	Glucosyltransferase	Sequestration of antibiotics, upregulation of ethanol oxidation genes	Thomas et al. (2009)
exaA, pqqC, erbR	Tobramycin	Ethanol oxidation players	Unknown	Kohanski et al. (2010)
PA1875–1877	Tobramycin, gentamicin, ciprofloxacin	Biofilm-specific antibiotic efflux pump	Efflux of antibiotics out of the cell	Zheng and Stewart (2004)
tssC1, hcp1	Tobramycin, gentamicin, ciprofloxacin	Type VI secretion components	Unknown	Webb et al. (2003)
PA0756–0757	Tobramycin, gentamicin	Two-component system	Unknown	Kohanski et al. (2010)
PA2070	Tobramycin, gentamicin	TonB-dependent receptor	Unknown	Lechner et al. (2012)
PA5033	Tobramycin, gentamicin	Hypothetical proteins	Unknown	Webb et al. (2003)
pslABCDEFGHIJKLMNO	Colistin, polymyxin B, tobramycin, ciprofloxacin	Psl biosynthetic enzymes	Unknown	Zheng and Stewart (2004)
pelABCDEFG	Tobramycin, gentamicin	Pel biosynthetic enzymes	Unknown	Webb et al. (2003)
relA, spoT	Ofloxacin, meropenem, colistin, gentamicin	Players in the stringent response	Upregulate antioxidant defenses and downregulate pro-oxidants	Zheng and Stewart (2004)

rapA	Penicillin G, norfloxacin, chloramphenicol, gentamicin	Helicase-like protein	Upregulation of YhcQ and of exopolysaccharide synthesis	Van Acker and Coenye (2016)
yafQ	Tobramycin, cefazolin	Toxin	Persister cell formation	Kohanski et al. (2010)
epaOX	Gentamicin	Glycosyltransferase	Maintenance of cell wall integrity	Whiteley et al. (2001)
epaI	Daptomycin	Glycosyltransferase	Unknown	Whiteley et al. (2001)
gelE	Gentamicin, daptomycin, linezolid	Gelatinase	Unknown	Spoering and Lewis (2001)
fsrA, *fsrC*	Gentamicin, daptomycin, linezolid	Quorum-sensing players	Unknown	Thomas et al. (2009)
dltABCD	Gentamicin	Enzymes involved in D-alanylation of teichoic acid	Decrease in the negative charge of the cell wall	Van Acker and Coenye (2016)

was the F plasmid. The FII plasmid R100 carries a class I In2 inside Tn21 which is in turn present inside IS-mediated Tn carrying chloramphenicol-resistant gene (*catA1*) (Liebert et al. 1999).

Different strains of *Staphylococci* consist of plasmids that provide resistance to different antibiotics, disinfectants, and heavy metal ions that are known to be plasmid-borne. The three major classes of staphylococci-resistant plasmids are small plasmids, multiresistance plasmids, and conjugative multiresistance plasmids (Firth and Skurray 2006; Fischetti et al. 2006).

8.6 Effects of Antibiotic Resistance

As bacterial biofilm shows increased tolerance to disinfectant chemicals and antibiotics and other components of the body's defense system, it is considered to be the root cause of chronic infection. The protective environment created by the biofilm makes the bacteria tolerant to various stresses and resistant to antimicrobials. The extracellular matrix present in the biofilms prevents the antibiotic diffusion by forming a mechanical barrier, and as a result the antibiotics cannot reach the microorganisms. The bacterial metabolism which includes gradients of oxygen, metabolites, and nutrients plays a crucial part in antibiotic tolerance in the biofilm (Olivares et al. 2020). For example, the tenacity of *Staphylococcus* infections in relation to the foreign bodies is because of the biofilm formation, and an another example is chronic *Pseudomonas aeruginosa* lung infection by cystic fibrosis patients which occurs due to the biofilm-growing mucoid strains. According to the studies, from the entire biofilm there remain gradients of nutrient and oxygen, and these gradients are responsible for reduction in bacterial metabolic activity and also for the increase in the doubling time of the bacterial cells. Growth of biofilm is related to rise in level of mutations and quorum-sensing-regulated mechanisms. Standard resistance mechanisms like chromosomal β-lactamase, upregulated efflux pumps, and mutations in antibiotic target molecules in bacteria also help in the survival of biofilms (Høiby et al. 2010).

Acinetobacter baumannii is a pathogen of great importance for its multiple resistance to antibiotics and ability to survive, which directly links to biofilm formation. In order to characterize the contribution of AdeABC, AdeFGH, and AdeIJK resistance-nodulation-cell division (RND)-type efflux systems to acquired and intrinsic resistance, a set of isogenic mutants overexpressing each system following introduction of a point mutation in their cognate regulator or a deletion for the pump by allelic replacement is formed from a completely sequenced susceptible strain of *A. baumannii*. When every derivative is pairwise compared with parental strain, it indicated that when AdeABC and AdeFGH are overproduced, they are regulated tightly and contribute to antibiotic resistance. AdeABC had a wide substrate range that includes β-lactams, fluoroquinolones, tetracycline-tigecycline, macrolides-lincosamides, and chloramphenicol, and it provides clinical resistance to aminoglycosides. On the other hand, AdeIJK was responsible after being expressed for intrinsic resistance to the major drug classes, same as AdeABC and also antifolates and fusidic acid. Studies and researches have shown that composition of bacterial membrane has been altered, resulting in reduction of biofilm formation but not motility with overproduction of AdeABC and AdeIJK. Recipient that overproduces AdeABC has shown decrease in natural transformation and transfer of plasmid. If there is alteration in the expression of efflux systems, it then results in different changes in the relation between the host and its environment, along with antibiotic resistance.

The Gram-negative pathogen, *Klebsiella pneumoniae*, creates different types of community-acquired infections and infects patients with indwelling medical devices, especially urinary catheters, where the microorganism is able to produce biofilm. The highly frequent acquisition of antibiotic resistance by *K. pneumoniae* strains has resulted in a global spread of this multidrug-resistant pathogen, mainly at the hospital level. The circumstances worsen due to the development of biofilm by *K. pneumoniae*. Initially only small amounts of chromosomal penicillinase enzyme were possessed by *K. pneumoniae strain*; it is a well-known "collector" of multidrug-resistant plasmids that usually encoded resistance to aminoglycosides, till the end of the 1980s, while, later, encoding extended-spectrum β-lactamases (ESBLs), mostly temoniera (TEMs) and sulfhydryl variable (SHVs) active against last-generation cephalosporins, and a variety of genes conferring resistance to drugs besides β-lactams (Vuotto et al. 2014).

8.7 Conclusion

From researches it is found that 75% of bacterial infections includes biofilms that are attached onto the surface colonies present within a protective layer of extracellular matrix. Biofilm-forming bacteria are more antibiotic resistant (up to 1000 times) compared to free-floating bacteria (planktonic). Now this resistant property of bacteria against antibiotics severely affects the treatment option. There are several mechanisms that explain the high resistance of biofilms which include restricted penetration of antimicrobial agents into biofilms, slow growth owing to nutrient limitation, expression of genes involved in the general stress response, and an emergence of a biofilm-specific phenotype, expression of genes involved in the general stress response, and an emergence of a biofilm-specific phenotype. From the experimental data, it is observed that the resistance of biofilms can be attributed to formation of ampicillin-resistant subpopulations in the deeper layers of mature biofilms but not in young colony biofilms and that the production and resistance of the subpopulations were aided by biofilm-specific phenotypes, like slow growth. The combination of biofilm-controlling molecules along with antibiotics can help in treatment of biofilm infections. Various newly discovered innovative and effective antibiotic strategies are introduced which include quorum-sensing inhibitors using antibiotic combinations and dispersion of biofilms. 2-Amino-imidazole are small molecules, developed by researchers, that disrupt biofilms which makes antibiotic-resistant bacteria more exposed to the known traditional drugs. It was also seen that the antibiotics increases the ability of 2-amino-imidazoles to disrupt biofilms which shows that the new antibiotics are not the only way to combat biofilm infections if one could make an effective older antibiotics active again. Antimicrobial resistance in bacterial pathogens is a challenge in association with high morbidity and mortality. In Gram-positive and Gram-negative bacteria, multidrug resistance is sometimes difficult to treat, and also they become untreatable with the traditional antibiotics. Recently there is a scarcity of effective therapies, dearth of successful preventive measures, and only very few new antibiotics that require development of novel treatment options and other possible antimicrobial therapies. Biofilms are associated with multidrug resistance and can present challenges for control of infection. Another bacteria *Acinetobacter baumannii* is also a cause of various infections due to its increasing resistance against antibiotics and due to increasing virulence too. The capability of *Acinetobacter baumannii* to form biofilms helps it to survive at unfavorable environmental conditions that include hospital environments and medical devices. The interplay among microbial physicochemistry, changes in the phenotype and genotypic determinants, and the impact of existing ecological niche and the chemistry of antimicrobial agents has resulted in enhanced biofilm formation leading to limited access of drugs to their specific targets. Biomedical researchers got propelling interest in *A. baumannii* for its wide range of varied infections, mostly in the intensive care units in the hospital.

References

Ahiwale S, Tamboli N, Thorat K, Kulkarni R, Ackermann H, Kapadnis B (2011) In vitro management of hospital Pseudomonas aeruginosa biofilm using indigenous T7-like lytic phage. Curr Microbiol 62:335–340

Alekshun MN, Levy SB (1999) The mar regulon: multiple resistances to antibiotics and othertoxic chemicals. Trends Microbiol 7:410–413

Allesen-Holm M, Barken KB, Yang L et al (2006) A characterization of DNA release in Pseudomonas aeruginosa cultures and biofilms. Mol Microbiol 59:1114–1128

Anderl JN, Franklin MJ, Stewart PS (2000) Role of antibiotic penetration limitation in Klebsiella pneumoniae biofilm resistance to ampicillin and ciprofloxacin. Antimicrob Agents Chemother 44:1818–1824

Bagge N, Ciofu O, Skovgaard LT, Høiby N (2000) Rapid development in vitro and in vivo of resistance to ceftazidime in biofilm-growing Pseudomonas aeruginosa due to chromosomal beta-lactamase. APMIS 108:589–600

Bagge N, Hentzer M, Andersen JB, Ciofu O, Givskov M, Høiby N (2004) Dynamics and spatial distribution of beta-lactamase expression in Pseudomonas aeruginosa biofilms. Antimicrob Agents Chemother 48:1168–1174

Balaban NQ, Merrin J, Chait R, Kowalik L, Leibler S (2004) Bacterial persistence as a phenotypic switch. Science 305:1622–1625

Beauclerk AAD, Cundliffe E (1987) Site of action of two ribosomal RNA methylases responsible for resistance to aminoglycoside. J Mol Biol 193:661–671

Bennett PM, Richmond MH (1976) Translocation of a discrete piece of deoxyribonucleic acid carrying an amp gene between replicons in *Escherichia coli*. J Bacteriol 126:1–6

Bertram J, Strätz M, Dürre P (1991) Natural transfer of conjugative transposon tn916 between gram-positive and gram- negative bacteria. J Bacteriol 173:443–448

Billings N, Millan M, Caldara M et al (2013) The extracellular matrix component Psl provides fast-acting antibiotic defense in Pseudomonas aeruginosa biofilms. PLoS Pathog 9:e1003526

Bjarnsholt T, Kirketerp-Møller K, Kristiansen S, Phipps R, Nielsen AK, Jensen PØ et al (2007) Silver against Pseudomonas aeruginosa biofilms. APMIS 115:921–928

Bjarnsholt T, Jensen PØ, Fiandaca MJ, Pedersen J, Hansen CR, Andersen CB et al (2009) Pseudomonas aeruginosa biofilms in the respiratory tract of cystic fibrosis patients. Pediatr Pulmonol 44:547–558

Blair JM, Webber MA, Baylay AJ et al (2015) Molecular mechanisms of antibiotic resistance. Nat Rev Microbiol 13:42–51

Boles BR, Singh PK (2008) Endogenous oxidative stress produces diversity and adaptability in biofilm communities. Proc Natl Acad Sci U S A 105:12503–12508

Borriello G, Werner E, Roe F et al (2004) Oxygen limitation contributes to antibiotic tolerance of Pseudomonas aeruginosa in biofilms. Antimicrob Agents Chemother 48:2659–2664

Breidenstein Elena BM, de la Fuente Núñez C, Hancock EW (2011) *Pseudomonas aeruginosa*: all roads lead to resistance

Byrd MS, Sadovskaya I, Vinogradov E et al (2009) Genetic and biochemical analyses of the Pseudomonas aeruginosa Psl exopolysaccharide reveal overlapping roles for polysaccharide synthesis enzymes in Psl and LPS production. Mol Microbiol 73:622–638

Chiang W-C, Nilsson M, Jensen PØ et al (2013) Extracellular DNA shields against aminoglycosides in Pseudomonas aeruginosa biofilms. Antimicrob Agents Chemother 57:2352–2361

Ciofu O (2003) Pseudomonas aeruginosa chromosomal-lactamase in patients with cystic fibrosis and chronic lung infection—mechanism of antibiotic resistance and target of the humoral immune response. APMIS 111:4–47

Ciofu O, Giwercman B, Pedersen SS, Hoiby N (1994) Development of antibiotic resistance in Pseudomonas aeruginosa during two decades of antipseudomonal treatment at the Danish CF Center. APMIS 102:674–680

Ciofu O, Riis B, Pressler T, Poulsen HE, Høiby N (2005) Occurrence of hypermutable Pseudomonas aeruginosa in cystic fibrosis patients is associated with the oxidative stress caused by chronic lung inflammation. Antimicrob Agents Chemother 49:2276–2282

Colvin KM, Gordon VD, Murakami K et al (2011) The Pel polysaccharide can serve a structural and protective role in the biofilm matrix of Pseudomonas aeruginosa. PLoS Pathog 7:e1001264

Conibear TC, Collins SL, Webb JS (2009) Role of mutation in Pseudomonas aeruginosa biofilm development. PLoS One 14:e6289

Costerton JW, Lewandowski Z (1995) Microbial biofilms. Annu Rev Microbiol 49:711–745

Costerton JW, Stewart PS, Greenberg EP (1999) Bacterial biofilms: a common cause of persistent infections. Science 284:1318–1322

Costerton JW, Montanaro L, Arciola CR (2005) Biofilm in implant infections: its production and regulation. Int J Artif Org 28:1062–1068

Cox G, Wright GD (2013) Intrinsic antibiotic resistance: mechanisms, origins, challenges and solutions. Int J Med Microbiol 303:287–292

de la Fuente-Núnez C, Reffuveille F, Fernandez L, Hancock REW (2013) Bacterial biofilm development as a multicellular adaptation: antibiotic resistance and new therapeutic strategies. Curr Opin Microbiol 16:580–589

Del Pozo JL, Patel R (2009) Infection associated with prostetic joints. N Engl J Med 361:787–794

Domingues S, Harms K, Fricke WF, Johnsen PJ, da Silva GJ, Nielsen KM (2012) Natural transformation facilitates transfer of transposons, integrons and gene cassettes between bacterial species. PLoS Pathog 8:e1002837

Doroshenko N, Tseng BS, Howlin RP et al (2014) Extracellular DNA impedes the transport of vancomycin in Staphylococcus epidermidis biofilms preexposed to subinhibitory concentrations of vancomycin. Antimicrob Agents Chemother 58:7273–7282

Driffield K, Miller K, Bostock M, O'Neill AJ, Chopra I (2008) Increased mutability of Pseudomonas aeruginosa in biofilms. J Antimicrob Chemother 61:1053–1056

Dunne WM, Mason EO, Kaplan SL (1993) Diffusion of rifampin and vancomycin through a Staphylococcus epidermidis biofilm. Antimicrob Agents Chemother 37:2522–2526

Falagas ME, Bliziotis IA (2007) Pandrug-resistant Gram-negative bacteria: the dawn of the post-antibiotic era? Int J Antimicrob Agents 29:630–636

Fernández L, Breidenstein EBM, Hancock REW (2011) Creeping baselines and adaptive resistance to antibiotics. Drug Resist Updat 14:1–21

Firth N, Skurray RA (2006) The Staphylococcus—genetics: accessory elements and genetic exchange, p 413–426

Fischetti VA, Novick RP, Ferretti JJ, Portnoy DA, Rood JI (2006) Gram-positive pathogens. ASM Press, Washington, DC

Flemming HC, Neu TR, Wozniak DJ (2007) The EPS matrix: the "house of biofilm cells". J Bacteriol 189(22):7945–7947

Franklin MJ, Nivens DE, Weadge JT et al (2011) Biosynthesis of the Pseudomonas aeruginosa extracellular polysaccharides, alginate, Pel, and Psl. Front Microbiol 2:167

Fux CA, Costerton JW, Stewart PS, Stoodley P (2005) Survival strategies of infectious biofilms. Trends Microbiol 13:34–40

Giwercman B, Lambert PA, Rosdahl VT, Shand GH, Hoiby N (1990) Rapid emergence of resistance in Pseudomonas aeruginosa in cystic fibrosis patients due to in vivo selection of stable partially derepressed-lactamase producing strains. J Antimicrob Chemother 26:247–259

Gregory PD, Lewis RA, Curnock SP, Dyke KGH (2003) Studies of the repressor (BlaI) of β-lactamase synthesis in *Staphylococcus aureus*. Wiley Publications, New York

Gristina AG, Oga M, Webb LX, Hobgood CD (1985) Adherent bacterial colonization in the pathogenesis of osteomyelitis. Science 228:990–993

Haagensen JAJ, Klausen M, Ernst RK et al (2006) Differentiation and distribution of colistin- and sodium dodecyl sulfate tolerant cells in Pseudomonas aeruginosa biofilms. J Bacteriol 189:28–37

Hall-Stoodley L, Hu FZ, Gieseke A, Nistico A, Nguyen L, Hayes J et al (2006) Direct detection of bacterial biofilms on the middle-ear mucosa of children with chronic otitis media. JAMA 296:202–211

Hancock REW (1981) Aminoglycoside uptake and mode of action-with special ref-erence to streptomycin and gentamicin II. Effects of aminoglycosides on cells. J Antimicrob Chemother 8:429–445

Hassett DJ, Ma JF, Elkins JG, McDermott TR, Ochsner UA, West SEH et al (1999) Quorum sensing in Pseudomonas aeruginosa controls expression of catalase and superoxide dismutase genes and mediates biofilm susceptibility to hydrogen peroxide. Mol Microbiol 34:1082–1093

Hauryliuk V, Atkinson GC, Murakami KS et al (2015) Recent functional insights into the role of (p)ppGpp in bacterial physiology. Nat Rev Microbiol 13:298–309

Hawkey PM, Jones AM (2009) The changing epidemiology of resistance. J Antimicrob Chemother i3:64

Høiby N, Johansen HK, Moser C, Song ZJ, Ciofu O, Kharazmi A (2001) Pseudomonas aeruginosa and the biofilm mode of growth. Microbes Infect 3:1–13

Høiby N, Bjarnsholt T, Givskov M, Molin S, Ciofu O (2010) Antibiotic resistance of bacterial biofilms, vol 35. Elsevier Publications, Amsterdam, p 4

Olivares E et al (2020) Clinical impact of antibiotics for the treatment of *Pseudomonas aeruginosa* biofilm infections. Front Microbiol 10:2894

Hoiby N, Ciofu O, Johansen HK, Song ZJ, Moser C, Jensen PØ, Molin S, Givskov M, Tolker-Nielsen T, Bjarnsholt T (2011) The clinical impact of bacterial biofilms. Int J Oral Sci 3:55

Holloway BW (1969) Genetics of *Pseudomonas*. Bacteriol Rev 33:419–443

Hu Y, Yang X, Li J, Lv N, Liu F, Wu J, Lin IYC, Wu N, Weimer BC, Gao GF, Liu Y, Zhu B (2016) The bacterial mobile resistome transfer network connecting the animal and human microbiomes. Appl Environ Microbiol 82:6672–6681

Jefferson KK, Goldmann DA, Pier GB (2005) Use of confocal microscopy to analyze the rate of vancomycin penetration through Staphylococcus aureus biofilms. Antimicrob Agents Chemother 49:2467–2473

Jensen PØ, Briales A, Brochmann RP et al (2014) Formation of hydroxyl radicals contributes to the bactericidal activity of ciprofloxacin against Pseudomonas aeruginosa biofilms. Pathog Dis 70:440–443

Kamruzzaman M, Patterson JD, Shoma S, Ginn AN, Partridge SR, Iredell JR (2015) Relative strengths of promoters provided by common mobile genetic elements associated with resistance gene expression in Gram-negative bacteria. Antimicrob Agents Chemother 59:5088–5091. https://doi.org/10.1128/AAC.00420-15

Kaplan JB, Izano EA, Gopal P et al (2012) Low levels of β-lactam antibiotics induce extracellular DNA release and biofilm formation in Staphylococcus aureus. MBio 3:e00198–e00112

Keren I, Kaldalu N, Spoering A, Wang YP, Lewis K (2004) Persister cells and tolerance to antimicrobials. FEMS Microbiol Lett 230:13–18

Khakimova M, Ahlgren HG, Harrison JJ et al (2013) The stringent response controls catalases in Pseudomonas aeruginosa and is required for hydrogen peroxide and antibiotic tolerance. J Bacteriol 195:2011–2020

Khan W, Bernier SP, Kuchma SL et al (2010) Aminoglycoside resistance of Pseudomonas aeruginosa biofilms modulated by extracellular polysaccharide. Int Microbiol 13:207–212

Klumper U, Riber L, Dechesne A, Sannazzarro A, Hansen LH, Sorensen SJ, Smets BF (2015) Broad host range plasmids can invade an unexpectedly diverse fraction of a soil bacterial community. ISME J 9:934–945

Kohanski MA, Dwyer DJ, Collins JJ (2010) How antibiotics kill bacteria: from targets to networks. Nat Rev Microbiol 8:423–435

Kolenbrander PE, Palmer RJ Jr (2004) Human oral bacterial biofilms. In: Ghannoum MA, O'Toole GA (eds) Microbial biofilms. ASM Press, Washington, DC

Kolpen M, Appeldorff CF, Brandt S et al (2016) Increased bactericidal activity of colistin on Pseudomonas aeruginosa biofilms in anaerobic conditions. Pathog Dis 74:ftv086

Kumar B, Sharma D, Sharma P, Katoch VM, Venkatesan K, Bisht D (2013) Proteomic analysis of Mycobacterium tuberculosis isolates resistant to kanamycin and amikacin. J Proteome 94:68–77

Lata M, Sharma D, Deo N, Tiwari PK, Bisht D, Venkatesan K (2015) Proteomic analysis of ofloxacin-mono resistant Mycobacterium tuberculosis isolates. J Proteome 127:114–121

Lechner S, Lewis K, Bertram R (2012) Staphylococcus aureus persisters tolerant to bactericidal antibiotics. J Mol Microbiol Biotechnol 22:235–244

Lewenza S (2013) Extracellular DNA-induced antimicrobial peptide resistance mechanisms in Pseudomonas aeruginosa. Front Microbiol 4:21

Lewis K (2008) Multidrug tolerance of biofilms and persister cells. Curr Top Microbiol Immunol 322:107–131

Liebert CA, Hall RM, Summers AO (1999) Transposon Tn*21*, flagship of the floating genome. Microbiol Mol Biol Rev 63:507–522

López M, Sáenz Y, Álvarez-Martínez MJ, Marco F, Robredo B, Rojo-Bezares B, Ruiz-Larrea F, Zarazaga M, Torres C (2010) Tn*1546* structures and multilocus sequence typing of *vanA*-containing enterococci of animal, human and food origin. J Antimicrob Chemother 65:1570–1575

Macia MD, Blanquer D, Togores B, Sauleda J, Perez JL, Oliver A (2005) Hypermutation is a key factor in development of multiple-antimicrobial resistance in Pseudomonas aeruginosa strains causing chronic lung infections. Antimicrob Agents Chemother 49:3382–3386

Mai-Prochnow A, Lucas-Elio P, Egan S, Thomas T, Webb J, Sanchez-Amat A et al (2008) Hydrogen peroxide linked to lysine oxidase activity facilitates biofilm differentiation and dispersal in several Gram-negative bacteria. J Bacteriol 190:5493–5501

Mandsberg LF, Ciofu O, Kirkby N, Christiansen LE, Poulsen HE, Høiby N (2009) Antibiotic resistance in Pseudomonas aeruginosa strains with increased mutation frequency due to inactivation of the DNA oxidative repair system. Antimicrob Agents Chemother 53:2483–2491

McCarty SM, Cochrane CA, Clegg PD, Percival SL (2012) The role of endogenous and exogenous enzymes in chronic wounds: a focus on the implications of aberrant levels of both host and bacterial proteases in wound healing. Wound Repair Regen 20:125–136

McPhee JB, Bains M, Winsor G et al (2006) Contribution of the PhoPPhoQ and PmrA-PmrB two-component regulatory systems to Mg2+-induced gene regulation in Pseudomonas aeruginosa. J Bacteriol 188:3995–4006
McPhee JB, Lewenza S, Hancock REW (2003) Cationic antimicrobial peptides activate a two-component regulatory system, PmrA-PmrB, that regulates resistance to polymyxin B and cationic antimicrobial peptides in Pseudomonas aeruginosa. Mol Microbiol 50:205–217
Messi P, Guerrieri E, De Niederhäusern S, Sabia C, Bondi M (2006) Vancomycin-resistant enterococci (VRE) in meat and environmental samples. Int J Food Microbiol 107:218–222
Mijnendonckx K, Leys N, Mahillon J, Silver S, Van Houdt R (2013) Antimicrobial silver: uses, toxicity, and potential for resistance
Molin S, Tolker-Nielsen T (2003) Gene transfer occurs with enhanced efficiency in biofilms and induces enhanced stabilisation of the biofilm structure. Curr Opin Biotechnol 14:255–261
Mulcahy H, Charron-Mazenod L, Lewenza S (2008) Extracellular DNA chelates cations and induces antibiotic resistance in Pseudomonas aeruginosa biofilms. PLoS Pathog 4:e1000213
Musken M, Di Fiore S, Dötsch A, Fischer R, Haussler S (2010) Genetic determinants of Pseudomonas aeruginosa biofilm establishment. Microbiology 156:431–441
Musovic S, Oregaard G, Kroer N, Sørensen SJ (2006) Cultivation independent examination of horizontal transfer and host range of an IncP-1 plasmid among gram-positive and gram-negative bacteria indigenous to the barley rhizosphere. Appl Environ Microbiol 72:6687
Neuhaus FC, Baddiley J (2003) A continuum of anionic charge: structures and functions of D-alanyl-teichoic acids in Gram positive bacteria. Microbiol Mol Biol Rev 67:686–723
Nguyen D, Joshi-Datar A, Lepine F et al (2011) Active starvation responses mediate antibiotic tolerance in biofilms and nutrient-limited bacteria. Science 334:982–986
Nilsson M, Rybtke M, Givskov M et al (2016) The dlt genes play a role in antimicrobial tolerance of Streptococcus mutans biofilms. Int J Antimicrob Agents 48:298–304
Oliver A, Sanchez JM, Blazquez J (2002) Characterization of the GO system of Pseudomonas aeruginosa. FEMS Microbiol Lett 217:31–35
Pamp SJ, Gjermansen M, Johansen HK et al (2008) Tolerance to the antimicrobial peptide colistin in Pseudomonas aeruginosa biofilms is linked to metabolically active cells, and depends on the pmr and mexAB-oprM genes. Mol Microbiol 68:223–240
Partridge SR, Kwong SM, Firth N, Jensen SO (2018) Mobile genetic elements associated with antimicrobial resistance. Clin Microbiol Rev 31:e00088–e00017
Partridge SR, Hall RM (2005) Evolution of transposons containing bla_{TEM}. Genesis 49:1267–1268
Perry JA, Wright GD (2013) The antibiotic resistance "Mobilome": searching for the link between environment and clinic. Front Microbiol 4:1–7
Ragan MA, Beiko RG (2009) Lateral genetic transfer: open issues. Philos Trans R Soc Lond Ser B Biol Sci 364:2241–2251
Schaible B, Taylor CT, Schaffer K (2012) Hypoxia increases antibiotic resistance in Pseudomonas aeruginosa through altering the composition of multidrug efflux pumps. Antimicrob Agents Chemother 56:2114–2118
Schuster CF, Bertram R (2016) Toxin-antitoxin systems of Staphylococcus aureus. Toxins 8:140
Shah D, Zhang Z, Khodursky AB, Kaldalu N, Kurg K, Lewis K (2006) Persisters: a distinct physiological state of E. coli. BMC Microbiol 6:53
Singh R, Ray P, Das A et al (2010) Penetration of antibiotics through Staphylococcus aureus and Staphylococcus epidermidis biofilms. J Antimicrob Chemother 65:1955–1958
Solano C, Echeverz M, Lasa I (2014) Biofilm dispersion and quorum sensing. Curr Opin Microbiol 18:96–104
Spoering AL, Lewis K (2001) Biofilms and planktonic cells of Pseudomonas aeruginosa have similar resistance to killing by antimicrobials. J Bacteriol 183:6746–6751
Stewart PS, Davison WM, Steenbergen JN (2009) Daptomycin rapidly penetrates a Staphylococcus epidermidis biofilm. Antimicrob Agents Chemother 53:3505–3507
Stewart PS, Franklin MJ, Williamson KS et al (2015) Contribution of stress responses to antibiotic tolerance in Pseudomonas aeruginosa biofilms. Antimicrob Agents Chemother 59:3838–3847

Stewart PS, William Costerton J (2001) Antibiotic resistance of bacteria in biofilms. Lancet 358:135–138

Stokes HW, Gillings MR (2011) Gene flow, mobile genetic elements and the recruitment of antibiotic resistance genes into Gram-negative pathogens. FEMS Microbiol Rev 35(5):790–819

Suci PA, Mittelman MW, Yu FP et al (1994) Investigation of ciprofloxacin penetration into Pseudomonas aeruginosa biofilms. Antimicrob Agents Chemother 38:2125–2133

Taber HW, Mueller JP, Miller PF et al (1987) Bacterial uptake of aminoglycoside antibiotics. Microbiol Rev 51:439–457

Taconelli E, Smith G, Hieke K, Lafuma A, Bastide P (2009) Epidemiology, medical outcomes and costs of catheter-related bloodstream infection in intensive care units of four European countries: literature- and registry-based estimates. J Hosp Infect 72:97–103

Tata M, Wolfinger MT, Amman F et al (2016) RNASeq based transcriptional profiling of Pseudomonas aeruginosa PA14 after short and long-term anoxic cultivation in synthetic cystic fibrosis sputum medium. PLoS One 11:e0147811

Thomas CM, Nielsen KM (2005) Mechanisms of, and barriers to, horizontal gene transfer between bacteria. Nat Rev Microbiol 3:711–721

Thomas VC, Hiromasa Y, Harms N et al (2009) A fratricidal mechanism is responsible for eDNA release and contributes to biofilm development of Enterococcus faecalis. Mol Microbiol 72:1022–1036

Trampuz A, Zimmerli W (2005) Prosthetic joint infections: update in diagnosis and treatment. Swiss Med Wkly 135:243–251

Tseng BS, Zhang W, Harrison JJ et al (2013) The extracellular matrix protects Pseudomonas aeruginosa biofilms by limiting the penetration of tobramycin. Environ Microbiol 15:2865–2878

Van Acker H, Coenye T (2016) The role of efflux and physiological adaptation in biofilm tolerance and resistance. J Biol Chem 291(24):12565–12572

Van Acker H, Gielis J, Acke M et al (2016) The role of reactive oxygen species in antibiotic-induced cell death in Burkholderia cepacia complex bacteria. PLoS One 11:e0159837

Van Acker H, Sass A, Bazzini S et al (2013) Biofilm-grown Burkholderia cepacia complex cells survive antibiotic treatment by avoiding production of reactive oxygen species. PLoS One 8: e58943

Von Wintersdorff CJH, Penders J, van Niekerk JM, Mills ND, Majumder S, van Alphen LB, Savelkoul PHM, Wolffs PFG (2016) Dissemination of antimicrobial resistance in microbial ecosystems through horizontal gene transfer. Front Microbiol 7:1–10

Vuotto C, Longo F, Balice MP, Donelli G, Varaldo PE (2014) Antibiotic Resistance Related to Biofilm Formation in *Klebsiella pneumonia*. Pathog Publ 3(3):743–758

Walters MC, Roe F, Bugnicourt A et al (2003) Contributions of antibiotic penetration, oxygen limitation, and low metabolic activity to tolerance of *Pseudomonas aeruginosa* biofilms to ciprofloxacin and tobramycin. Antimicrob Agents Chemother 47:317–323

Webb JS, Thompson LS, James S et al (2003) Cell death in Pseudomonas aeruginosa biofilm development. J Bacteriol 185:4585–4592

Whitchurch CB, Tolker-Nielsen T, Ragas PC, Mattick JS (2002) Extracellular DNA required for bacterial biofilm formation. Science 295:1487

Whiteley M, Bangera MG, Bumgarner RE et al (2001) Gene expression in Pseudomonas aeruginosa biofilms. Nature 413:860–864

Wilkins M, Hall-Stoodley L, Allan RN, Faust SN (2014) New approaches to the treatment of biofilm-related infections. J Inf Secur 69:S47–S52

Wilton M, Charron-Mazenod L, Moore R et al (2015) Extracellular DNA acidifies biofilms and induces aminoglycoside resistance in Pseudomonas aeruginosa. Antimicrob Agents Chemother 60:544–553

Wingender J, Strathmann M, Rode A, Leis A, Flemming HC (2001) Isolation and biochemical characterization of extracellular polymeric substances from Pseudomonas aeruginosa. Methods Enzymol 336:302–314

Wright GD (2010) Antibiotic resistance in the environment: a link to the clinic? Curr Opin Microbiol 13:589–594

Yamamoto M, Matsumura Y, Gomi R, Matsuda T, Tanaka M, Nagao M, Takakura S, Uemoto S, Ichiyama S (2016) Interspecies dissemination of a mobilizable plasmid harboring BlaIMP-19 and the possibility of horizontal gene transfer in a single patient. Antimicrob Agents Chemother 60:5412–5419

Zahller J, Stewart PS (2002) Transmission electron microscopic study of antibiotic action on *Klebsiella pneumoniae* biofilm. Antimicrob Agents Chemother 46:2679–2683

Zegans ME, Wozniak D, Griffin E et al (2012) Pseudomonas aeruginosa exopolysaccharide Psl promotes resistance to the biofilm inhibitor polysorbate 80. Antimicrob Agents Chemother 56:4112–4122

Zheng Z, Stewart PS (2002) Penetration of rifampin through Staphylococcus epidermidis biofilms. Antimicrob Agents Chemother 46:900–903

Zheng Z, Stewart PS (2004) Growth limitation of Staphylococcus epidermidis in biofilms contributes to rifampin tolerance. Biofilms 1:31–35

Inhibition of Biofilm Formation

9

Bhisma Narayan Ratha, Dibyajit Lahiri, and Rina Rani Ray

Abstract

Biofilm formation increases the survival chances of the microbial species. Increase in microbial cell density triggers quorum sensing and leads to colony formation. The emergence of multidrug resistance (MDR) strains with biofilm formation ability of human pathogen has limited effective antibiotic-based treatment. Hence it is immediately required to identify novel molecules for the treatment of such microbial infections forming a biofilm. Here, traditional medicine has a lot to offer in search of a novel therapeutic molecule. Researchers from the field of microbiology, phytochemistry, and pharmacognosy have screened several plant extracts and purified molecules, which have shown promising results. Here, we have covered druggable targets in a biofilm, interacting members in a biofilm, and utilize those as targets. In the subsequent part, we have discussed the phytochemicals and herbal drugs in inhibiting biofilm. It will assist in having an overview of current research progress and understanding the druggable targets.

B. N. Ratha (✉)
Department of Phytopharmaceuticals, School of Agriculture and Bioengineering, Centurion University of Technology and Management, Parlakhemundi, Odisha, India

D. Lahiri
Department of Biotechnology, University of Engineering & Management, Kolkata, West Bengal, India

R. R. Ray (✉)
Department of Biotechnology, Maulana Abul Kalam Azad University of Technology, West Bengal, Kolkata, West Bengal, India

R. R. Ray et al. (eds.), *Biofilm-Mediated Diseases: Causes and Controls*,
https://doi.org/10.1007/978-981-16-0745-5_9

9.1 Introduction

The bacterial biofilm is a matrix-enclosed, self-synthesized, adherent group of cells that provides anchorage to a large population of microbial cells. These are the groups of sessile communities that attach irreversibly to the biotic or abiotic surface using pili and secrete extracellular polymeric substances (EPS) by changing their phenotype with respect to transcription factor and growth rate (Donlan 2002). The cells forming the biofilm become enveloped in a matrix comprised of glycocalyx that protects them from surfactant, biocides, and phagocytic cells (Donlan and Costerton 2002). The growth rates and antimicrobial resistance of the planktonic (freely suspended) form of microorganisms differ widely from the respective biofilm-bound (sessile) cells (Bogino et al. 2013). The development of pathogenicity within the biofilm occurs by a density-dependent communication system known as quorum sensing (QS) which are the groups of diffusible signal molecules known as autoinducers (AI) possessing the ability of regulating gene expressions. The mechanism of QS helps in shifting the cells from its planktonic to sessile forms forming consortia (Kaufmann et al. 2008). Often biofilm forming pathogenic bacteria develop resistance to antibiotics. Drug resistance in a biofilm occurs because of the extracellular polymeric substances (EPS). EPS comprises of polysaccharide, proteins, nucleic acid, and lipids, these macromolecular ensamble assist in sequestering antibiotics. Moreover, the environment persisting within the biofilm lacks suitable amount of nutrients and oxygen and thus make them resistant against any types of environmental stresses.

The ill effect of biofilm is fatal since they develop antimicrobial resistance as the polymeric substances that formed around the bacterial cells do not allow the penetration of smaller molecules and as a result numerous bacterial infections become difficult to get treated by antibiotics (Costerton et al. 1999). As a consequence, the growth of microbial biofilm causes the development of nosocomial diseases. For example, the growth of microbial biofilm on medical devices causes pathogenesis of numerous bacterial infections that are difficult to get treated by antibiotics (Costerton et al. 1999). Microbial biofilms may occur on indwelling medical devices from the skin of the patients or healthcare personalities, tap water to which the devices are exposed, and other environmental sources. Medical devices like mechanical heart valves, central venous catheters, contact lenses, endotracheal tubes, uro-catheters, intra-uterine devices, pacemakers, prosthetic joints, peritoneal dialysis catheters, and teeth surfaces are more prone to the development of microbial biofilms, and thus biofilms pose a serious threat to public health (Donlan 2002).

The biofilm formation on medical devices comprises Gram-positive as well as Gram-negative bacterial cell. Gram-positive bacteria commonly isolated from devices are *Enterococcus faecalis*, *Streptococcus viridans*, *Staphylococcus epidermidis*, *Bacillus subtilis*, and *Staphylococcus aureus*, and the Gram-negative strains include *Klebsiella pneumoniae*, *Escherichia coli*, *Proteus mirabilis*, and *Pseudomonas aeruginosa* (Kwakman et al. 2006). Biofilm provides resistance to different microbes, thereby giving protection against invaders like protozoan and defenses offered by the hosts (Mah and O'Toole 2001). The researches in this field proved that antibiotic treatment can only eliminate planktonic forms of bacteria but the sessile form develops resistance toward antibiotics and keeps on propagating inside the biofilm (Moser et al. 2015). The persistent bacterial cells develop antibiotic and multidrug resistance by unregulated use of antibiotics and genetic evolution of the species (Rizzello and Pompa 2014). The extracellular polymeric substance (EPS) matrix prevents the diffusion of antibiotic compounds into the biofilm. This prevents the conventional approach of medicines (Ro et al. 2006). Thus, biofilm is considered as a target for pharmacological development. Recent studies have shown that natural agents having plant secondary metabolites can disrupt biofilm. Moreover, plant extracts are considered to be the safest as they are naturally derived and do not harm the host tissues surrounding the biofilm while acting upon it. This chapter would focus on various phytocompounds and natural compounds which play a potent role in the mechanism of inhibiting the biofilm.

9.2 Inhibition of Quorum Sensing and Signalling Pathways

Quorum sensing (QS) is the ability of the bacterial cells to sense bacterial density by cell-to-cell signalling using autoinducers. These autoinducers cause aggregation of the biofilm-forming cells by extracellular polymeric substances. QS is the mechanism by which both Gram-positive and Gram-negative bacterial cells develop biofilm by producing signals that are comprised of small peptides, acyl homoserine lactones (AHL), and quinolones. The AHL-dependent QS has three components: synthase protein, LUXI family homoserine lactone signal, and LuxR family. The Lux R is AHL-responsive DNA binding transcriptional regulator. In contrast to this, conserved rings linked by amide bonds are found in acyl homoserine lactone signal (Papenfort and Bassler 2016). The production of AHL with the help of LUX I protein occurs from S-adenosyl methionine (SAM) and acetylated acyl carrier protein (ACP) from fatty acid biosynthesis pathway (Schaefer et al. 1996). The AHL produced by LUX I are autoinducers which increase with the increase in cell population density. The binding of LUX-R proteins with AHL causes increase in the concentration of autoinducers resulting in the activation of target gene transcription (Engebrecht et al. 1983). The Gram-negative group of bacteria can efficiently couple gene expression during fluctuations in cell population density. The density-dependent quorum sensing in Gram-positive bacteria occurs with the help of peptide being secreted via ATP-binding cassette (ABC) transporter. It also has two component adaptive response proteins for detecting autoinducers. The signalling in these groups of organisms usually occurs with the help of phosphorylation or dephosphorylation cascade. The peptide signals are detected with the help of sensor kinases. In Gram-positive bacteria, AIP are released out of the cell and upon reaching a suitable concentration, interacts with the kinase receptors. Kinase receptor activation results in phosphorylation events. (Miller and Bassler 2001) (Table 9.1).

The mechanisms of QS result in the development of virulence, non-virulence, and survival virtues of the bacteria. Hence, it attracts the attention of global research to dissect the probable underlying mechanism of quorum signalling, which may help to

Table 9.1 Various compounds acting as quorum sensing inhibitors

Anti-quorum sensing agent	Mode of action	References
Halogenated furanones from *Delisea pulchra*	Inhibit the quorum sensing pathway by competitively inhibiting the LuxR protein responsible for QS and thus lead to proteolytic degradation without killing the bacterial cells leading to the diminishing of the biofilm layer	Choo et al. (2006)
Anacardic acid mixture from *Amphipterygium adstringens*	It decreases the elastase activity, thus reducing QS activity causing inhibition of biofilm	Castillo et al. (2013)
Piericidin A and glucopiericidin A isolated from *Streptomyces xanthocidicus*	It suppresses the activity of the virulent genes like pelC, pehA, nip, and celIV and thus inhibits the process of QS in organisms like *Erwinia carotovora* which is a plant pathogen	Kang et al. (2016)
Malabaricone C obtained from *Myristica cinnamomea*	It prevents the synthesis of violacein in *Chromobacterium violaceum*. It also prevents the QS in *Pseudomonas aeruginosa* by inhibiting the production of phycocyanin	Chong et al. (2011)
Flavone a group of flavonoid compound	It inhibits the formation of QS-dependent biofilm formation with *P. aeruginosa*, performs proteolytic degradation within *Agrobacterium tumefaciens*, and also inhibits the TraR receptor within the *E. coli*	Davies et al. (1998)
Flavanones, taxifolins, naringenin, and eriodictyol from *Combretum albiflorum*	It inhibits QS by inhibiting the production of phycocyanin and elastase in *P. aeruginosa*. The compounds like naringenin and taxifolin inhibit the QS gene like lasI, lasR, rhlR, lasA, lasB, and rhlA	Vandeputte et al. (2011)
Quercetin	It inhibits the QS by competitively inhibiting the LasR receptor pathway	Ouyang et al. (2016)
Catechin isolated from *Combretum albiflorum*	It inhibits the expression of QS regulating genes like lasI, lasR, rhlI, and rhlR. It has the ability to inhibit *N*-butanoyl-L-homoserine lactone by RhlR leading to the presentation of quorum sensing	Vandeputte et al. (2010)
Chestnut honey	Reduces QS by preventing the secretion of acyl homoserine lactone (AHL)	Asfour (2018)
Clove oil	It acts as anti-quorum sensing therapeutics by inhibiting virulence gene LasB, total protease, total chitinase, and swimming mortality	Husain et al. (2013)

understand the mechanism of autoinducer-mediated cell-to-cell communication and support in designing modulators of the quorum signalling. For example, Gram-positive bacteria use peptides processed from precursors as autoinducers (Sturme et al. 2002). There are two types of autoinducing peptides (Sifri 2008), listed in

Table 9.1 (Sifri 2008). On the other hand, Gram-negative bacteria use *N*-acyl homoserine lactone (AHL) autoinducer signal molecules for quorum sensing (Eberl 1999; Lade et al. 2014). Three core components of all AHL-based quorum sensing systems are *N*-(3-oxooctanoyl)-L-homoserine lactone (OOHL), *N*-hexanoyl-L-homoserine lactone (HHL), and *N*-decanoyl-L-homoserine lactone (DHL) in *E. chrysanthemi* (Eberl 1999).

9.2.1 Quorum Quenching (QQ)

It is a unique mechanism of disrupting the communication among the various microcolonies, i.e., by inhibiting the mechanism of quorum signalling that varies among various species of bacteria (Williams 2002; Lazar 2011). This can be accomplished by (1) inhibiting the production of autoinducers or introduction of AI antagonist (Smith et al. 2003; Suga and Smith 2003), (2) inhibiting the binding of autoinducers with receptors (Shiner et al. 2006), and (3) degradation of autoinducers (Fuentes et al. 2016) (Table 9.2).

1. **Inhibiting the production of AI**—The three types on enzymes which possess the ability of degrading AHL are AHL-lactonases which possess the ability of hydrolysing the lactone moiety, AHL-acylase that possesses the ability of hydrolysing the amide bond, and AHL oxidoreductase that brings about oxidation of the third carbon being present within the AHL molecules (Dong and Zhang 2005; Dong et al. 2002). QQ based on AHL inhibition is found to be used by Gram-negative bacteria, where the enzymes like acylase, lactonase, and oxidoreductase can preferably inhibit AHL, which prevents the accumulation of AHL in the extracellular environment, and subsequent suppression of quorum sensing-related gene expression [35]. In *Bacillus* species, AiiA enzyme assists in the hydrolysis of AHL; similarly, AhlK protein in *Klebsiella pneumoniae* is responsible for the degradation of AHL [36–38].
2. **Inhibition of autoinducer receptors**—Interference of QS by blocking signal production is not very common (Hentzer and Givskov 2003). In another approach, preventing interaction of AHL with its receptor will also disrupt the signalling cascade. Inhibition can be competitive inhibitor that competes with AHL for the same binding site on the receptor surface (Petersen 2016). Molecules that mimic the structure of AHL by alteration in the acyl side chain or lactone ring can be considered for this purpose (Lucas et al. 1993). These molecules should ideally have a higher binding affinity on the receptor interface as compared to AHL molecule and should not trigger the signalling cascade upon binding.
3. **Degradation of the autoinducers**—Certain types of bacteria and mammalian cells possess the ability to degrade AHL in the presence of various enzymes. It has been observed that various phylogenetically diverse AHL-inactivating bacteria like *Klebsiella*, *Alteromonas*, *Bosea*, *Brevundimonas*, *Achromobacter*, *Alcaligenes*, *Shewanella*, *Sphingomonas*, *Stenotrophomonas*, etc. are grouped as QQ bacteria (Dong and Zhang 2005).

Table 9.2 Inhibition of quorum sensing

Type of input signal	Mechanism of action	Organisms	References
Host signalling molecules			
AHL (Las)	Inhibiting the formation of EPS subsequent to a repression of the DGC TpbB in the presence of phosphatase TpbA	*P. aeruginosa*	Ueda and Wood (2009)
AHL (LasI/LasR and RhlI/RhlR)	The synthesis of rhamnolipid reduces the surface tension	*P. aeruginosa*	Davey et al. (2003)
DSF (*RpfF* synthesis)	Degrading the endo-β-1,4-mannanase being present within the matrix	*Xanthomonas campestris*	Dow et al. (2003)
AIP (*agr* system)	Phenol-soluble modulins (PSM) help in reducing the surface tension	*S. aureus* and *S. epidermidis*	Wang et al. (2011)
AIP (*agr* system)	Proteolytic activity of aureolysin that brings about protein degradation of the matrix	*S. aureus*	Boles and Horswill (2008)
Nutrient starvation			
Carbon source	Cleaves the surface adhesion LapA by proteinase cysteine and LapG brings about the reducing of c-di-GMP, thus regulating the formation of the biofilm Reducing the surface tension by the production of the lipopeptide viscosin	*Pseudomonas putida*, *Pseudomonas fluorescens*	Gjermansen et al. (2010) Bonnichsen et al. (2015)
Oxygen, glucose, nitrogen	Decreases the level of c-di-GMP	*P. aeruginosa*	Schleheck et al. (2009)
Nitrogen and glucose	Enhances the hydrophobicity of the cells	*P. fluorescens*	Delaquis et al. (1989)
Iron	Enhances the production of rhamnolipid, thus reducing surface tension	*P. aeruginosa*	Glick et al. (2010)
Excess amount of nutrient availability			
Heavy metals and carbohydrate source	Reduction in c-di-GMP concentration by the activation of BdlA by DGC GcbA	*P. aeruginosa*	Petrova et al. (2015)
Mercury chloride, ammonium chloride, and glutamate	Enhances the swarming motility and by the activation of PDE DipA	*P. aeruginosa*	Roy et al. (2012)
Succinate, silver, glutamate, and mercury arsenate	Enhances the swarming motility by the activation of RbdA and DipA	*P. aeruginosa*	Morgan et al. (2006)
Iron	Inhibition of biofilm by downregulating the biofilm-forming genes	*P. aeruginosa*	Musk et al. (2005)

4. **Inhibition of surface adhesion initiation**—The mechanism of alteration in the topography of the surface or the release of nitric oxide (NO) brings about inhibition of bacterial adhesion and formation of biofilm (Xu et al. 2017). The alteration of the topography of the surface prevents the mechanism of bacterial attachment to the surface and thus reduces the chances of the formation of the biofilm (Estrela et al. 2009).
   (a) **Targeting and modulating second messenger signalling molecule**—The formation of biofilm is being controlled by the group of secondary messenger molecules like guanosine tetraphosphate and guanosine pentaphosphate which predominantly participates in the mechanism of QS (Ghosh et al. 2020). Another type of messenger molecule that plays an important role in the mechanism of biofilm formation is cyclic bis(3′-5′)diguanylic acid (c-di-GMP) that regulates the transition of the bacterial cells from the planktonic to the sessile forms (Ghosh et al. 2020).
   (b) **Inhibition of biofilm maturation**—Various types of bioactive compounds like linezolid play an important role against the methicillin-resistant *Staphylococcus epidermidis* (Reiter et al. 2013). It has been further observed that the use of zinc-oxide nanoparticles along with titanium implants prevents the formation and maturation of the biofilm. Copper ions have also showed their efficacy as antibacterial agents and also play an important role in preventing bacterial adhesion upon the surface.
   (c) **Disruption of the mature biofilm**—There are various mechanisms by which various intracellular signals are inhibited, thus preventing the formation of the biofilm. Various conditions express the inhibition of the secondary messengers and other signal molecules which in contrary prevents the formation of the biofilm.

9.3 Other Interacting Members in a Bacterial Biofilm

Biofilm is a complex organization of microbial cell along with extracellular polymeric substances (EPS); recently, some are using the term "matrixome" for EPS synonymously (Karygianni et al. 2020). Among the EPS proteins, DNA, polysaccharide, and RNA are the significant components (Sutherland 2001). EPS ensures nutrient flow into the biofilm and survival of the microbes forming the biofilm (Prakash et al. 2003). Destabilization of the EPS can be used as a drug target; for that, we have to understand the complex organization of the biofilm matrixome. EPS is regarded as one of the vital components of biofilm, extensively characterized in the field of dental medicine studies (Jain et al. 2013; Saini et al. 2011).

EPS composition may vary from organism to organism local shear stress availability of nutrients and the growth environment. Furthermore, the nature of constituent microbes in a biofilm is an essential determinant of the composition of biofilm

EPS (Di Martino 2018). With the advancement in biophysical techniques, molecules of diverse nature and characteristics have been identified in the last few decades. Majorly these can be classified into two groups according to their location in the biofilm.

1. **Cell Surface-Attached EPS Molecules**
   Cell surface-attached EPS molecules are allied with cell motility functions, evident from their close association with cell motility appendages like flagella and pili (Wood et al. 2006; Guttenplan and Kearns 2013). Moreover, evidence supports the role of these molecules in functional amyloid formation, which provide mechanical stability by modulating the cell adhesion in biofilm (Erskine et al. 2018).
2. **Molecules Secreted to EPS**
   Other groups of molecules which are secretory such as the exopolysaccharides, proteins, RNA, and DNA act as a matrix scaffold (Flemming et al. 2016; Saxena et al. 2019). Besides bacterial molecules, host proteins are present in the EPS, which assist in microbial attachment and are also a source of the nutrient. The EPS components are essential to characterize the biofilm, both physically and chemically. The nucleic acid is a vital constituent of the majority of bacterial and fungal biofilm, and the extracellular DNA (eDNA) is scaffolding in nature and used for communication among the constituent microbes in the biofilm (Dominiak et al. 2011; Das et al. 2013). The eDNA is often used as a medium of gene transfer and modulates the interaction with other matrix components. Lipid components of EPS also play a significant role in pathogen survival and virulence; e.g., lipopolysaccharide (LPS) found in Gram-negative bacteria acts as the endotoxin. LPS molecules in Gram-negative bacteria also have these key roles (Emody et al. 2003; Bandara et al. 2010). Apart from LPS, other lipids which play a vital role in the EPS organization, such as teichoic acid in *S. aureus*, have an essential role in adhesion-cohesion, immune evasion, and protection of the biofilm (Gross et al. 2001; Kohler et al. 2009). Major constituents are protein and polysaccharides, and a list of identified protein and polysaccharide EPS components in different model bacterial biofilm is provided in Table 9.3.

It is evident from Table 9.3 that EPS molecules provide mechanical strength, adhesion, and cohesion to the growth surface. In addition to the mechanical properties, these molecules are responsible for the spatial organization, physical and social interactions, enhanced immune evasion, chemical heterogeneity, and antimicrobial tolerance. Specifically, these molecules can be targeted for therapeutic interventions. As EPS provides mechanical and nutrient support along with immune evasion and antimicrobial tolerance, degrading the EPS with enzyme or phytochemicals, it may disassemble the matrix of the biofilm, which will result in weak biofilm spatial organization, easing the vulnerability toward antimicrobial agents and the immune system. It may also hamper the continuous nutrient support to the constituting microbe in the biofilm and slow down the spread of biofilm. For

Table 9.3 Biofilm model organism with their extracellular polymeric substance (EPS) components and the role of those molecules in the EPS

Pathogen	EPS molecule	Class of EPS molecule	Location and function	References
Bacillus subtilis	Exopolysaccharide	Polysaccharide	Extracellular, adhesion, and scaffolding role	Harimawan and Ting (2016)
	γ-PGA (poly-γ-glutamate)		Extracellular, adhesion, scaffolding, and provide nutrition	Yu et al. (2016)
	Biofilm surface layer protein (BslA)	Protein	Extracellular, increases surface hydrophobicity, plays a protective role	Kobayashi and Iwano (2012)
	TasA anchoring and assembly protein (TapA)		Found on the cell wall, plays a scaffolding role	Romero et al. (2011)
	Flagellum		Associated with cell, role in motility, and mechanosensing	Houry et al. (2010)
Pseudomonas aeruginosa	Psl	Polysaccharide	Cell associated and extracellular, protection against the immune response	Ma et al. (2006)
	Pel		Cell associated and extracellular, protection against antibiotics	Colvin et al. (2011); Colvin et al. (2012)
	Alginate		Cell associated and extracellular, protects biofilm	Leid et al. (2005)
	T4P (Type IV pilins)	Protein	Associated with cell, role in motility, and mechanosensing	Giltner et al. (2006)
	Lectins		Cell associated and extracellular, role in cell-to-cell connectivity, stability, and cytotoxin	Diggle et al. (2006)
Staphylococcus aureus	Poly-β(1–6)-*N*-acetylglucosamine (PNAG)	Polysaccharide	Extracellular, role in cohesion and adhesion, provide stability and protection against antibiotics	Lin et al. (2015)

(continued)

Table 9.3 (continued)

Pathogen	EPS molecule	Class of EPS molecule	Location and function	References
	Fibronectin-binding proteins (FnBPs)	Protein	Cell associated and extracellular, role in cell-to-cell connectivity	Houston et al. (2011)
	S. aureus surface protein G (SasG)		Cell associated and extracellular, role in cell-to-cell connectivity	Geoghegan et al. (2010)
	Staphylococcal protein A (SpA)		Cell associated and extracellular, role in immune evasion	Chippaux and Puchelle (1994)
	Biofilm-associated protein (BAP)		Extracellular, scaffolding	Merino et al. (2009)
	Phenol-soluble modulins (PSMs)		Extracellular, act as proinflammatory, toxic to host cell, assist in spreading biofilm	Cucarella et al. (2004)
Candida albicans	α-Mannans and β-glucans	Polysaccharide	Extracellular, role in the formation of mannan-glucan complex, protection, antifungal resistance, and scaffolding	Schwartz et al. (2012)
Mycobacterium tuberculosis	Complex sugars	Polysaccharides	Extracellular, role in anchoring	Pierce et al. (2017)
	Mycolic acid	Lipids	Cell wall lipid, role in attachment and virulence	Trivedi et al. (2016)

targeting the EPS molecules in a biofilm, it is necessary to understand the mechanism of formation of biofilm.

9.3.1 Role of EPS Molecules in the Formation of Biofilm

The primary role of EPS molecules in a biofilm is to establish the cell-to-cell communication and nutrient supply along with providing mechanical support, which is emphasized in existing reports (Huber et al. 2001). Adhesion and cohesion properties imparted by EPS molecules lead to the formation of cell clusters in a characteristic spatial arrangement. Present knowledge of EPS molecules of biofilm has developed from extensive studies on biofilm model organisms like *Staphylococcus aureus*, *Streptococcus mutans*, *Escherichia coli*, *Candida albicans*, *Vibrio cholerae*, *Bacillus subtilis*, and *Pseudomonas aeruginosa*. Here we emphasize on the precise role of EPS molecules in the formation of biofilm to understand the process and utilization of specific step as an antibiofilm target.

Initially, adhesion of the bacterial species to the surface occurs via an adhesin-receptor interaction, surface scanning, and sensing phenomena (Beachey et al. 1988). Some of the prominent examples are the attachment of *B. subtilis* and *V. cholerae* through mechanosensing with flagella (Cairns et al. 2013). Interestingly, protein BAP1 in *V. cholerae* is secreted by the founder cell and remains bound to the attachment surface in mature biofilm as demonstrated by the single-cell imaging technique (Yan et al. 2016), while pioneering *S. mutans* cells sense the external adhesion forces through quorum sensing (Wang et al. 2019).

With the attachment of pioneering cells to the substratum, cohesive EPS entities facilitate cell aggregation to proceed with the biofilm formation further. In contrast, other sensing mechanisms mediate interspecies interactions. In general, the biofilm formation with EPS proceeds by adhesion of pioneering cells to the substratum, followed by the formation of EPS at the cell surface of attached cells and by secretion to the adjacent substratum. Formation of EPS and release of AIs facilitate cell clustering around the attachment surface. With the increase in the cell population, the EPS production also increases and leads to the formation of the 3D matrix. Within the 3D matrix, the bacterial cell cluster is formed. Growing 3D matrix supports attachment of more microbes and their clustering, which are termed as microcolonies (Flemming et al. 2007). It has been observed that the process of biofilm microcolony formation is conserved in various biofilm-producing microbial models, which indicates an evolutionary relationship. Enhanced survival of microbes in biofilm is of ecological and evolutionary importance as it is found in diverse microbial species. Shreds of evidence exist from genetic studies, supporting parallel evolution which enhances bacterial survival and productivity (Steenackers et al. 2016).

EPS molecules like glucosyltransferases of *S. mutans* and TasA of *B. subtilis* mediate interspecies colonization in microbial biofilm (Liu et al. 2017; Duanis-Assaf et al. 2018). The occurrence of multiple species in a biofilm is advantageous in the survival of the colonizing species, as it helps to recruit microbial species lacking EPS formation machinery. Some of the proteins in EPS are capable of forming amyloid fibers such as phenol-soluble modulins (PSMs) (Tayeb-Fligelman et al. 2020). The amyloid fibers provide spatial support for the growth and progression of microbial colonization. In general, all known amyloid fibers adopt a cross-β sheet structure, but recently an exception found in the case of *S. aureus* PSMα3 shows a cross-α amyloid-like fibril (Tayeb-Fligelman et al. 2020). Examples for other functional amyloid associated with biofilms can be observed in Bap fibers of *S aureus*, Fap fibers of *P aeruginosa*, and P1 adhesin of *S. mutans* (Erskine et al. 2018). Characterization of *P. aeruginosa* biofilm with the assistance of super-resolution microscopy revealed that prophage endolysin-mediated explosive cell lysis assists in the release of eDNA and other EPS molecule of biofilm (Turnbull et al. 2016).

Apart from the mechanical support to growing microbial colonies in biofilm, EPS also provides the nutrients and retards the penetration of antimicrobial agents into the biofilm. The antimicrobial protection is achieved by hindrance in diffusion and limiting more in-depth entry into the biofilm. Moreover, the interaction of antimicrobial agents with the EPS components further slows down the process of penetration, which aids in the enzymatic degradation of the applied therapeutic agent. β-Glucans of fungus *C. albicans* protect co-colonizing bacteria *S. aureus* in the biofilm against an antibacterial drug, which demonstrates a mixed-culture crossed-kingdom protection through extracellular EPS molecule (Kean et al. 2017). On the other hand, glucans from *S. mutans* are shown to sequester the antifungal agents like fluconazole (Kim et al. 2018). Several other examples can be found in literature where EPS material secreted by one species is protecting other species in the mixed-culture biofilms. However, a detailed mechanism for the process of antimicrobial drug neutralization still lacks that needs attention to thorough exploration for better understanding of the mechanism. Reports exist supporting the involvement of the EPS components of biofilm in enhancing the virulence of the pathogen. As listed in Table 9.3, *S. aureus* polysaccharide PNAG and protein EPS component SpA are involved in enhanced virulence of the bacteria. Cytokine elevation in cystic fibrosis patient is caused by EPS component and not by the *P. aeruginosa* itself (Bielen et al. 2017). Collectively the mutual protection in a mixed cell biofilm, nutrient supply, and EPS mediated-virulence require special attention to design and explore antibiofilm agents.

It is a daunting task to treat a biofilm infection with the conventional therapeutic approach. Treatment requires multidrug therapy with consideration to various targets. It is because of the complex physical, chemical, and biological nature of EPS. Targeting the vital functional role of EPS in a biofilm can help in the removal of the biofilm or weaken to make it susceptible to the host immune system. Also, EPS production can be targeted, which is regulated by molecules involved in signalling or by non-signalling mechanisms. Thereby targeting such molecules either disables them to carry out their vital function or may cause their degradation. Degradation of the signalling molecule will prevent further proceeding of the signalling cascade. For example, inhibition or degradation of adhesion molecules will prevent the biofilm formation itself. Here it is challenging to completely inhibit the biofilm formation, as a single species utilizes more than one adhesion molecule or receptor-mediated surface attachment (Richards and Melander 2009). Hence it becomes crucial to understand the surface attachment process completely and target more than one adhesion process. Further, multispecies biofilm formation complicates the case more.

As discussed in the earlier section, apart from the attachment machinery, the quorum sensing molecules can also be targeted. So far, numerous phytochemicals and natural antimicrobial agents have been tested on bacterial biofilm to evaluate their potency and recent advancements on potential antibiofilm agents from natural sources and different phytochemicals.

9.4 Natural Antibiofilm Compounds

The ability of the microbe to resist the efficacy of drug molecules is termed as antimicrobial resistance (AMR) (Prestinaci et al. 2015). Bacteria like *Streptococcus mutans*, *Enterococcus faecalis*, *Actinomyces naeslundii*, *Lactobacillus fermentum*, *Prevotella buccae*, *Actinomyces viscosus*, and *Propionibacterium acidifaciens* characterized as biofilm-forming bacteria are also sensitive toward antibiotics like amoxicillin, penicillin G, clindamycin, vancomycin, doxycycline, fosfomycin, gentamicin, rifampicin, ciprofloxacin, tetracycline, and moxifloxacin. Antimicrobial resistance arises via three mechanisms: (a) The first is by a specific gene product (enzyme) that protects the bacterial cell from antibiotic (Abraham and Chain 1988). The enzyme β-lactamase, first identified in *Escherichia coli* (Lautenbach et al. 2001), inhibits the activity of a group of antibiotics like penicillin by denaturing its β-lactam ring. Another group of antibiotics of aminoglycoside-aminocyclitol family is also inhibited by an antimicrobial-resistant element present in bacteria (Wright 1999). (b) The second is by a protein with altered structure synthesized due to the chromosomal mutation that leads to the changes in protein synthesis and forms mosaic protein through transformation. (c) The third is by the incorporation of a new genetic material into the same or different species or genera (Spratt 1994), caused by the horizontal as well as vertical gene transfer. Mutation results in the generation of antibiotic resistance genes, as, for example, mutation through base substitution of 23S rRNA gene or within the ribosomal protein L4 and L22 results in macrolide resistance (Vester and Douthwaite 2001). Random mutations in microbial gene sequences result in random mutations. It is also observed that prolonged exposure to antimicrobials encourages selective mutations providing resistance and spontaneous occurrence of random mutations and extended use of antimicrobials promotes selection for mutations rendering antimicrobials ineffective (Long et al. 2016).

Since the identification of biofilm and its link to human diseases, the scientific community has explored the natural resources to find a suitable antibiofilm compound for a specific type of disease-associated biofilm. Microbes in a typical biofilm exhibit elevated adaptive resistance to antibiotics in comparison to their planktonic form, which makes developing a drug more challenging. Herbs are used for the treatment of several ailments since the early days of human civilization, and exploring the traditional herbal knowledge base for finding an antibiofilm drug appears a more feasible approach. Traditional medicine from herbs probably provides a drug for fighting biofilm-associated infections. Recently extracts from a variety of plant species have demonstrated antibiofilm activity.

All the stages of biofilm development can be targeted, i.e., molecules inhibiting the polymeric matrix or cell surface proteins having a role in attachment can inhibit the initial attachment and the later irreversible attachment. Different molecules targeting the matrix components such as nucleic acid and carbohydrate can inhibit the quorum sensing, and inhibition of autoinducers can inhibit the biofilm maturation. Such activities have been reported from several plant products from in vitro culture. Primarily the activity of such plant-derived drugs is observed due to the prevention of matrix formation and attachment, suppressing quorum sensing, or by inhibiting the synthesis of virulence factor. Plant extracts of *Cocculus trilobus*, *Ginkgo biloba*, *Coptis chinensis*, and many others have demonstrated antibiofilm activity. Plant extracts or purified molecules and their precise mechanism of inhibition in microbial species are listed in Table 9.4.

N-(Heptylsulfanylacetyl)-L-homoserine lactone (HepS-AHL) from garlic was found to have potent quorum sensing inhibition by transcription regulation of LuxR and LasR genes in several clinical isolates (Lu et al. 2019). Commonly used food additives like garlic have several other antimicrobial molecules. A study in mice model of infection with *P. aeruginosa* demonstrated that garlic extract could increase sensitivity toward the host immune system as well as antibiotics (Bjarnsholt et al. 2005). Garlic extract has also exhibited antibiofilm activity in *P. aeruginosa* UTI infection mice model (Harjai et al. 2010), while mixed extracts and loading garlic extract to nanoparticle have revealed enhanced activity against biofilm and microbes. Upon loading the garlic extract to silver-based nanoparticles, it shows enhanced antibiofilm activity in comparison to free garlic extract (Girish et al. 2019). In a similar finding, garlic clove extract-loaded silver nanoparticle has also shown enhanced antibiofilm activity, while it was found non-toxic toward human HEK293 embryonic cell line (Vijayakumar et al. 2019). The mechanism of action of nanoparticle loaded with garlic extract is still unknown.

While inhibition activity in specific plant extracts such as *C. trilobus*, *Coptis chinensis*, and *Humulus lupulus* prevents the adherence to substratum in biofilm-forming bacterial models (Rozalski et al. 2013), adhesion prevention of these plant extract is mediated by suppressing the membrane-associated sortase enzyme. Sortase group of enzymes are mostly found in Gram-positive bacteria, which are linked covalently with the peptidoglycan membrane (Cossart and Jonquières 2000). Alcoholic leaf extract of *Hymenocallis littoralis* has shown antibiofilm activity. It is rich in apigenin 7-(4″,6″-diacetylalloside)-4′-alloside, emodic acid, and catechin 7-*O*-apiofuranoside along with several other alkaloids, terpenes, terpenoids, flavonoids, and phenolic compounds (Nadaf et al. 2018). The computational study indicated anti-adhesion, sortase inhibition, and agglutinin inhibition in *S. aureus* and *C. albicans*, while histological experiments revealed the extract was non-toxic to mice skin (Nadaf et al. 2018) (Table 9.4).

In *P. aeruginosa*, six essential biofilm-associated genes were studied with a designed luxCDABE-based reporter system. Among several plant extracts, the *H. patriniae* extract exhibited inhibition of most of the biofilm-associated gene (Kim et al. 2004). Interestingly the study has shown activity on mature biofilm and also reduces EPS polysaccharide formation (Kim et al. 2004). *Ginkgo biloba* extract has demonstrated antibiofilm activity for *Salmonella*, *Listeria*, *E. coli* O157: H7, and *S. aureus* (Fu et al. 2017). In *E. coli*, it was shown the ginkgolic acid in the extract repressed prophage and curli genes, resulting in reduced fimbriae production and ultimately preventing biofilm formation (Wu et al. 2016). Antioxidant compound phloretin, which is abundantly found in apple, has shown similar fimbria production inhibition in *E. coli* O157:H7, along with prevention of attachment on the colon epithelium in a rat model. Molecular-level study on phloretin-based inhibition showed gene repression of AI importer gene, curli genes, and toxin production

Table 9.4 List of some plant extract and molecules tested against biofilm-forming microbe with the mechanism of inhibition

Molecule or extract	Natural source	Target species	Mechanism	References
HepS-AHL (*N*-(heptylsulfanylacetyl)-L-homoserine lactone)	*Allium sativum*	*P. aeruginosa*	Protease inhibition, transcription regulation	Rasch et al. (2007)
Ethyl acetate extract of *Cocculus trilobus*	*Cocculus trilobus*	Gram-positive bacteria	Sortase activity prevents adhesion	Kim et al. (2004)
Herba patriniae extract	*Herba patriniae*	*P. aeruginosa*	Inhibit biofilm by regulating EPS production at the molecular level	Fu et al. (2017)
Ginkgolic acid	*Ginkgo biloba*	*E. coli*	Prevents attachment to certain substratum	Lee et al. (2014a)
Phloretin	*Malus domestica*	*E. coli* and *S. aureus*	Repression of vital genes for biofilm formation	Lee et al. (2011)
Cinnamaldehyde	*Cinnamomum zeylanicum*	*E. coli* and *V. cholerae*	Reduces stress response, motility, and virulence	Amalaradjou et al. (2010)
Hordenine	*Hordeum vulgare*	*P. aeruginosa*	Quorum sensing inhibition	Zhou et al. (2018)
Quercetin	*Raphanus sativus*	*P. aeruginosa*	Molecular inhibition of biofilm formation and virulence factor production	Vipin et al. (2019)
Epigallocatechin-3-gallate (EGCG)	*Camellia sinensis*	*Stenotrophomonas* sp.	Reduced cell viability	Adegoke et al. (2017)
Erianin	*Dendrobium chrysotoxum*	*S. aureus*	Inhibition of sortase prevents cell adhesion	Ouyang et al. (2018)
Parthenolide	*Tanacetum parthenium*	*P. aeruginosa*	Quorum sensing inhibition	Kalia et al. (2018)
Chloroform extract	*Aspergillus fumigatus*	Various microbes	Inhibition of initial cell attachment and disruption of mature biofilm	Kaur et al. (2020)
Methanolic extract	*Zingiber officinale*	*S. mutans*	Inhibition of surface antigen	Giriraju and Yunus (2013)

genes. Phloretin also inhibited biofilm in *S. aureus* RN4220 strain and *Salmonella typhimurium* (Shuai-Cheng et al. 2016).

Reduced stress response, swarming motility, and virulence are reported in the case of *E. coli*, *P. aeruginosa*, and *V. cholerae* upon treatment with cinnamaldehyde, which is abundantly present in cinnamon (Vasconcelos et al. 2018; Niu and Gilbert 2004; Topa et al. 2018; Albano et al. 2019). Hordenine extracted from germinating barley has shown potent activity against *P. aeruginosa* PAO1 biofilm by quorum sensing inhibition. In comparison, loading hordein on anisotropic gold nanoparticles has potentiated the antibiofilm activity of hordein (Rajkumari et al. 2017).

Quercetin isolated from the hairy root extract of radish has demonstrated antibiofilm activity against *P. aeruginosa* (Muthusamy and Shanmugam 2020). Quercetin and other flavonoids have shown inhibitory activity in various microbial biofilms (reviewed in details elsewhere) (Memariani et al. 2019). Parthenolide extracted from feverfew plant exhibited potent quorum sensing inhibition by downregulating AI synthase and AI-associated receptor molecules. Organic solvent extracts of a fungus *Aspergillus fumigatus* reported initial inhibition of attachment to substratum in various biofilm-forming species, while that of zinger inhibited surface antigen in *S. mutans* leading to biofilm formation inhibition.

List of natural products exhibiting antibiofilm activity is quite long, and a mechanistic overview is shown in this chapter. However, the traditional knowledge of herbal and natural drugs is far more exhaustive, and it is needed to examine their antibiofilm activity to find promising drug candidates.

9.4.1 Antimicrobial Resistance Versus Phytotherapy

The biggest problem threatening the global public health is the development of antimicrobial resistance due to the rampant use of antibiotics, adaptations to antibiotics, and genetic evolution. This resulted in the exploration to alternative therapies to combat against antibiotic-resistant organisms, and hence the development of drugs from natural sources becomes a major interest. Although the majority of the existing antibiofilm agents are non-biocidal, some bactericidal molecules could be considered as antibiofilm agents. In the following section, we will focus on naturally derived compounds having therapeutic activities and their effects on biofilm formation.

9.4.2 Phytochemicals and Biofilm

The emergence of multidrug resistance (MDR) bacteria has initiated the post-antibiotic era in clinical infectious disease treatment. Current antibiotic resistance scenario has compelled the scientific community to identify a potent alternative to antibiotic-based therapeutics. Moreover, the evolution of bacteria exhibiting biofilm forming new clinical isolates makes it more complicated (Vidal et al. 1998). It becomes complicated to treat MDR infection, which is in biofilm form (Drenkard 2003). Because of growing antibiotic resistance, plant secondary metabolites can become an alternative for treatment, which appears promising from in vitro tests. Many of the plant extracts and their constituents are studied for antibiofilm activity. Secondary metabolites such as terpenes, phenolics, and nitrogen−/sulfur-containing compounds have shown antibiofilm activity.

Terpenes and terpenoids are known for their antimicrobial activity against both Gram-positive and Gram-negative bacteria. In the context of antibiofilm phytochemical, phenol monoterpenes like carvacrol have shown activity against *S. aureus* and *S. typhimurium* (Knowles et al. 2005), while acyclic sesquiterpene alcohol farnesol demonstrated moderated antibiofilm activity against *Streptococcus* sp. biofilm (Koo et al. 2002). Another study showed enhanced activity of farnesol when added with xylitol against *S. aureus* biofilm (Katsuyama et al. 2005). Xanthorrhizol, which is a phenol sesquiterpene, showed in vitro anti-adherence activity against *S. mutans* biofilm (Rukayadi and Hwang 2006). It was reported that the presence of sesquiterpenes potentiated antibiotic activity against MDR *S. aureus* under in vitro condition (Gonçalves et al. 2011). Diterpene isolated from the root of *Salvia sclarea*, namely, salvipisone and aethopinone, demonstrated antibiofilm activity against *S. aureus* and *S. epidermidis* (Kuźma et al. 2007). When treated to methicillin-resistant *S. aureus* (MRSA) and methicillin-resistant *S. epidermidis* (MRSE), these diterpenes potentiated the activity of antibiotics, such that the minimum inhibitory concentration (MIC) of antibiotics like oxacillin, vancomycin, and linezolid almost reduced by 50% (Walencka et al. 2007). Triterpenes like olenic acid and ursolic acid have shown biofilm inhibition against *S. mutans*, *S. sobrinus*, and *Listeria monocytogenes* (Liu et al. 2015; Kurekci et al. 2013; Cunha et al. 2010).

Apart from the terpenes class of phytochemicals, plant phenols also have shown antimicrobial and antibiofilm activity. Plant phenols are a large group of compounds having bioactive properties; it includes tannins, flavonoids, lignans, anthocyanins, coumarins, phenolic acids, and stilbenes. These molecules have a role in plant defense mechanism and also have shown activity against microbes. In the context of antibacterial activity, phenolic phytochemicals have exhibited interaction with membrane proteins and membrane disruption and reduced membrane fluidity, are involved in the inhibition of nucleic acid synthesis and cell wall synthesis, and also demonstrated quorum sensing inhibition to impede the biofilm formation (Gyawali and Ibrahim 2014).

A study showed anthocyanin group of polyphenols like malvidin, petunidin, and cyanidin were showing potent antibiofilm activity against *K. pneumoniae* by regulating EPS production (Gopu et al. 2015). Coumarins are another subgroup of polyphenols which are chromenone with a keto group at 2-position. Some of the coumarins are FDA-approved marketed drugs such as warfarin and calanolide A, which shows the drug potential of coumarins. Studies in the context of antibiofilm activity of coumarins have shown similar potential results, for example, coumarin inhibits biofilm of *E. coli*, *S. aureus*, *Vibrio anguillarum*, and *Edwardsiella tarda* (Lee et al. 2014b). Other coumarins like umbelliferone and coladonin have shown antibiofilm activity against *E. coli* O157:H7 by reducing biofilm formation, while esculetin, psoralen, and esculin reported inhibiting biofilm formation of *P. aeruginosa* (Zeng et al. 2008). Flavonoids also exhibit antibiofilm activity, for example, quercetin (discussed in the earlier section of this chapter) has demonstrated antibiofilm activity on *E. coli*, *S. mutans*, and *S. aureus* (Zhang et al. 2014). Some of the flavonoids like nobiletin and sinensitin inhibit swarming motility in *E. coli* biofilm (Vikram et al. 2010). Moreover, tannins like catechin, EGCG, gallic acid, methyl gallate, tannic acid, and rosmarinic acid also show antibiofilm activity on different biofilm-forming species (Rendeková et al. 2016).

With the emergence of multidrug resistance and biofilm-forming strains of microbial species, a safe and effective antimicrobial drug is much needed to reduce the burden on human healthcare. This book chapter has provided glimpses of potential antimicrobial agents taken from nature's library. Concerning biofilm, most of the phytochemicals discussed have shown anti-adhesion or quorum quenching activity. As of now, it is required to examine the efficacy of individual components of plant extract after separating it from the whole extract. In the twenty-first century, it has become possible because of the advancement in separation technology and biophysical tool for characterization of the purified molecules (Table 9.5).

Interestingly, many of these identified molecules are filed for a patent, and few of the clinical studies are ongoing. Further application of cheminformatics to modify the side chain structure of the natural compounds and their activity assessment may provide a potent antibiofilm compound for treating MDR and biofilm-forming strains. Once the safety and efficacy of these natural antibiofilm molecules are established, it can be introduced as a solo or combination therapy to get rid of microbial infections forming a biofilm (Table 9.5).

9.5 Conclusion

Biofilm-associated bacterial infections frequently caused by *Staphylococcus epidermidis, Staphylococcus aureus, Escherichia coli, and Pseudomonas aeruginosa* are found in majority of human diseases. Effectiveness of many antimicrobial drugs has been lost due to the evolution of pathogenic resistance. Many of the microorganisms are no longer susceptible to most of the existing antibiotics and therapeutic agents. Various groups of natural compounds having different medicinal properties have a potential to cure these diseases. In this review, it has been described that compounds isolated from natural sources have the ability to inhibit biofilm formation of different bacterial strains that were isolated from different infection sources such as wound, uro-catheter infections, and septicemia.

Table 9.5 List of phytochemicals having antibiofilm activity

Bioactive compound	Phytochemicals	Mechanism of action	References
Flavonoid	Amentoflavone	Induction of fungal apoptosis	Lee and Lee (2015)
	Isoquercitrin	Disruption of cell membrane	Lee and Lee (2015)
	Catechin	Damage in cell wall	Lee and Lee (2015)
Terpenoid	Carvacrol	Inhibition of cell adhesion, calcium stress	Lee and Lee (2015)
	Eugenol	Disintegration of cell membrane, ergosterol biosynthesis inhibition, influences cytoplasmic permeases	Lee and Lee (2015)
	Caffeic acids	Biofilm inhibition	Lee and Lee (2015)
	Quercetin	Inhibition of cell adhesion	Lee and Lee (2015)
	Galangin	Biosynthesis inhibition	Lee and Lee (2015)
	Apigenin	Damage in cell wall	Lee and Lee (2015)
	Kaempferol	Induction of fungal apoptosis	Lee and Lee (2015)
	O-coumaric acid	Induction of fungal apoptosis	Lee and Lee (2015)
Carotenoid Saponin	Lycopene	Deformation of membrane, induction of fungal apoptosis	Lee and Lee (2015)
	α-Tomatine	Induction of fungal apoptosis, disruption of cell membrane	Lee and Lee (2015)
Polyphenol	Curcumin	Formation of pore on membrane, ROS, inhibition of morphogenetic switch	Moshe et al. (2011)
	Resveratrol	Induction of fungal apoptosis	Lee and Lee (2015)
Lectins and polypeptides Coumarins		Interaction with cell adhesion receptors, formation of ion channels in the membrane	Cavalcante et al. (2014)
	Hydroxycoumarins	Cell membrane damage, antioxidation	Upadhyay et al. (2014)
	Xanthotol, esculetin, daphnetin	Free radical formation	Upadhyay et al. (2014)
Essential oils	Eugenol	Membrane potential disbalance, disintegration of cell membrane	Zhang et al. (2016)
	Thymol	Pore formation on cell wall	Faleiro (2011)

(continued)

Table 9.5 (continued)

Bioactive compound	Phytochemicals	Mechanism of action	References
Tannin	Cyanidin	Prevent biofilm formation on the polystyrene surface	Upadhyay et al. (2014)
	Catechin	Interaction with cell adhesion receptors, formation of ion channels in the membrane	Upadhyay et al. (2014)
Other phenolic compounds	Epicachin	Pore formation on cell wall	Upadhyay et al. (2014)
	Rosmarinic acid	Disintegration of cell membrane	Upadhyay et al. (2014)
	Isoflavonoids	Cell membrane damage, antioxidation	Upadhyay et al. (2014)

References

Abraham EP, Chain E (1988) An enzyme from bacteria able to destroy penicillin. Rev Infect Dis 10:677–678

Adegoke AA, Stenström TA, Okoh AI (2017) Stenotrophomonas maltophilia as an emerging ubiquitous pathogen: looking beyond contemporary antibiotic therapy. Front Microbiol 8:2276

Albano M, Crulhas BP, Alves FCB, Pereira AFM, Andrade B, Barbosa LN, Furlanetto A, Lyra L, Rall VLM, Junior AF (2019) Antibacterial and anti-biofilm activities of cinnamaldehyde against S. epidermidis. Microb Pathog 126:231–238. https://doi.org/10.1016/j.micpath.2018.11.009

Amalaradjou MAR, Narayanan A, Baskaran SA, Venkitanarayanan K (2010) Antibiofilm effect of trans-cinnamaldehyde on uropathogenic Escherichia coli. J Urol 184(1):358–363

Asfour HZ (2018) Anti-Quorum Sensing Natural Compounds. J Microsc Ultrastruct 6(1):1–10

Bandara H, Lam O, Watt R, Jin L, Samaranayake L (2010) Bacterial lipopolysaccharides variably modulate in vitro biofilm formation of Candida species. J Med Microbiol 59(10):1225–1234

Beachey EH, Giampapa CS, Abraham SN (1988) Bacterial adherence. Adhesin receptor-mediated attachment of pathogenic bacteria to mucosal surfaces. Am Rev Respir Dis 138(6 Pt 2):S45–S48. https://doi.org/10.1164/ajrccm/138.6_Pt_2.S45

Bielen K, s Jongers B, Boddaert J, Raju TK, Lammens C, Malhotra-Kumar S, Jorens PG, Goossens H, Kumar-Singh S (2017) Biofilm-Induced Type 2 Innate Immunity in a Cystic Fibrosis Model of Pseudomonas aeruginosa. Front Cell Infect Microbiol 7:274. https://doi.org/10.3389/fcimb.2017.00274

Bjarnsholt T, Jensen P, Rasmussen TB, Christophersen L, Calum H, Hentzer M, Hougen HP, Rygaard J, Moser C, Eberl L, Høiby N, Givskov M (2005) Garlic blocks quorum sensing and promotes rapid clearing of pulmonary Pseudomonas aeruginosa infections. Microbiology (Reading, England) 151(Pt 12):3873–3880. https://doi.org/10.1099/mic.0.27955-0

Bogino PC, de las M, Oliva M, Sorroche FG, Giordano W (2013) The role of bacterial biofilms and surface components in plant-bacterial associations. Int J Mol Sci 30:15838–15859

Boles BR, Horswill AR (2008) Agr-mediated dispersal of staphylococcus aureus biofilms. PLoS Pathog 4:e1000052

Bonnichsen L, Bygvraa Svenningsen N, Rybtke M, de Bruijn I, Raaijmakers JM, Tolker-Nielsen T et al (2015) Lipopeptide biosurfactant viscosin enhances dispersal of *Pseudomonas fluorescens* SBW25 biofilms. Microbiology 161:2289–2297

Cairns LS, Marlow VL, Bissett E, Ostrowski A, Stanley-Wall NR (2013) A mechanical signal transmitted by the flagellum controls signalling in Bacillus subtilis. Mol Microbiol 90(1):6–21. https://doi.org/10.1111/mmi.12342

Castillo J, García-Contreras R, Velázquez-Guadarrama N, Soto-Hernández M, Martínez-Vázquez M (2013) Amphypterygium adstringens anacardic acid mixture inhibits quorum sensing-controlled virulence factors of Chromobacterium violaceum and Pseudomonas aeruginosa. Arch Med Res 44:488–494

Cavalcante TTA, Sá Firmino N, Tajra FS, de Andrade CR, Costa RA (2014) Plant lectins as alternative tools against bacterial biofilms. Afr J Microbiol Res 8:2555–5264

Chippaux C, Puchelle E (1994) Cytotoxicity of Pseudomonas aeruginosa internal lectin PA-I to respiratory epithelial cells in primary culture. Infect Immun 62(10):4481–4487

Chong YM, Yin WF, Ho CY, Mustafa MR, Hadi AH, Awang K, Narrima P, Koh CL, Appleton DR, Chan KG (2011) Malabaricone C from Myristica cinnamomea exhibits anti-quorum sensing activity. J Nat Prod 74(10):2261–2264

Choo JH, Rukayadi Y, Hwang JK (2006) Inhibition of bacterial quorum sensing by vanilla extract. Lett Appl Microbiol 42:637–641

Colvin KM, Gordon VD, Murakami K, Borlee BR, Wozniak DJ, Wong GC, Parsek MR (2011) The pel polysaccharide can serve a structural and protective role in the biofilm matrix of Pseudomonas aeruginosa. PLoS Pathog 7(1):e1001264

Colvin KM, Irie Y, Tart CS, Urbano R, Whitney JC, Ryder C, Howell PL, Wozniak DJ, Parsek MR (2012) The Pel and Psl polysaccharides provide Pseudomonas aeruginosa structural redundancy within the biofilm matrix. Environ Microbiol 14(8):1913–1928

Cossart P, Jonquières R (2000) Sortase, a universal target for therapeutic agents against gram-positive bacteria? Proc Natl Acad Sci U S A 97(10):5013–5015. https://doi.org/10.1073/pnas.97.10.5013

Costerton JW, Stewart PS, Greenberg EP (1999) Bacterial biofilms: a common cause of persistent infections. Science 284:1318–1229

Cucarella C, Tormo MA, Ubeda C, Trotonda MP, Monzón M, Peris C, Amorena B, Lasa Í, Penadés JR (2004) Role of biofilm-associated protein bap in the pathogenesis of bovine Staphylococcus aureus. Infect Immun 72(4):2177–2185

Cunha WR, de Matos GX, Souza MG, Tozatti MG, Andrade e Silva ML, Martins CH, da Silva R, Da Silva Filho AA (2010) Evaluation of the antibacterial activity of the methylene chloride extract of Miconia ligustroides, isolated triterpene acids, and ursolic acid derivatives. Pharm Biol 48(2):166–169. https://doi.org/10.3109/13880200903062648

Das T, Sehar S, Manefield M (2013) The roles of extracellular DNA in the structural integrity of extracellular polymeric substance and bacterial biofilm development. Environ Microbiol Rep 5 (6):778–786

Davey ME, Caiazza NC, O'Toole GA (2003) Rhamnolipid surfactant production affects biofilm architecture in Pseudomonas aeruginosa PAO1. J Bacteriol 185:1027–1036

Davies DG, Parsek MR, Pearson JP, Iglewski BH, Costerton JW, Greenberg EP (1998) The involvement of cell-to-cell signals in the development of a bacterial biofilm. Science 280 (5361):295–298

Delaquis PJ, Caldwell DE, Lawrence JR, McCurdy AR (1989) Detachment of Pseudomonas fluorescens from biofilms on glass surfaces in response to nutrient stress. Microb Ecol 18:199–210

Di Martino P (2018) Extracellular polymeric substances, a key element in understanding biofilm phenotype. AIMS Microbio 4(2):274–288. https://doi.org/10.3934/microbiol.2018.2.274

Diggle SP, Stacey RE, Dodd C, Cámara M, Williams P, Winzer K (2006) The galactophilic lectin, LecA, contributes to biofilm development in Pseudomonas aeruginosa. Environ Microbiol 8 (6):1095–1104

Dominiak DM, Nielsen JL, Nielsen PH (2011) Extracellular DNA is abundant and important for microcolony strength in mixed microbial biofilms. Environ Microbiol 13(3):710–721

Dong YH, Zhang LH (2005) Quorum sensing and quorum-quenching enzymes. J Microbiol 43:101–109

Dong YH, Gusti AR, Zhang Q, Xu JL, Zhang LH (2002) Identification of quorum-quenching N-acyl homoserine lactonases from Bacillus species. Appl Environ Microbiol 68:1754–1759. https://doi.org/10.1128/AEM.68.4.1754-1759.2002

Donlan RM (2002) Biofilms: microbial life on surfaces. Emerg Infect Dis 8:881–890

Donlan RM, Costerton JW (2002) Biofilms: survival mechanisms of clinically relevant microorganisms. Clin Microbiol Rev 15:167–193

Dow JM, Crossman L, Findlay K, He Y-Q, Feng J-X, Tang J-L (2003) Biofilm dispersal in Xanthomonas campestris is controlled by cell–cell signaling and is required for full virulence to plants. Proc Natl Acad Sci U S A 100:10995–11000

Drenkard E (2003) Antimicrobial resistance of Pseudomonas aeruginosa biofilms. Microbes Infect 5(13):1213–1219

Duanis-Assaf D, Duanis-Assaf T, Zeng G, Meyer RL, Reches M, Steinberg D, Shemesh M (2018) Cell wall associated protein TasA provides an initial binding component to extracellular polysaccharides in dual-species biofilm. Sci Rep 8(1):9350. https://doi.org/10.1038/s41598-018-27548-1

Eberl L (1999) N-acyl homoserinelactone-mediated gene regulation in gram-negative bacteria. Syst Appl Microbiol 22(4):493–506

Emody L, Kerenyi M, Nagy G (2003) Virulence factors of uropathogenic Escherichia coli. Int J Antimicrob Agents 22:29–33

Engebrecht J, Nealson K, Silverman M (1983) Bacterial bioluminescence: isolation and genetic analysis of functions from Vibrio fischeri. Cell 32(3):773–781. https://doi.org/10.1016/0092-8674(83)90063-6

Erskine E, MacPhee CE, Stanley-Wall NR (2018) Functional amyloid and other protein fibers in the biofilm matrix. J Mol Biol 430(20):3642–3656

Estrela AB, Heck MG, Abraham WR (2009) Novel approaches to control biofilm infections. Curr Med Chem 16(12):1512–1530. https://doi.org/10.2174/092986709787909640. PMID: 19355904

Faleiro ML (2011) The mode of antibacterial action of essential oils. In: Méndez-Vilas A (ed) Science against microbial pathogens: communicating current research and technological advances. World Scientific, Singapore, pp 1143–1156

Flemming HC, Neu TR, Wozniak DJ (2007) The EPS matrix: the "house of biofilm cells". J Bacteriol 189(22):7945–7947. https://doi.org/10.1128/jb.00858-07

Flemming H-C, Wingender J, Szewzyk U, Steinberg P, Rice SA, Kjelleberg S (2016) Biofilms: an emergent form of bacterial life. Nat Rev Microbiol 14(9):563

Fu B, Wu Q, Dang M, Bai D, Guo Q, Shen L, Duan K (2017) Inhibition of Pseudomonas aeruginosa biofilm formation by traditional Chinese medicinal herb Herba patriniae. Biomed Res Int 2017:9584703

Fuentes JL, Garbayo I, Cuaresma M, Montero Z, González-Del-Valle M, Vílchez C (2016) Impact of microalgae-bacteria interactions on the production of algal biomass and associated compounds. Mar Drugs 14(5):100

Geoghegan JA, Corrigan RM, Gruszka DT, Speziale P, O'Gara JP, Potts JR, Foster TJ (2010) Role of surface protein SasG in biofilm formation by Staphylococcus aureus. J Bacteriol 192 (21):5663–5673

Ghosh A, Jayaraman N, Chatterji D (2020) Small-molecule inhibition of bacterial biofilm. ACS Omega 5(7):3108–3115. https://doi.org/10.1021/acsomega.9b03695. PMID: 32118127; PMCID: PMC7045314

Giltner CL, Van Schaik EJ, Audette GF, Kao D, Hodges RS, Hassett DJ, Irvin RT (2006) The Pseudomonas aeruginosa type IV pilin receptor binding domain functions as an adhesin for both biotic and abiotic surfaces. Mol Microbiol 59(4):1083–1096

Giriraju A, Yunus G (2013) Assessment of antimicrobial potential of 10% ginger extract against Streptococcus mutans, Candida albicans, and Enterococcus faecalis: an in vitro study. Indian J Dent Res 24(4):397

Girish VM, Liang H, Aguilan JT, Nosanchuk JD, Friedman JM, Nacharaju P (2019) Anti-biofilm activity of garlic extract loaded nanoparticles. Nanomedicine 20:102009. https://doi.org/10.1016/j.nano.2019.04.012

Gjermansen M, Nilsson M, Yang L, Tolker-Nielsen T (2010) Characterization of starvation-induced dispersion in Pseudomonas putida biofilms: genetic elements and molecular mechanisms. Mol Microbiol 75:815–826

Glick R, Gilmour C, Tremblay J, Satanower S, Avidan O, Déziel E et al (2010) Increase in rhamnolipid synthesis under iron-limiting conditions influences surface motility and biofilm formation in Pseudomonas aeruginosa. J Bacteriol 192:2973–2980

Gonçalves O, Pereira R, Gonçalves F, Mendo S, Coimbra MA, Rocha SM (2011) Evaluation of the mutagenicity of sesquiterpenic compounds and their influence on the susceptibility towards antibiotics of two clinically relevant bacterial strains. Mutat Res 723(1):18–25. https://doi.org/10.1016/j.mrgentox.2011.03.010

Gopu V, Kothandapani S, Shetty PH (2015) Quorum quenching activity of Syzygium cumini (L.) Skeels and its anthocyanin malvidin against Klebsiella pneumoniae. Microb Pathog 79:61–69. https://doi.org/10.1016/j.micpath.2015.01.010

Gross M, Cramton SE, Götz F, Peschel A (2001) Key role of teichoic acid net charge in staphylococcus aureus colonization of artificial surfaces. Infect Immun 69(5):3423–3426

Guttenplan SB, Kearns DB (2013) Regulation of flagellar motility during biofilm formation. FEMS Microbiol Rev 37(6):849–871

Gyawali R, Ibrahim SA (2014) Natural products as antimicrobial agents. Food Control 46:412–429

Harimawan A, Ting Y-P (2016) Investigation of extracellular polymeric substances (EPS) properties of P. aeruginosa and B. subtilis and their role in bacterial adhesion. Colloids Surf B: Biointerfaces 146:459–467

Harjai K, Kumar R, Singh S (2010) Garlic blocks quorum sensing and attenuates the virulence of Pseudomonas aeruginosa. FEMS Immunol Med Microbiol 58(2):161–168. https://doi.org/10.1111/j.1574-695X.2009.00614.x

Hentzer M, Givskov M (2003) Pharmacological inhibition of quorum sensing for the treatment of chronic bacterial infections. J Clin Invest 112:1300–1307

Houry A, Briandet R, Aymerich S, Gohar M (2010) Involvement of motility and flagella in Bacillus cereus biofilm formation. Microbiology 156(4):1009–1018

Houston P, Rowe SE, Pozzi C, Waters EM, O'Gara JP (2011) Essential role for the major autolysin in the fibronectin-binding protein-mediated Staphylococcus aureus biofilm phenotype. Infect Immun 79(3):1153–1165

Huber B, Riedel K, Hentzer M, Heydorn A, Gotschlich A, Givskov M, Molin S, Eberl L (2001) The cep quorum-sensing system of Burkholderia cepacia H111 controls biofilm formation and swarming motility. Microbiology (Reading, England) 147(Pt 9):2517–2528. https://doi.org/10.1099/00221287-147-9-2517

Husain FM, Ahmad I, Asif M, Tahseen Q (2013) Influence of clove oil on certain quorum-sensing-regulated functions and biofilm of Pseudomonas aeruginosa and Aeromonas hydrophila. J Biosci 38(5):835–844

Jain K, Parida S, Mangwani N, Dash HR, Das S (2013) Isolation and characterization of biofilm-forming bacteria and associated extracellular polymeric substances from oral cavity. Ann Microbiol 63(4):1553–1562

Kalia M, Yadav VK, Singh PK, Sharma D, Narvi SS, Agarwal V (2018) Exploring the impact of parthenolide as anti-quorum sensing and anti-biofilm agent against Pseudomonas aeruginosa. Life Sci 199:96–103. https://doi.org/10.1016/j.lfs.2018.03.013

Kang JE, Han JW, Jeon BJ, Kim BS (2016) Efficacies of quorum sensing inhibitors, piericidin A and glucopiericidin A, produced by Streptomyces xanthocidicus KPP01532 for the control of potato soft rot caused by Erwinia carotovora subsp. atroseptica. Microbiol Res 184:32–41

Karygianni L, Ren Z, Koo H, Thurnheer T (2020) Biofilm matrixome: extracellular components in structured microbial communities. Trends Microbiol 28:668–681

Katsuyama M, Kobayashi Y, Ichikawa H, Mizuno A, Miyachi Y, Matsunaga K, Kawashima M (2005) A novel method to control the balance of skin microflora Part 2. A study to assess the effect of a cream containing farnesol and xylitol on atopic dry skin. J Dermatol Sci 38 (3):207–213. https://doi.org/10.1016/j.jdermsci.2005.01.003

Kaufmann GF, Park J, Janda KD (2008) Bacterial quorum sensing: a new target for anti-infective immunotherapy. Expert Opin Biol Ther 8:719–724

Kaur N, Arora DS, Kalia N, Kaur M (2020) Antibiofilm, antiproliferative, antioxidant and antimutagenic activities of an endophytic fungus Aspergillus fumigatus from Moringa oleifera. Mol Biol Rep 47(4):2901–2911. https://doi.org/10.1007/s11033-020-05394-7

Kean R, Rajendran R, Haggarty J, Townsend EM, Short B, Burgess KE, Lang S, Millington O, Mackay WG, Williams C, Ramage G (2017) Candida albicans mycofilms support *Staphylococcus aureus* colonization and enhances miconazole resistance in dual-species interactions. Front Microbiol 8:258. https://doi.org/10.3389/fmicb.2017.00258

Kim S-H, Shin D-S, Oh M-N, Chung S-C, Lee J-S, Oh K-B (2004) Inhibition of the bacterial surface protein anchoring transpeptidase sortase by isoquinoline alkaloids. Biosci Biotechnol Biochem 68(2):421–424

Kim D, Liu Y, Benhamou RI, Sanchez H, Simón-Soro Á, Li Y, Hwang G, Fridman M, Andes DR, Koo H (2018) Bacterial-derived exopolysaccharides enhance antifungal drug tolerance in a cross-kingdom oral biofilm. ISME J 12(6):1427–1442. https://doi.org/10.1038/s41396-018-0113-1

Knowles JR, Roller S, Murray DB, Naidu AS (2005) Antimicrobial action of carvacrol at different stages of dual-species biofilm development by Staphylococcus aureus and Salmonella enterica serovar Typhimurium. Appl Environ Microbiol 71(2):797–803. https://doi.org/10.1128/AEM.71.2.797-803.2005

Kobayashi K, Iwano M (2012) BslA (YuaB) forms a hydrophobic layer on the surface of Bacillus subtilis biofilms. Mol Microbiol 85(1):51–66

Kohler T, Weidenmaier C, Peschel A (2009) Wall teichoic acid protects Staphylococcus aureus against antimicrobial fatty acids from human skin. J Bacteriol 191(13):4482–4484

Koo H, Pearson SK, Scott-Anne K, Abranches J, Cury JA, Rosalen PL, Park YK, Marquis RE, Bowen WH (2002) Effects of apigenin and tt-farnesol on glucosyltransferase activity, biofilm viability and caries development in rats. Oral Microbiol Immunol 17(6):337–343. https://doi.org/10.1034/j.1399-302x.2002.170602.x

Kurekci C, Padmanabha J, Bishop-Hurley SL, Hassan E, Al Jassim RA, McSweeney CS (2013) Antimicrobial activity of essential oils and five terpenoid compounds against Campylobacter jejuni in pure and mixed culture experiments. Int J Food Microbiol 166(3):450–457. https://doi.org/10.1016/j.ijfoodmicro.2013.08.014

Kuźma Ł, Różalski M, Walencka E, Różalska B, Wysokińska H (2007) Antimicrobial activity of diterpenoids from hairy roots of Salvia sclarea L.: salvipisone as a potential anti-biofilm agent active against antibiotic resistant Staphylococci. Phytomedicine 14(1):31–35. https://doi.org/10.1016/j.phymed.2005.10.008

Kwakman PHS, te Velde AA, Grauls CMJEV, van Deventer SJH, Zaat SAJ (2006) Treatment and prevention of staphylococcus epidermidis experimental biomaterial-associated infection by bactericidal peptide 2. Antimicrob Agents Chemother 50:3977–3983

Lade H, Paul D, Kweon JH (2014) N-acyl homoserine lactone-mediated quorum sensing with special reference to use of quorum quenching bacteria in membrane biofouling control. Biomed Res Int 2014:162584

Lautenbach E, Patel JB, Bilker WB, Edelstein PH, Fishman NO (2001) Extended-spectrum beta-lactamase-producing Escherichia coli and Klebsiella pneumoniae: risk factors for infection and impact of resistance on outcomes. Clin Infect Dis 32:1162–1171

Lazar V (2011) Quorum sensing in biofilms–how to destroy the bacterial citadels or their cohesion/power? Anaerobe 17(6):280–285

Lee H, Lee DG (2015) Mode of action of bioactive phytochemicals, plant secondary metabolites, possessing antimicrobial properties. In: Méndez-Vilas A (ed) The battle against microbial pathogens: basic science, technological advances and educational programs. World Scientific, Singapore

Lee JH, Regmi SC, Kim JA, Cho MH, Yun H, Lee CS, Lee J (2011) Apple flavonoid phloretin inhibits Escherichia coli O157:H7 biofilm formation and ameliorates colon inflammation in rats. Infect Immun 79(12):4819–4827. https://doi.org/10.1128/IAI.05580-11

Lee JH, Kim YG, Cho HS, Ryu SY, Cho MH, Lee J (2014a) Coumarins reduce biofilm formation and the virulence of Escherichia coli O157:H7. Phytomedicine 21(8–9):1037–1042. https://doi.org/10.1016/j.phymed.2014.04.008

Lee J-H, Kim Y-G, Ryu SY, Cho MH, Lee J (2014b) Ginkgolic acids and Ginkgo biloba extract inhibit Escherichia coli O157:H7 and Staphylococcus aureus biofilm formation. Int J Food Microbiol 174:47–55

Leid JG, Willson CJ, Shirtliff ME, Hassett DJ, Parsek MR, Jeffers AK (2005) The exopolysaccharide alginate protects Pseudomonas aeruginosa biofilm bacteria from IFN-γ-mediated macrophage killing. J Immunol 175(11):7512–7518

Lin MH, Shu JC, Lin LP, Yu Chong K, Cheng YW, Du JF, Liu S-T (2015) Elucidating the crucial role of poly N-acetylglucosamine from Staphylococcus aureus in cellular adhesion and pathogenesis. PLoS One 10(4):e0124216

Liu Q, Niu H, Zhang W, Mu H, Sun C, Duan J (2015) Synergy among thymol, eugenol, berberine, cinnamaldehyde and streptomycin against planktonic and biofilm-associated food-borne pathogens. Lett Appl Microbiol 60(5):421–430. https://doi.org/10.1111/lam.12401

Liu S, Qiu W, Zhang K, Zhou X, Ren B, He J, Xu X, Cheng L, Li M (2017) Nicotine enhances interspecies relationship between Streptococcus mutans and Candida albicans. Biomed Res Int 2017:7953920. https://doi.org/10.1155/2017/7953920

Long H, Miller SF, Strauss C, Zhao C, Cheng L, Ye Z, Griffin K, Te R, Lee H, Chen CC, Lynch M (2016) Antibiotic treatment enhances the genome-wide mutation rate of target cells. Proc Natl Acad Sci U S A 113:E2498–E2505

Lu L, Hu W, Tian Z, Yuan D, Yi G, Zhou Y, Cheng Q, Zhu J, Li M (2019) Developing natural products as potential anti-biofilm agents. Chin Med 14:11. https://doi.org/10.1186/s13020-019-0232-2

Lucas REB, Wheeler D, Hettige H (1993) Economic development, environmental regulation, and the international migration of toxic industrial pollution, 1960–1988. Washington, DC (1818 H, NW, Washington 20433): Office of the Vice President, Development Economics, World Bank

Ma L, Jackson KD, Landry RM, Parsek MR, Wozniak DJ (2006) Analysis of Pseudomonas aeruginosa conditional psl variants reveals roles for the psl polysaccharide in adhesion and maintaining biofilm structure postattachment. J Bacteriol 188(23):8213–8221

Mah TF, O'Toole GA (2001) Mechanisms of biofilm resistance to antimicrobial agents. Trends Microbiol 9:34–39

Memariani H, Memariani M, Ghasemian A (2019) An overview on anti-biofilm properties of quercetin against bacterial pathogens. World J Microbiol Biotechnol 35(9):143. https://doi.org/10.1007/s11274-019-2719-5

Merino N, Toledo-Arana A, Vergara-Irigaray M, Valle J, Solano C, Calvo E, Lopez JA, Foster TJ, Penadés JR, Lasa I (2009) Protein A-mediated multicellular behavior in Staphylococcus aureus. J Bacteriol 191(3):832–843

Miller MB, Bassler BL (2001) Quorum sensing in bacteria. Annu Rev Microbiol 55:165–199. https://doi.org/10.1146/annurev.micro.55.1.165

Morgan R, Kohn S, Hwang S-H, Hassett DJ, Sauer K (2006) BdlA, a chemotaxis regulator essential for biofilm dispersion in Pseudomonas aeruginosa. J Bacteriol 188:7335–7343

Moser H, Wu C, Wang HZ, Høiby N, Song ZJ (2015) Strategies for combating bacterial biofilm infections. Int J Oral Sci 7:1–7

Moshe M, Lellouche J, Banin E (2011) Curcumin: a natural antibiofilm agent. Sci Technol Against Microb Pathog 2011:89–93

Musk DJ, Banko DA, Hergenrother PJ (2005) Iron salts perturb biofilm formation and disrupt existing biofilms of Pseudomonas aeruginosa. Chem Biol 12:789–796

Muthusamy B, Shanmugam G (2020) Analysis of flavonoid content, antioxidant, antimicrobial and antibiofilm activity of in vitro hairy root extract of radish (Raphanus sativus L.). Plant Cell Tiss Organ Cult (PCTOC) 140(3):619–633

Nadaf NH, Parulekar RS, Patil RS, Gade TK, Momin AA, Waghmare SR, Dhanavade MJ, Arvindekar AU, Sonawane KD (2018) Biofilm inhibition mechanism from extract of Hymenocallis littoralis leaves. J Ethnopharmacol 222:121–132. https://doi.org/10.1016/j.jep.2018.04.031

Niu C, Gilbert ES (2004) Colorimetric method for identifying plant essential oil components that affect biofilm formation and structure. Appl Environ Microbiol 70(12):6951–6956. https://doi.org/10.1128/AEM.70.12.6951-6956.2004

Ouyang J, Sun F, Feng W, Sun Y, Qiu X, Xiong L, Liu Y, Chen Y (2016) Quercetin is an effective inhibitor of quorum sensing, biofilm formation and virulence factors in Pseudomonas aeruginosa. J Appl Microbiol 120(4):966–974

Ouyang P, He X, Yuan ZW, Yin ZQ, Fu H, Lin J, He C, Liang X, Lv C, Shu G, Yuan ZX, Song X, Li L, Yin L (2018) Erianin against Staphylococcus aureus infection via inhibiting Sortase A. Toxins (Basel) 10(10):385. https://doi.org/10.3390/toxins10100385

Papenfort K, Bassler BL (2016) Quorum sensing signal-response systems in Gram-negative bacteria. Nat Rev Microbiol 14(9):576–588. https://doi.org/10.1038/nrmicro.2016.89. PMID: 27510864; PMCID: PMC5056591.

Petersen J (2016) Heap leaching as a key technology for recovery of values from low-grade ores—a brief overview. Hydrometallurgy 165:206–212

Petrova OE, Cherny KE, Sauer K (2015) The diguanylate cyclase GcbA facilitates *Pseudomonas aeruginosa* biofilm dispersion by activating BdlA. J Bacteriol 197:174–187

Pierce CG, Vila T, Romo JA, Montelongo-Jauregui D, Wall G, Ramasubramanian A, Lopez-Ribot JL (2017) The Candida albicans biofilm matrix: composition, structure and function. J Fungi 3 (1):14

Prakash B, Veeregowda B, Krishnappa G (2003) Biofilms: a survival strategy of bacteria. Curr Sci 2003:1299–1307

Prestinaci F, Pezzotti P, Pantosti A (2015) Antimicrobial resistance: a global multifaceted phenomenon. Pathog Glob Health 109:309–318

Rajkumari J, Meena H, Gangatharan M, Busi S (2017) Green synthesis of anisotropic gold nanoparticles using hordenine and their antibiofilm efficacy against Pseudomonas aeruginosa. IET Nanobiotechnol 11(8):987–994. https://doi.org/10.1049/iet-nbt.2017.0069

Rasch M, Kastbjerg VG, Bruhn JB, Dalsgaard I, Givskov M, Gram L (2007) Quorum sensing signals are produced by Aeromonas salmonicida and quorum sensing inhibitors can reduce production of a potential virulence factor. Dis Aquat Org 78(2):105–113. https://doi.org/10.3354/dao01865

Reiter KC, Villa B, Paim TGDS, de Oliveira CF, d'Azevedo PA (2013) Inhibition of biofilm maturation by linezolid in methicillin-resistant Staphylococcus epidermidis clinical isolates: comparison with other drugs. J Med Microbiol 62(Pt 3):394–399. https://doi.org/10.1099/jmm.0.048678-0

Rendeková K, Fialová S, Jánošová L, Mučaji P, Slobodníková L (2016) The activity of Cotinus coggygria Scop. leaves on Staphylococcus aureus strains in planktonic and biofilm growth forms. Molecules (Basel, Switzerland) 21(1):50

Richards JJ, Melander C (2009) Controlling bacterial biofilms. ChemBioChem 10(14):2287–2294. https://doi.org/10.1002/cbic.200900317

Rizzello L, Pompa PP (2014) Nanosilver-based antibacterial drugs and devices: mechanisms, methodological drawbacks, and guidelines. Chem Soc Rev 43:1501–1518

Ro DK, Paradise EM, Ouellet M, Fisher KJ, Newman KL, Ndungu JM, Ho KA, Eachus RA, Ham TS, Kirby J, Chang MC, Withers ST, Shiba Y, Sarpong R, Keasling JD (2006) Production of the antimalarial drug precursor artemisinic acid in engineered yeast. Nature 440(7086):940–943

Romero D, Vlamakis H, Losick R, Kolter R (2011) An accessory protein required for anchoring and assembly of amyloid fibres in B. subtilis biofilms. Mol Microbiol 80(5):1155–1168

Roy AB, Petrova OE, Sauer K (2012) The phosphodiesterase DipA (PA5017) is essential for pseudomonas aeruginosa biofilm dispersion. J Bacteriol 194:2904–2915

Rozalski M, Micota B, Sadowska B, Stochmal A, Jedrejek D, Wieckowska-Szakiel M, Rozalska B (2013) Antiadherent and antibiofilm activity of Humulus lupulus L. derived pro ducts: new pharmacological properties. Biomed Res Int 2013:101089. https://doi.org/10.1155/2013/101089

Rukayadi Y, Hwang JK (2006) Effect of coating the wells of a polystyrene microtiter plate with xanthorrhizol on the biofilm formation of Streptococcus mutans. J Basic Microbiol 46 (5):410–415. https://doi.org/10.1002/jobm.200510088

Saini R, Saini S, Sharma S (2011) Biofilm: a dental microbial infection. J Natur Sci Biol Med 2 (1):71

Saxena P, Joshi Y, Rawat K, Bisht R (2019) Biofilms: architecture, resistance, quorum sensing and control mechanisms. Indian J Microbiol 59(1):3–12

Schaefer AL, Val DL, Hanzelka BL, Cronan JE Jr, Greenberg EP (1996) Generation of cell-to-cell signals in quorum sensing: acyl homoserine lactone synthase activity of a purified Vibrio fischeri LuxI protein. Proc Natl Acad Sci U S A 93(18):9505–9509

Schleheck D, Barraud N, Klebensberger J, Webb JS, McDougald D, Rice SA et al (2009) Pseudomonas aeruginosa PAO1 preferentially grows as aggregates in liquid batch cultures and disperses upon starvation. PLoS One 4:e5513

Schwartz K, Syed AK, Stephenson RE, Rickard AH, Boles BR (2012) Functional amyloids composed of phenol soluble modulins stabilize Staphylococcus aureus biofilms. PLoS Pathog 8(6):e1002744

Shiner E, Terentyev D, Bryan A, Sennoune S, Martinez-Zaguilan R, Li G, Gyorke S, Williams S, Rumbaugh K (2006) Pseudomonas aeruginosa autoinducer modulates host cell responses through calcium signalling. Cell Microbiol 8(10):1601–1610

Shuai-Cheng W, Ben-Dong F, Xiu-Ling C, Jian-Qing S, Yun-Xing F, Zhen-Qiang C, Dao-Xiu X, Zong-Mei W (2016) Subinhibitory concentrations of phloretin repress the virulence of *Salmonella typhimurium* and protect against Salmonella typhimurium infection. Antonie Van Leeuwenhoek 109(11):1503–1512. https://doi.org/10.1007/s10482-016-0752-z

Sifri CD (2008) Quorum sensing: bacteria talk sense. Clin Infect Dis 47(8):1070–1076

Smith KM, Bu Y, Suga H (2003) Library screening for synthetic agonists and antagonists of a Pseudomonas aeruginosa autoinducer. Chem Biol 10(6):563–571

Spratt BG (1994) Resistance to antibiotics mediated by target alterations. Science 264:388–393

Steenackers HP, Parijs I, Dubey A, Foster KR, Vanderleyden J (2016) Experimental evolution in biofilm populations. FEMS Microbiol Rev 40(6):980. https://doi.org/10.1093/femsre/fuw030

Sturme MH, Kleerebezem M, Nakayama J, Akkermans AD, Vaughan EE, De Vos WM (2002) Cell to cell communication by autoinducing peptides in gram-positive bacteria. Antonie Van Leeuwenhoek 81(1-4):233–243

Suga H, Smith KM (2003) Molecular mechanisms of bacterial quorum sensing as a new drug target. Curr Opin Chem Biol 7(5):586–591

Sutherland IW (2001) The biofilm matrix–an immobilized but dynamic microbial environment. Trends Microbiol 9(5):222–227

Tayeb-Fligelman E, Salinas N, Tabachnikov O, Landau M (2020) Staphylococcus aureus PSMα3 cross-α fibril polymorphism and determinants of cytotoxicity. Structure (London, England: 1993) 28(3):301–313. https://doi.org/10.1016/j.str.2019.12.006

Topa SH, Subramoni S, Palombo EA, Kingshott P, Rice SA, Blackall LL (2018) Cinnamaldehyde disrupts biofilm formation and swarming motility of Pseudomonas aeruginosa. Microbiology (Reading, England) 164(9):1087–1097. https://doi.org/10.1099/mic.0.000692

Trivedi A, Mavi PS, Bhatt D, Kumar A (2016) Thiol reductive stress induces cellulose-anchored biofilm formation in Mycobacterium tuberculosis. Nat Commun 7(1):11392. https://doi.org/10.1038/ncomms11392

Turnbull L, Toyofuku M, Hynen AL, Kurosawa M, Pessi G, Petty NK, Osvath SR, Cárcamo-Oyarce G, Gloag ES, Shimoni R, Omasits U, Ito S, Yap X, Monahan LG, Cavaliere R, Ahrens CH, Charles IG, Nomura N, Eberl L, Whitchurch CB (2016) Explosive cell lysis as a mechanism for the biogenesis of bacterial membrane vesicles and biofilms. Nat Commun 7:11220. https://doi.org/10.1038/ncomms11220

Ueda A, Wood TK (2009) Connecting Quorum Sensing, c-di-GMP, Pel Polysaccharide, and Biofilm Formation in Pseudomonas aeruginosa through Tyrosine Phosphatase TpbA (PA3885). PLoS Pathog 5:e1000483

Upadhyay A, Upadhyaya I, Johny AK, Venkitanarayanan K (2014) Combating pathogenic microorganisms using plant-derived antimicrobials: a minireview of the mechanistic basis. Biomed Res Int 2014:761741

Vandeputte OM, Kiendrebeogo M, Rajaonson S, Diallo B, Mol A, El Jaziri M, Baucher M (2010) Identification of catechin as one of the flavonoids from Combretum albiflorum bark extract that reduces the production of quorum-sensing-controlled virulence factors in *Pseudomonas aeruginosa* PAO1. Appl Environ Microbiol 76(1):243–253

Vandeputte OM, Kiendrebeogo M, Rasamiravaka T, Stévigny C, Duez P, Rajaonson S, Diallo B, Mol A, Baucher M, El Jaziri M (2011) The flavanone naringenin reduces the production of quorum sensing-controlled virulence factors in Pseudomonas aeruginosa PAO1. Microbiology (Reading) 157(Pt 7):2120–2132

Vasconcelos NG, Croda J, Simionatto S (2018) Antibacterial mechanisms of cinnamon and its constituents: a review. Microb Pathog 120:198–203. https://doi.org/10.1016/j.micpath.2018.04.036

Vester B, Douthwaite S (2001) Macrolide resistance conferred by base substitutions in 23S rRNA. Antimicrob Agents Chemother 45:1–12

Vidal O, Longin R, Prigent-Combaret C, Dorel C, Hooreman M, Lejeune P (1998) Isolation of an Escherichia coli K-12 mutant strain able to form biofilms on inert surfaces: involvement of a new ompR allele that increases curli expression. J Bacteriol 180(9):2442–2449

Vijayakumar S, Malaikozhundan B, Saravanakumar K, Duran-Lara EF, Wang MH, Vaseeharan B (2019) Garlic clove extract assisted silver nanoparticle—antibacterial, antibiofilm, antihelminthic, anti-inflammatory, anticancer and ecotoxicity assessment. J Photochem Photobiol B 198:111558. https://doi.org/10.1016/j.jphotobiol.2019.111558

Vikram A, Jayaprakasha GK, Jesudhasan PR, Pillai SD, Patil BS (2010) Suppression of bacterial cell-cell signalling, biofilm formation and type III secretion system by citrus flavonoids. J Appl Microbiol 109(2):515–527. https://doi.org/10.1111/j.1365-2672.2010.04677.x

Vipin C, Mujeeburahiman M, Ashwini P, Arun A, Rekha PD (2019) Anti-biofilm and cytoprotective activities of quercetin against Pseudomonas aeruginosa isolates. Lett Appl Microbiol 68(5):464–471

Walencka E, Rozalska S, Wysokinska H, Rozalski M, Kuzma L, Rozalska B (2007) Salvipisone and aethiopinone from Salvia sclarea hairy roots modulate staphylococcal antibiotic resistance and express anti-biofilm activity. Planta Med 73(6):545–551. https://doi.org/10.1055/s-2007-967179

Wang R, Khan BA, Cheung GYC, Bach T-HL, Jameson-Lee M, Kong K-F et al (2011) Staphylococcus epidermidis surfactant peptides promote biofilm maturation and dissemination of biofilm-associated infection in mice. J Clin Invest 121:238–248

Wang C, Hou J, van der Mei HC, Busscher HJ, Ren Y (2019) Emergent properties in streptococcus mutans biofilms are controlled through adhesion force sensing by initial colonizers. MBio 10:5. https://doi.org/10.1128/mBio.01908-19

Williams P (2002) Quorum sensing: an emerging target for antibacterial chemotherapy? Exp Opin Therap Target 6(3):257–274

Wood TK, Barrios AFG, Herzberg M, Lee J (2006) Motility influences biofilm architecture in Escherichia coli. Appl Microbiol Biotechnol 72(2):361–367

Wright GD (1999) Aminoglycoside-modifying enzymes. Curr Opin Microbiol 2:499–503

Wu Y, Park KC, Choi BG, Park JH, Yoon KS (2016) The antibiofilm effect of ginkgo biloba extract against salmonella and listeria isolates from poultry. Foodborne Pathog Dis 13(5):229–238. https://doi.org/10.1089/fpd.2015.2072

Xu LC, Wo Y, Meyerhoff ME, Siedlecki CA (2017) Inhibition of bacterial adhesion and biofilm formation by dual functional textured and nitric oxide releasing surfaces. Acta Biomater 51:53–65

Yan J, Sharo AG, Stone HA, Wingreen NS, Bassler BL (2016) Vibrio cholerae biofilm growth program and architecture revealed by single-cell live imaging. Proc Natl Acad Sci U S A 113 (36):E5337–E5343. https://doi.org/10.1073/pnas.1611494113

Yu Y, Yan F, Chen Y, Jin C, Guo J-H, Chai Y (2016) Poly-γ-glutamic acids contribute to biofilm formation and plant root colonization in selected environmental isolates of Bacillus subtilis. Front Microbiol 7:1811

Zeng Z, Qian L, Cao L, Tan H, Huang Y, Xue X, Shen Y, Zhou S (2008) Virtual screening for novel quorum sensing inhibitors to eradicate biofilm formation of Pseudomonas aeruginosa. Appl Microbiol Biotechnol 79(1):119–126. https://doi.org/10.1007/s00253-008-1406-5

Zhang J, Rui X, Wang L, Guan Y, Sun X, Dong M (2014) Polyphenolic extract from Rosa rugosa tea inhibits bacterial quorum sensing and biofilm formation. Food Control 42:125–131

Zhang J, Ye KP, Zhang X, Pan DD, Sun YY, Cao JX (2016) Antibacterial activity and mechanism of action of black pepper essential oil on meat-borne Escherichia coli. Front Microbiol 7:2094

Zhou JW, Luo HZ, Jiang H, Jian TK, Chen ZQ, Jia AQ (2018) Hordenine: a novel quorum sensing inhibitor and antibiofilm agent against pseudomonas aeruginosa. J Agric Food Chem 66 (7):1620–1628. https://doi.org/10.1021/acs.jafc.7b05035

Novel and Future Treatment Strategies for Biofilm-Associated Infections

10

Sougata Ghosh, Raymond J. Turner, Tanay Bhagwat, and Thomas J. Webster

Abstract

Biofilm formation by pathogenic bacteria is a matter of considerable concern in healthcare as it is a leading cause of emerging multidrug resistance in microbes. Microbial biofilms are often found on the surfaces of biomaterials such as contact lenses and medical devices including implants and urinary catheters. Several bacteria like *Enterococcus faecium*, *Staphylococcus aureus*, *Klebsiella pneumoniae*, *Acinetobacter baumannii*, *Pseudomonas aeruginosa*, and *Enterobacter* spp., together termed as ESKAPE pathogens, are responsible for nosocomial infections that result in increased morbidity and mortality. This imposes a significant financial burden on the healthcare system. Available antimicrobial therapeutics are rendered ineffective due to compact and complex biofilm architecture which is composed of mainly extracellular polymeric substances (EPS). This leads to persistent infections untreatable by conventional therapy. Thus, there is a high demand for novel strategies for inhibition as well as disruption of biofilms to control biofilm-associated infections. In this chapter, we provide an elaborate account of complementary and alternative therapeutic strategies for biofilm control that include isolation of quorum-sensing inhibitors, metal chelators, and biofilm efflux pump inhibitors from medicinal plants. Further, dispersion of a preformed biofilm can be achieved by ultrasonic disruption, enzyme-mediated degradation of EPS, acidic electrolyzed water-assisted

S. Ghosh (✉)
Department of Microbiology, School of Science, RK University, Rajkot, Gujarat, India

Department of Chemical Engineering, Northeastern University, Boston, MA, USA

R. J. Turner
Department of Biological Sciences, University of Calgary, Calgary, AB, Canada

T. Bhagwat · T. J. Webster
Department of Chemical Engineering, Northeastern University, Boston, MA, USA

R. R. Ray et al. (eds.), *Biofilm-Mediated Diseases: Causes and Controls*,
https://doi.org/10.1007/978-981-16-0745-5_10

disruption, and bacteriophage-assisted biofilm disruption. Another attractive approach includes modification of the surface of medical devices by antibiofilm nanoscale biomaterials. Gold, silver, iron oxide, and bimetallic nanoparticles either individually or multifunctionalized with polymeric substances or drugs have also been fabricated to control biofilm formation by the interruption of quorum sensing, cell-to-cell communication, and multidrug efflux pumps. Similarly, antibiofilm nanostructures may also induce oxidative stress by generating reactive oxygen species which can play an important role in biofilm inhibition. Additionally, the future scope of integrating therapeutics, employing drugs, targeting ligands, and nanomedicine is discussed as promising strategies for better biofilm control compared to conventional treatments.

10.1 Introduction

Biofilms are aggregated microbial cells embedded in a polymeric extracellular matrix and are firmly attached to surfaces like metals, plastics, plants, body tissue, medical devices, and implant materials. Various nosocomial (hospital borne) infections are attributed to biofilm formation by pathogens on indwelling medical devices and implants such as heart valves, pacemakers, vascular grafts, catheters, prosthetic joints, intrauterine devices, sutures, and contact lenses and pose a critical problem of infection. Intravascular catheters for patient care are often the main cause for the rise in central line-associated blood stream infections (CLABSI), and approximately 250,000 cases of primary blood stream infections are reported each year in the USA. Biofilm formation results in multidrug resistance and challenges for available therapeutic strategies causing significant morbidity and mortality apart from huge increases in healthcare costs (Subhadra et al. 2018).

The predominant biofilm-forming bacteria include the ESKAPE pathogens which include six pathogens, namely, *Enterococcus faecium*, *Staphylococcus aureus*, *Klebsiella pneumoniae*, *Acinetobacter baumannii*, *Pseudomonas aeruginosa*, and *Enterobacter* spp. The indiscriminate use of high doses of antibiotics has provoked the emergence of multidrug resistant (MDR) and extensively drug-resistant (XDR) ESKAPE pathogens, which render even the most effective drugs ineffective. Biofilm-associated multidrug resistance is increasing in pathogens at an alarming rate, and the World Health Organization (WHO) has also recently listed ESKAPE12 bacteria against which new and more effective biocidal agents are immediately required. Carbapenem-resistant *A. baumannii* and *P. aeruginosa* along with extended-spectrum β-lactamase (ESBL) or carbapenem-resistant *K. pneumoniae* and *Enterobacter* spp. are considered in the critical priority list of pathogens; whereas, vancomycin-resistant *E. faecium* (VRE) and methicillin- and vancomycin-resistant *S. aureus* (MRSA and VRSA) are in the high priority group list (Mulani et al. 2019). These bacteria exhibit a potent biofilm-forming ability that challenges existing therapeutic agents.

Biofilm formation is initiated when the free-floating (planktonic) bacterial cells attach and adhere to surfaces which is facilitated by factors such as bacterial motility, increased shear forces, and hydrodynamic and electrostatic interactions between the microorganism and surface. Multiple factors like cell surface proteins, capsular polysaccharide/adhesin, protein autolysin, etc. play an important role in cell surface-material surface interactions. Staphylococcal surface protein-1 and staphylococcal surface protein-2 (SSP-1 and SSP-2) are essential for the adhesion of *S. epidermidis* to polystyrene (Veenstra et al. 1996).

Additionally, host factors also determine the extent of adherence of bacterial cells to implants which are usually covered by host plasma and other extracellular components. On attaching to the surfaces, the bacterial cells proliferate, aggregate, and differentiate into mature biofilm structures that concomitantly allow bacterial cells to detach and spread to other organs, thereby resulting in persistent chronic infections (Fey and Olson 2010).

Biofilm-based infections are attributed to the increased tolerance of biofilm cells to biocides compared to planktonic bacteria which is mainly associated with plasmids containing multidrug-resistant genes. Biofilms serve as an ideal niche for plasmid exchange between the closely associated immobilized bacterial cells in the biofilm matrix. The mechanisms of biofilm-associated increased drug resistance includes slow or incomplete penetration of antimicrobial agents through the extracellular polymeric matrix, the formation of persister or dormant cells in a spore-like non-dividing state, and slow growth rate of cells in the biofilm, thereby reducing the number of targets for antimicrobial molecules, etc. (Ghosh and Chopade 2017).

Thus, there is a high demand for alternative strategies to control biofilm-based infections other than antibiotic therapy. In this chapter, the most successful complementary and alternative biofilm control strategies are discussed in detail that includes phytomedicine, bacteriophages, and nanomedicines. Further, the development of new methods for biofilm disruption using ultrasound, enzymes, and acidic electrolyzed water is also elaborated. Finally, the scope of including these strategies in the therapeutic regimes and further investigations required in this aspect in order to ensure the safety and efficiency during medical applications are also presented.

10.2 Biofilm Formation in Infections

Chronic wounds have the characteristic of being colonized with pathogenic bacteria growing as a biofilm. Microbial biofilms affect the management of mostly all infections due to their high tolerance to antimicrobials lending them to be extremely recalcitrant (overviewed by Lebeaux et al. 2014). Their multifactorial mechanisms of high tolerance and resistance toward antibiotics lead these chronic issues and device-related infection to have cyclic recurrence. Infection biofilms can influence antimicrobial resistance, modulate immune responses, as well as develop and maintain chronic infections. Biofilms of infectious organisms on medical devices and surfaces are also responsible for the source of infections (Veerachamy et al. 2014). Reports from the National Institutes of Health and the Center for Disease Control estimate between 65% and 80% of microbial infections involve biofilms. Examples of biofilm-driven infections include dental caries and root canal infections (Walsh 2020), orthopedic (Zimmerli and Sendi 2017) and medical implant materials (Arciola et al. 2018), urinary track and prostatitis (Delcaru et al. 2016), bacterial vaginosis (Jung et al. 2017), diabetic foot infections (Malik et al. 2013), and arterial plaques leading to enhanced cardiovascular disease (Lanter et al. 2014).

The process of infection follows hierarchically ordered pathways that see the free-form "planktonic" bacteria invading the wound site followed soon after by the bacterium cell attaching to the tissue cells and transitioning to a state of acute biofilm consisting of early-stage biofilms of cells in microcolonies. This is followed by genetic signaling changes directing these microcolonies to bloom into mature chronic state biofilm communities (Monds and O'Toole 2008). It is this final chronic biofilm state of infection that is notoriously difficult to treat. The mature chronic biofilms in a wound infection may be comprised of a single species or a polymicrobial community. Mixed-species community biofilms can lead to further increased antimicrobial tolerance and antimicrobial resistance changes (e.g., see Lemire et al. 2017; Schwering et al. 2013) due to social biological processes (Nadell et al. 2009).

The **attachment** stage of biofilm formation is the interaction of an individual bacterial cell to a surface. This adhesion process has been nicely reviewed by Berne et al. (2018) and briefly described here. Getting to the target surface may be through passive or active movement. Bacterial self-propelled movement is through swimming via flagellar-based motility or twitch motility mediated by pili (Eruardt 2016). Flow dynamics and shear force are important characteristics that influence the ability to attach. This may be facilitated by air flow or the aquatic habitats the organism is inhabiting as well as the surface moisture and oils on the human skin. In the case of infection, these physical characteristics would be dictated by the biological fluid flows from peristaltic and physical forces. Once the "target" surface is reached, its physicochemical characteristics of polarity and charge density can influence attachment. The bacterial cell attachment is through chemical bonding interactions including van der Waals, electrostatic, H-bonding, and hydrophobicity of molecules associated with abiotic and biotic surfaces and biochemical associated with the bacterium surface. Carbohydrates (glycocalyx), lipids, proteins (membrane proteins or S-layer proteins), and even nucleic acids can be involved; thus, basic physiological conditions of ionic strength and pH of the interacting environment are critical variables in attachment. Many bacteria have surface proteins with functional structures that are also employed for attachment and referred to as "adhesions," which include surface membrane proteins, fimbriae, pili, or flagellum. These adhesions are key to the initial surface attachment contact and immobilization of individual cells.

Not all initial surface contacts between the bacterium and the target surface are productive, and thus such nonproductive interactions are referred to as the "reversible adhesion" stage. If a bacterium can overcome any repulsion forces, it can soon become "irreversibly attached." The action of attachment leads to signaling of further adhesion molecules and the beginning of the production of the extra-polymeric substance (EPS) molecules mediated by complex signaling by the regulator cyclic-di-GMP (Jenal et al. 2017). Depending on the species, additional adhesion proteins may be produced, but for the most part, most adhesions that are involved in initial attachment, such as those involved also in motility (pili and flagella), are disassembled allowing for closer/tighter cell contact with the surface through cell-cell repositioning or cell shape changes (Petrova and Sauer 2012).

Beyond c-di-GMP, other **signaling** molecules play a role and influence attachment. Cyclic-AMP is involved in irreversible attachment conversion and pili modulation (Ono et al. 2014). Cyclic-di-AMP can control polysaccharide production mediating further biofilm formation (Peng et al. 2016; Corrigan and Grundling 2013). Also, the alarmone guanosine pentaphosphate or tetraphosphate, (p)ppGpp, which is the regulator of the stringent response, also regulates steps of adhesion (Hauryliuk et al. 2015). Finally, noncoding small RNAs (sRNAs) were found to regulate adhesions and EPS (Wolska et al. 2016). All these signaling events lead to considerable changes in gene expression, protein expression, and subsequent metabolic changes. Booth et al. (2011) used metabolomics to show that *Pseudomonas fluorescens* metabolism was distinct between the planktonic and biofilm states and that these states responded differently to an antimicrobial challenge. A study on *P. aeruginosa* demonstrated that the global gene expression pattern was distinct in biofilms and that the pattern showed that the biofilm-associated cells are not just surface-attached cells in the stationary phase (Dotsch et al. 2012). Jia et al. (2017) found that 71% of the transcriptome was differentially expressed between biofilm and planktonic cells for *Salmonella enterica*. This study also observed 15 differentially expressed noncoding sRNAs. For these studies, the most notable biofilm-specific genes were those involved in EPS molecule production, repression of various secretion systems, and adaptation to microaerophilic and stress conditions (Dotsch et al. 2012). This illustrates the complexity occurring in gene regulation moving from the early attachment phase to biofilm formation.

The next stage of biofilm formation after initial cell attachment is the growth and division of cells leading to **microcolonies**. During this stage, one sees the initial signaling changes leading to physiological changes. It is at this stage that extracellular molecules begin to be produced to generate the biofilm **extracellular matrix (ECM)**. The matrix of EPS molecules can be the majority of the biomass of a biofilm (Flemming and Wingender 2010). The EPS molecules are key to the biofilm architecture and generate spatial physiological differentiation. The biofilm structure and matrix lead to localized nutrient and waste gradients, sorption abilities, secreted biomolecule retention, antimicrobial tolerance, and resistance properties. The concept of the matrix is expanded as the EPS components go beyond a mixture of polysaccharides to contain proteins, lipid vesicles, and extracellular DNA (eDNA). The matrix is not completely impermeable and has pores and channels providing a level of a circulatory system (Wilking et al. 2013). The matrix is complex and provides a "home" for polymicrobial consortia with species syntrophic and competition activities (Nadell et al. 2009).

The proximity and density of cells in a biofilm allow for efficient cell-to-cell communications either through quorum-sensing molecules, electrical signals (ions), or other small metabolites (reviewed in Flemming et al. 2016). Many of the biofilm properties have been found to be regulated by these extracellular signals. The biofilm development, maturation rate, and proliferation extent are dependent on a multitude of environmental signals. Such signals are provided by the physicochemical environment and the microbial community within the biofilm. However, multiple hormones and host-derived factors have also been shown to function as exogenous signaling that further influences biofilm physiology (Feraco et al. 2016).

The outcome of the matrix development and signaling leads to a **mature chronic biofilm** state. With time, further genomic changes occur leading toward a mature biofilm. This is seen in *P. aeruginosa* moving from acute biofilm at 24 to a 48 h chronic biofilm showing a considerably different gene expression profile from the early-stage biofilm (Dotsch et al. 2012). This mature biofilm contains diverse microniches and physiologically distinct cell states (Harrison et al. 2007). Studies on GacA/S-regulated phase variants demonstrated that these lead to not only metabolically distinct cell states (Workentine et al. 2010) but also that these variants show a different special organization in mature biofilms (Workentine et al. 2013). Additionally, metabolically dormant "persister" cells (Lewis 2010) become more prevalent in mature biofilms and are suspected players in the recalcitrance of chronic infections. This diversity leads to a highly antimicrobial tolerant consortium where, no matter what weapon is chosen to attack the community, a subpopulation is in a state to survive the attack.

Vestby and coworker's recent review (Vestby et al. 2020) on biofilm-related diseases highlights that the presence of a biofilm exasperates the severity and prognosis of a disease. With a mature matrix protecting the biofilm, the bacteria directly attached to host cells can trigger signals in the host cell that activate processes that allow for the entrance of some bacteria via bacterial surface proteins call invasins. Their close association can facilitate the injection of toxin proteins into host cells through various secretion systems. The various disease pathologies and processes are beyond the context of the discussion here, but it is important to note that the biofilm becomes a source of bacterium cells for direct attack to the host as well as a reservoir for the secreted toxins. Furthermore, release of free cells by sloughing or bursting of motile bacteria provides large infectious dosing to invade other wound sites or propagate into the blood stream leading to sepsis. This seeding is prevalent in both tissue- and device-based biofilm infections.

A critical next step is to appreciate mixed-species biofilms and their synergistic interactions as well as understand the complete infection wound ecosystem. This will require reliable in vivo cutaneous wound models (Zhao et al. 2013).

10.3 Biofilm Control Strategies

Attachment of bacteria to surfaces using their surface structures such as pili, fimbriae, flagella, and glycocalyx facilitates biofilm formation. Further, nonspecific factors like surface tension, hydrophobicity, and electrostatic forces are also involved in biofilm formation. Rougher and more hydrophobic surfaces accumulate biofilms more rapidly which can grow to reach thicknesses of 100 mm. On depletion of nutrients, the biofilm-associated cells enter a slow-growing stationary state that can render the biofilm up to 1000 times more resistant to available antimicrobial agents compared to planktonic cells. Various strategies have been developed recently for eradicating these complicated slow-growing biofilms. In this section, promises of phytomedicine, ultrasonic disruption, biofilm matrix-degrading enzymes, acidic electrolyzed water, bacteriophages, and nanotechnology-driven solutions are discussed in details.

10.3.1 Phytomedicine

The field of antibiofilm agents is increasing dramatically, and the exploration of naturally occurring complementary and alternative antibiofilm agents has attracted wide attention. Medicinal plants have a plethora of bioactive principles with antimicrobial and antibiofilm functions in vitro (Ghosh et al. 2012a, 2013). Figure 10.1 illustrates the diverse groups of phytochemicals that were extracted and that can inhibit the formation of polymer matrix, suppress cell adhesion and attachment, interrupt extracellular matrix (ECM) generation, and decrease virulence factor

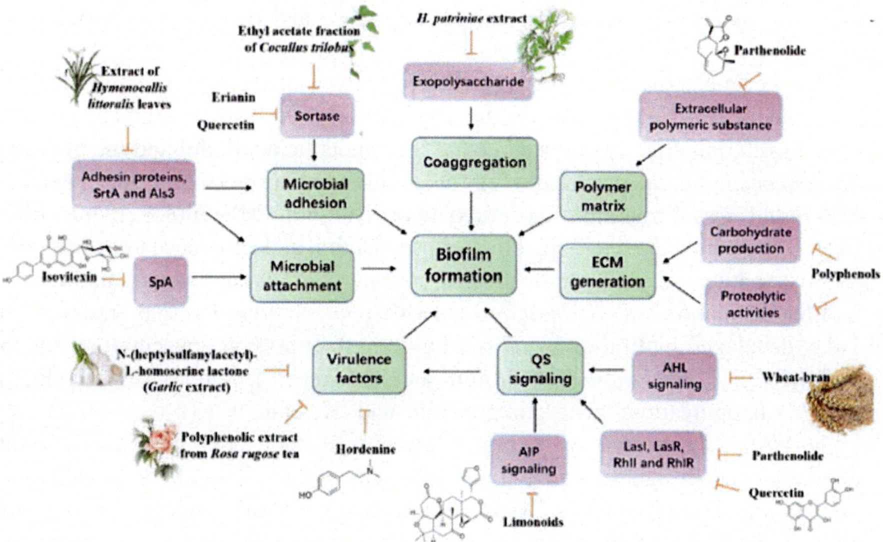

Fig. 10.1 Antibiofilm agents derived from natural plants and their potential mechanisms of action. The inhibition of biofilm formation mainly due to several aspects including the suppression of microbial adhesion and attachment, the inhibition of a polymer matrix and ECM generation, and interference with bacterial coaggregation and the QS network. *QS* quorum sensing, *SrtA* sortase A, *SpA* staphylococcal protein A, *ECM* extracellular matrix, *AIP* autoinducer peptide, and *AHL* acyl-homoserine lactone. (Reprinted from Lu L, Hu W, Tian Z, Yuan D, Yi G, Zhou Y, Cheng Q, Zhu J, Li M (2019) Developing natural products as potential anti-biofilm agents. Chin Med 14:11. Copyright © 2019 The Author(s))

production, thereby blocking the quorum-sensing (QS) network and biofilm development (Lu et al. 2019).

Table 10.1 shows various phytochemicals extracted from medicinal plants with their underlying mechanism behind their biofilm inhibitory activity. *Allium sativum* (garlic) is considered as a rich source of many antimicrobial compounds which can exhibit a potent quorum-sensing inhibitory effect. Biological screening of bioactive principals from garlic resulted in the identification of *N*-(heptylsulfanylacetyl)-L-homoserine lactone that efficiently interrupted quorum-sensing signaling by competitively inhibiting the transcriptional regulators LuxR and LasR (Persson et al. 2005).

Ethyl acetate and aqueous extracts of *Cocculus trilobus* and *Coptis chinensis* blocked the adherence of bacteria to fibronectin-coated surfaces by exerting anti-adhesin effects at the adhesion stage of biofilm formation. Also, suppression of a membrane enzyme named sortase which is responsible for catalyzing the covalent anchoring of surface proteins to peptidoglycan in gram-positive bacteria played a major role in biofilm control (Kim et al. 2002). *Vaccinium oxycoccos* (cranberry) fruit extracts rich in polyphenols affected key events of biofilm formation that include glucan-binding protein activity, the activity of enzymes that cause the destruction of the ECM polymers, carbohydrate production, bacterial hydrophobicity, proteolytic activities, and coaggregation. Thus, these bioactive molecules were speculated to work as promising agents for the prevention and/or treatment of oral diseases, including dental caries and periodontitis (Bodet et al. 2008).

Another medicinal plant, *Ginkgo biloba*, is extensively used in traditional Chinese medicine. Ginkgolic acids are the predominant phytochemicals found in extracts of Ginkgo leaves or seed coats. Ginkgoneolic acid inhibited the growth of *Streptococcus mutans* planktonic cells at a minimum inhibitory concentration (MIC) of 4 μg/mL and a minimum bactericidal concentration (MBC) of 8 μg/mL. Moreover, ginkgoneolic acid inhibited the acid production and adherence to saliva-coated hydroxyapatite of *S. mutans* at sub-MIC concentration. It significantly inhibited the biofilm formation of *S. mutans* ($MBIC_{50}$ = 4 μg/mL) and reduced the 1-day-developed biofilm of *S. mutans* by more than 50% at concentrations as low as $MBRC_{50}$ = 32 μg/mL. The biofilm architecture of *S. mutans* severely lost its integrity upon treatment with ginkgoneolic acid (He et al. 2013).

In *Vibrio* spp., three types of QS systems include acyl-homoserine lactone (AHL)-mediated signaling in the LuxM/N QS system, (S)-3-hydroxytridecan-4-one ("cholera autoinducer 1," CAI-1)-mediated signaling in the CqsA/S system, and a third QS system based on a mixture of interconvertible molecules collectively referred to as autoinducer-2 (AI-2). Halogenated furanones disrupted AHL- and AI-2-based quorum sensing in *Vibrio* spp. by decreasing the DNA-binding activity of the response regulator LuxR thus attenuating the virulence of several *Vibrio* spp. in gnotobiotic brine shrimp *Artemia franciscana.* Cinnamaldehyde reduced the total biofilm biomass with 26 ± 7% and 27 ± 13% in *Vibrio anguillarum* LMG 4411 and *Vibrio vulnificus* LMG 16867, respectively. 2-NO_2-cinnamaldehyde and 4-MeO-cinnamaldehyde resulted in a significant decrease in biomass of *Vibrio*

Table 10.1 Natural antibiofilm agents and their molecular mechanisms in antibiofilm effects

Plant extract/ compounds	Mechanism/ molecular addresses	Target bacteria	Antibiofilm effect
N-(Heptylsulfanylacetyl)-L-homoserine lactone (garlic extract)	Transcriptional regulators LuxR and LasR	*P. aeruginosa*	Decreased elaboration of virulence factors and reduced production of QS signals
Ethyl acetate fraction of *Cocculus trilobus*	Sortase	Gram-positive bacteria	Exerted anti-adhesin effects at the adhesion stage of biofilm formation
Polyphenols (cranberry)	Glucan-binding proteins, enzymes involved in biofilm formation	Cariogenic and periodontopathogenic bacteria	Affected the destruction of the extracellular matrix, carbohydrate production, bacterial hydrophobicity, proteolytic activities, and coaggregation which are involved in biofilm formation
Patriniae	Biofilm-associated genes	*P. aeruginosa*	Inhibited biofilm formation and reduced exopolysaccharide production
Ginkgolic acids	Curli genes and prophage genes	*E. coli* O157:H7	Inhibited biofilm formation on the surfaces of glass, polystyrene, and nylon membranes
Cinnamaldehyde	DNA-binding ability of LuxR	*E. coli* and *Vibrio* spp.	Affected biofilm formation and structure, the swimming motility, stress response, and virulence
Phloretin	Toxin genes (*hlyE* and *stx(2)*), autoinducer-2 importer genes (*lsrACDBF*), curli genes (*csgA* and *csgB*), and prophage genes in *E. coli* O157:H7	*E. coli* O157:H7	Reduced biofilm formation and fimbria production

(continued)

Table 10.1 (continued)

Plant extract/ compounds	Mechanism/ molecular addresses	Target bacteria	Antibiofilm effect
Phloretin	Efflux protein genes	*S. aureus* RN4220 and SA1199B	Antibiofilm formation at low concentrations (1–256 μg/mL)
Isolimonic acid	LuxO and AI-3/ epinephrine activated cell-cell signaling pathway	*Vibrio harveyi*	Interfered with cell-cell signaling and biofilm formation
Hordenine	QS-related genes	*P. aeruginosa*	Blocked QS-controlled phenotypes like biofilm formation and reduced virulence factors
Quercetin	SrtA	*Streptococcus pneumoniae*	Blocked function of SrtA, affect sialic acid production, and impair biofilm formation
Quercetin	LasI, LasR, RhlI, and RhlR	*P. aeruginosa*	Inhibited biofilm formation and production of virulence factors
Quercetin	pH	*S. mutans*	Disrupted the pH in a biofilm
Quercetin	Glycolytic, protein translation elongation, and protein-folding pathways	*Enterococcus faecalis*	Blocked glycolytic, protein translation elongation, and protein-folding pathways
Methanolic fraction of *Zingiber officinale*	The virulence genes, F-ATPase activity, surface protein antigen SpaP	*S. mutans*	Inhibition of surface protein antigen SpaP and inhibitory effect on cell-surface hydrophobicity index of *S. mutans*
Ethanolic extract of *P. betle* leaf (PbLE)	Pyocyanin	*P. aeruginosa strain* PAO1	Inhibition of pyocyanin production and reduction of swarming, swimming, and twitching ability of the bacteria by PbLE extract
Bergenia crassifolia (L.) leaf extract	Gtfs, EPSs	*S. mutans*	Decreased the adherence property

(continued)

Table 10.1 (continued)

Plant extract/ compounds	Mechanism/ molecular addresses	Target bacteria	Antibiofilm effect
			of *S. mutans* through inhibiting Gtfs to synthesize EPSs
Ethanol extract from *Rhodomyrtus tomentosa*	Not investigated	*S. aureus, Staphylococcus epidermidis*	Inhibited staphylococcal biofilm formation and killed mature biofilm
Extract of *Hymenocallis littoralis* leaves	Adhesin proteins, SrtA and Als3	*S. aureus* NCIM 2654 *and C. albicans* NCIM 3466	Antimicrobial, antibiofilm formation and antioxidant activities
Polyphenolic extract (epigallocatechin-3-gallate) from *Camellia sinensis*	Not investigated	*Stenotrophomonas maltophilia* (sm) isolated from cystic fibrosis (CF)	Reduced bacterial cell viability in biofilms in vitro and significantly reduced Sm bacterial counts in an acute infection model with wild-type and CF mice
Polyphenolic extract from *Rosa rugosa* tea	QS-controlled violacein factors	*Chromobacterium violaceum* 026, *E. coli* K-12, and *P. aeruginosa* PAO1	Inhibited swarming motility and biofilm formation
Erianin	SrtA	*S. aureus*	Downregulated SrtA, thereby inhibiting cell adhesion
Isovitexin	SpA	USA300	Reduced SpA and inhibited biofilm formation
Parthenolide	LasI, RhlI, LasR, RhlR, and extracellular polymeric substance	*P. aeruginosa* PAO1	Inhibited QS-related genes expression including *LasI/LasR* and *RhlI/RhlR* and downregulated extracellular polymeric substance
Extract of *Chamaemelum nobile* flowers	Not investigated	*P. aeruginosa* PAO1 and strains isolated from patients	Inhibition of bacteria swarming and biofilm formation

(continued)

Table 10.1 (continued)

Plant extract/ compounds	Mechanism/ molecular addresses	Target bacteria	Antibiofilm effect
Wheat bran	AHL	*S. aureus*	Inhibition of QS and biofilm formation through downregulating AHL level

Abbreviations: *QS* quorum sensing, *SrtA*, sortase A, *SpA* staphylococcal protein A, and *AHL* acyl-homoserine lactone

anguillarum LMG 4411 (decrease of 34 ± 16% and 20 ± 12%, respectively). However, no effect of the cinnamaldehyde derivatives was observed on the *Vibrio vulnificus* LMG 16867 biomass. The QS inhibition mechanism revealed an interference with AI-2 based QS in various *Vibrio* spp. by decreasing the DNA-binding ability of LuxR (Brackman et al. 2008).

Biofilm formation and fimbria production in an *E. coli* O157:H7 strain were also inhibited by phloretin which is an antioxidant, abundantly found in apples. It also restricted the attachment of *E. coli* O157:H7 to human colon epithelial cells and suppressed the tumor necrosis factor alpha-induced inflammatory response. The underlying mechanism of biofilm inhibition was found to be phloretin-mediated repression of toxin genes (*hlyE* and *stx(2)*), autoinducer-2 importer genes (*lsrACDBF*), curli genes (*csgA* and *csgB*), and prophage genes in *E. coli* O157:H7 biofilm cells (Lee et al. 2011). Similarly, citrus limonoids are unique secondary metabolites that can interfere with cell-cell signaling and biofilm formation in *Vibrio harveyi* owing to the modulation of *luxO* expression but not luxR promoter activity. Isolimonic acid and ichangin are potent modulators of bacterial cell-cell signaling and inhibitors of the type III secretion system. Furthermore, isolimonic acid interfered with the AI-3/epinephrine activated cell-cell signaling pathway in QseBC- and QseA-dependent fashion (Vikram et al. 2012). Similarly, many other biofilm inhibitory plant metabolites like hordenine, quercetin, erianin, isovitexin, and parthenolide are listed in Table 10.1 with their mode of action.

10.3.2 New Biofilm Control Strategies

Since most available antimicrobial agents are unable to penetrate the profound biofilm matrix, various new antimicrobial strategies have been implemented in order to both inhibit and disrupt the bacterial biofilms. In this section, the innovative biofilm control strategies are discussed in detail.

10.3.2.1 Ultrasonic Disruption

Bacterial biofilms can be removed more effectively using mechanical force, brushing of teeth being one of the most common examples. Nevertheless, this process is effective only for those surfaces which are easily accessible, thus limiting its efficiency. Hence, an alternative strategy employing ultrasound treatment was developed which could remove implant associated biofilms with greater efficiency. Several indwelling medical devices like catheters, artificial cardiac valves, pacemakers, and prosthetic joints are at high risk of biofilm formation. Associated risks of periprosthetic joint infections due to longer resident time of the implant in the body are also a common health hazard. These implants can become infected from a distant infectious focus by the hematogenous route.

One of the most convenient ways used to remove bacterial cells and disperse them was by sonication of the implants after removal. The dispersed pathogens could be further subjected to multiplex PCR for detection of microbial DNA for the rapid identification of potential pathogens that in one study included *Propionibacterium acnes*, *Corynebacterium* species, *Finegoldia magna*, and *Peptostreptococcus* species (Achermann et al. 2010). Similarly, biofilms of *Staphylococcus aureus*, *Enterococcus faecalis*, and *P. acnes* can be significantly dislodged from titanium and steel surfaces using sonication at 30 kHz with a power output of 300 W at 37 °C for 5 min. However, it should be noted that the biofilm disruption using sonication depends on multiple factors that include the type of equipment, the output power, oscillation frequency, reaction volume, fluid temperature, and duration of sonication (Bjerkan et al. 2009; Harrison et al. 2010).

Life-threatening conditions may arise due to severe biofilm-associated infections of electrophysiological cardiac devices. The use of cardiac pacemakers in patients with atrioventricular conduction block, sick sinus syndrome, and sinus bradycardia is severely challenged due to biofilm formation by bacterial pathogens like *P. acnes*, *S. aureus*, *Streptococcus mitis*, and coagulase-negative staphylococci. These bacteria may form biofilms in the generator pocket and the leads or both within pacemakers. Similar problems are also observed for implantable cardioverter/defibrillators that are used for patients with cardiac failure after myocardial infarction and ventricular arrhythmia (Rohacek et al. 2010). Bacterial biofilms were disrupted showing significant symptomatic improvements in chronic rhinosinusitis after a six-session course of pulsed ultrasound therapy. Formation of a cavity in/on bacterial cell surfaces with a simultaneous generation of peroxides was found to be the underlying mechanism behind the bactericidal effect of high level ultrasonic treatment. However, lower levels of ultrasonic power were incapable of killing bacteria but can play a role in biofilm dispersal to their planktonic state. This makes the bacterial pathogen more susceptible to both antibiotics and innate and adaptive immunity. Synergistic biocidal effect was achieved when ultrasound and gentamicin were applied simultaneously that caused a notable reduction in the viability of sessile *P. aeruginosa*. This mode of dual activity was referred to as the bioacoustical effect. In view of the background, a combined therapy comprised of ultrasonic treatment and antibiotics may prove to be an alternative noninvasive effective medical intervention for the treatment of various biofilm-associated infections (Young et al. 2010).

10.3.2.2 Enzyme-Mediated Disruption

Newly developed biofilm prevention and control approaches target the intrinsic cellular processes involved in biofilm establishment and maturation, such as motility, cell-to-cell aggregation, production of EPS, and intercellular communication (quorum sensing, QS). Several enzymes used for biofilm removal are described in Table 10.2 which are considered as nontoxic substitutes of harmful and ineffective chemical biocides (Meireles et al. 2016).

Biofilm-disrupting enzymes mostly target the matrix EPS molecules in which the cells are embedded. Mechanism of action includes (a) degradation of biofilm components by direct attack, (b) induction of lysis in bacterial cells, (c) hindering quorum sensing, and (d) inducing the formation of antimicrobial molecules. Physical integrity of bacterial biofilms can be compromised by the conversion of EPS matrix components like carbohydrates, polysaccharides, proteins (frequently exhibiting amyloid-like properties), glycoproteins, lipids, phospholipids, glycolipids, and nucleic acids to their monomers which are easy to degrade by metabolic processes. Cystic fibrosis, a disease of the respiratory tract, is associated with mucoid- and alginate-producing strains of *P. aeruginosa* which is responsible for high morbidity and mortality. Coadministration of a combination comprised of alginate lyase and antibiotics like gentamicin and ceftazidime disrupted bacterial biofilms very effectively. The exopolysaccharide produced by the mucoid strains of *P. aeruginosa* was degraded by alginate lyase which facilitated the penetration of antibiotics and, thus, enhanced the efficacy of the antibiotic in respiratory tract infections (Alkawash et al. 2006).

Various enzymes interfere with quorum sensing disrupting the intercellular communication used by many species of bacteria in response to an increase in cell density. QS is usually a complex gene regulatory system which relies on the production, release, and detection of small signaling molecules called autoinducers (AIs) like acyl homoserine lactones (AHLs), autoinducing peptides (AIPs), and autoinducer-2 (AI-2). QS inhibiting enzymes like *N*-acyl homoserine lactonases and acylases inhibit either of these signaling mechanisms. Lactonases hydrolyze the bond in the homoserine ring, avoiding the binding of AHLs to transcriptional regulators, thus disrupting the signaling during QS. Treatment with lactonase was reported to reduce biofilm formation in *P. aeruginosa* by 69–77% apart from decreasing its virulence. Likewise, acylase I was used to remove 9.0% of the adherent cells from a reverse osmosis membrane (Meireles et al. 2016).

Several oxidative enzymes can target the extracellular DNA (eDNA), a critical component for biofilm architecture. Treatment with deoxyribonuclease (DNase) significantly reduced biofilm accumulation of *Enterococcus faecalis*. Similarly, treatment with DNase could reduce biofilm formation in *Streptococcus pneumonia* and *Listeria monocytogenes* up to 85% and 50%, respectively (Meireles et al. 2016).

Numerous polysaccharide-degrading enzymes like amylase, alginate lyase, cellulase, and lysozymes are found to be very effective in disrupting the biofilm matrix.

Table 10.2 Antibiofilm applications of enzymes, their classification, and targets

Action	Enzyme class	Enzyme applied	Target biofilm producer	Surface material	Effect
Anti-QS enzymes	Hydrolase	Lactonase	*P. aeruginosa*	Polystyrene	69–77% biofilm removal
	Hydrolase	Acylase	Bacteria in a reverse osmosis membrane	Reverse osmosis membrane (material not specified)	9.0% biofilm removal
	Hydrolase	Lactonase (expressed by an engineered T7 bacteriophage)	*P. aeruginosa* and *E. coli*	Polyvinyl chloride	Biofilm formation inhibition
Oxidative enzymes	Hydrolase	DNase	*E. faecalis*	Polystyrene	Biofilm removal
	Hydrolase	DNase	*L. monocytogenes*	Polystyrene	50% biofilm removal
Polysaccharide-degrading enzymes	Hydrolase	Dispersin B	*S. epidermidis*	Glass	40% biofilm removal
	Hydrolase	α-amylase	*S. aureus*, *S. epidermidis*	Polystyrene	79% *S. aureus* biofilm removal; no biofilm removal for *S. epidermidis*
Proteolytic enzymes	Hydrolase	Pandion, resinase, spezyme, and paradigm used individually	*P. aeruginosa*	Polystyrene	4 log CFU/mL biofilm removal
	Hydrolase	Bacteriophage enzyme	*E. coli* O157:H7	Stainless steel	Removal of 2.8 log CFU per stainless steel coupon
	Hydrolase	Bacteriophage enzyme	*E. coli*	Plastic pegs	99.997% removal
	Hydrolase	Pronase	*P. fluorescens*	Borosilicate glass	30% biofilm removal
	Hydrolase	Savinase®	*Pseudoalteromonas* sp.	Polystyrene	Complete biofilm removal
	Hydrolase	Savinase®	*P. fluorescens*	Glass wool	80% biofilm removal
	Hydrolase	Endolysin (LysH5)	*S. aureus*	Polystyrene	1–3 log biofilm removal 34% biofilm removal
Anti QS + proteolytic enzymes	Hydrolase	Acylase I + proteinase K	Bacteria in a reverse osmosis membrane	Reverse osmosis membrane	34% biofilm removal

(continued)

Table 10.2 (continued)

Action	Enzyme class	Enzyme applied	Target biofilm producer	Surface material	Effect
				(material not specified)	
Oxidative + polysaccharide-degrading enzymes	Oxidoreductase + hydrolase	Glucose oxidase + lactoperoxidase	*S. aureus*, *S. epidermidis*, *P. aeruginosa*, *P. fluorescens*	Stainless steel	1–2 log CFU/disc biofilm removal of *Staphylococcus*;3 log CFU/disc biofilm removal of *Pseudomonas* spp.
Proteolytic + polysaccharide-degrading enzymes	Hydrolase	Cellulase + pronase	*P. fluorescens*	Borosilicate glass	94% of biofilm removal
Proteolytic enzyme + shear stress	Hydrolase	Savinase® + shear stress	*P. aeruginosa*	Polyethylene	90% biofilm removal
Proteolytic enzymes + ultrasounds	Hydrolase	Amyloglucosidase + US	*E. coli*	Stainless steel	96% biofilm removal
Polysaccharide-degrading enzymes + chemical treatment	Hydrolase	α-amylase + buffer withan anionic surfactant	*B. mycoides*	Stainless steel	2.98 log CFU/cm^2 biofilm removal
Polysaccharide-degrading enzymes + antibiotic	Lyase	Alginate lyase + gentamycin	*P. aeruginosa*	Cellulose fibers	Complete biofilm removal

Meireles A, Borges A, Giaouris E, Simões M (2016) The current knowledge on the application of anti-biofilm enzymes in the food industry. Food Res Int 86:140–146.

P. aeruginosa was inhibited significantly by cellulase which was responsible for the breakdown of complex polysaccharides to low molecular weight simple reducing sugars. Similarly, α-amylase inhibited biofilm formation in *S. aureus* up to 79%. Another active enzyme called dispersin B effectively hydrolyzed poly-*N*-acetylglucosamine present in the biofilms of *S. epidermidis* resulting in 40% of biofilm removal (Meireles et al. 2016).

Proteolytic enzymes, like Savinase®, were found as the most efficient enzymatic preparation for removing *Pseudomonas fluorescens* biofilms from glass wool. A similar effect of the enzyme was also found against *Pseudoalteromonas* sp. where bacterial adhesion was highly compromised inhibiting biofilm formation eventually causing complete biofilm removal. Upon treatment with different enzymatic agents (pandion, resinase, spezyme, and paradigm) for 30 min, a 4 log reduction of *P. aeruginosa* population was achieved. Another proteolytic enzyme, Pronase, removed *P. fluorescens* biofilms up to 30% (Meireles et al. 2016). Further, a combination of enzymes proved to be one of the most powerful strategies for biofilm control. Acylase I combined with proteinase K removed 33.7% of the cells present in a reverse osmosis membrane. Similarly, Orgaz et al. (2007) reported 94% removal of *P. fluorescens* biofilms due to treatment with cellulase and pronase.

Viruses infecting bacteria known as bacteriophages (or phages) have altogether different mechanisms of biofilm disruption predominantly using lysins (described in detail in subsequent section). However, certain genetically engineered phages producing enzymes have been very effective. Engineered T7 bacteriophage expressing a lactonase with broad range of QS inhibitory activity significantly reduced mixed biofilm of *P. aeruginosa* and *E. coli* (Meireles et al. 2016). Another example includes a phage with polysaccharide depolymerases that are able to disrupt the EPS matrix (Donlan 2009). These bacteriophages could reduce biofilms of *E. coli* O157:H7 by 2.8 log colony-forming units (CFU) per stainless steel (SS) coupon.

10.3.2.3 Electrolyzed Water

Electrolyzed water (EW) is considered as an alternative antibiofilm agent which is generally formed by electrolyzing a diluted salt (NaCl) solution. Acidic electrolyzed water (AEW) is obtained from the anode, and basic electrolyzed water (BEW) is simultaneously obtained from the cathode. Recently, AEW was introduced as a sanitizer. Sun et al. (2012) reported the use of AEW and BEW for the eradication of *S. aureus* biofilms. After biofilm formation, the medium was aspirated, and the wells were washed with phosphate-buffered saline followed by the addition of 200 μL of EW with different properties and further incubated at 37 °C for 2 min in order to check the biofilm viability. At pH 10.8 and pH 11.6, BEW could reduce *S. aureus* biofilm up to 42% and 78%, respectively. However, AEW did not show any biofilm reduction at pH 3.5 and 2.5, while it reduced biofilm viability significantly. As the pH of AEW dropped, the bactericidal efficacy of AEW increased. AEW reduced biofilm viability up to 80% and 95% at pH 3.5 and pH 2.5, respectively, after 2 min of treatment. Thus, AEW can be used as a bactericide for *S. aureus* imbedded in a biofilm, while BEW can be applied as a removing agent for an established *S. aureus* biofilm.

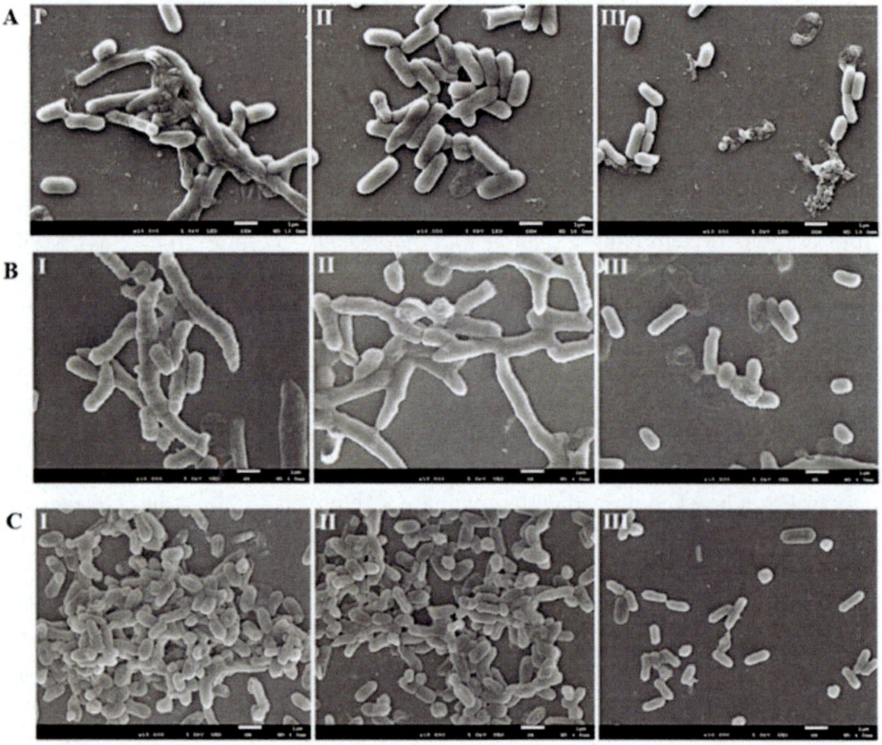

Fig. 10.2 Representative photomicrographs by SEM of biofilms formed by: (**a**) *E. coli*, (**b**) *V. parahaemolyticus*, and (**c**) *L. monocytogenes* for (*I*) untreated biofilms, (*II*) biofilms treated with SDW, and (*III*) biofilms treated with AEW-3 for 5 min. Scale bars represent 5 μm. Pictures were representative of three independent experiments with three replicates each. (Reprinted from Han Q, Song X, Zhang Z, Fu J, Wang X, Malakar PK, Liu H, Pan Y, Zhao Y (2017) Removal of foodborne pathogen biofilms by acidic electrolyzed water. Front Microbiol 8:988. (Open Access))

Even biofilms of foodborne pathogens were removed using acidic AEW which was monitored experimentally in green fluorescent protein-tagged *E. coli*. In this study, the efficiency of AEW for removing biofilms was indicated by a reduction in fluorescent signal of cells in the biofilm by more than 67%. Similarly, AEW treatment reduced the population of biofilm cells up to 82% and 52% in *L. monocytogenes* and *V. parahaemolyticus*, respectively. Biofilm disruption was found to be associated with triggered EPS disruption as revealed from deformation of the carbohydrate C-O-C bond and deformation of the aromatic rings in the amino acids tyrosine and phenylalanine. Figure 10.2 depicts the scanning electron microscopy (SEM) images of the deformed, nonuniform biofilm architecture induced by AEW treatment against *E. coli* (Han et al. 2017).

10.3.2.4 Bacteriophages

Bacteriophages are viruses that target bacteria and are classified according to their shape, size, and kind of nucleic acid. Most of the bacteriophages isolated belong to

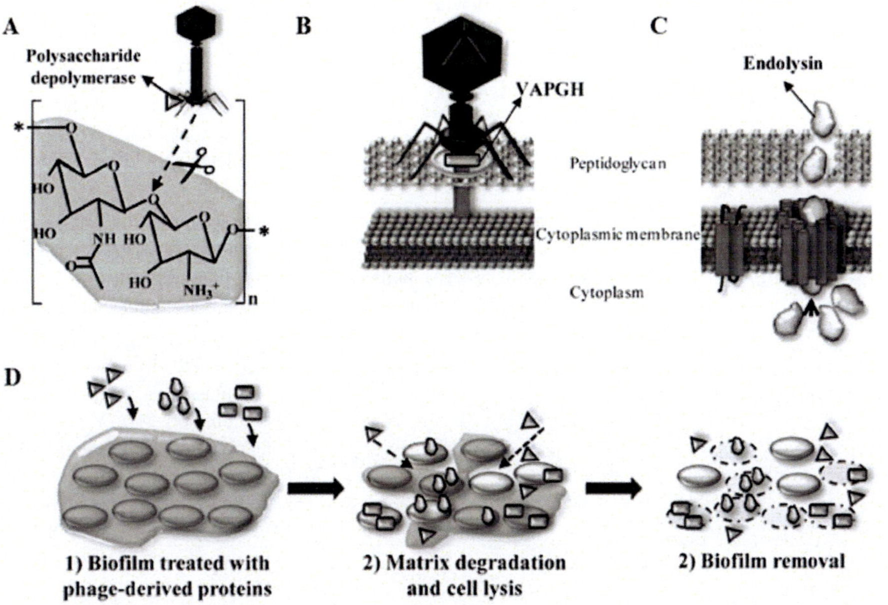

Fig. 10.3 Modes of biofilm inhibition by bacteriophages. (**a**) Location of exopolysaccharide depolymerase degrading b-(1,6) bonds of the biofilm extracellular matrix (PIA/PNAG) of the staphylococcal species in the phage particle and mode of action. (**b**) Location of virion-associated peptidoglycan hydrolase (VAPGH) at the phage particle and its role in the infection process. (**c**) Structure of gram-positive bacteria cell wall and the role of the endolysin during the bacterial lysis. (**d**) Activity of phage-derived proteins when added exogenously and their application as anti-biofilm agents degrading polysaccharide matrices (polysaccharide depolymerases) and lysing bacteria (VAPGHs and endolysins). (Reprinted from Gutiérrez D, Rodríguez-Rubio L, Martínez B, Rodríguez A, García P (2016) Bacteriophages as weapons against bacterial biofilms in the food industry. Front Microbiol 7: 825)

the *Caudovirales* order (tailed bacteriophages), which is further divided into three families (*Myoviridae*, *Podoviridae*, and *Siphoviridae*) according to the microscopic features of the tail morphology. Bacteriophages belonging to the *Siphoviridae* family are the most abundant (57.3%). Although both lytic and lysogenic life cycles are present in bacteriophages, the antimicrobial properties are associated to specifically the lytic cycle where the bacteriophage infects a bacterial cell, multiplies within it, and eventually bursts the host bacterial cells (lysis) releasing the progeny bacteriophages. On the contrary, the lysogenic cycle followed by temperate bacteriophages implies the survival and establishment of the phage genome into the bacterial chromosome (prophage) until environmental signals trigger a lytic cycle, thereby killing only a part of the infected population (Gutiérrez et al. 2016).

Figure 10.3 shows several mechanisms developed in bacteriophages for overcoming the barriers of the bacterial biofilms, the most significant being the production of polysaccharide depolymerases. Table 10.3 lists that various hydrolytic enzymes of bacteriophages can degrade polysaccharides of the biofilm matrix, thus enhancing invasion and dispersion. Polysaccharide depolymerase activity is

Table 10.3 Application of bacteriophages and phage proteins for biofilm removal

Phage or phage protein	Scope of application	Bacteria	Efficacy of the treatment
Phages LiMN4L, LiMN4p, and LiMN17	Stainless steel	*L. monocytogenes*	Phage cocktail reduced biofilm cell counts to undetectable levels after 75 min
Phage P100	Stainless steel	*L. monocytogenes*	Reduction in the cell counts from 3.5 to 5.4 log units/cm^2
Phage P100	Stainless steel	*L. monocytogenes*	Reduction of the biofilm cell counts to undetectable levels after 48 h
Phage K and phage derivatives	Polystyrene	*S. aureus*	Complete elimination of the biomass after 72 h of incubation Complete inhibition of biofilm formation was achieved when co-culturing phage mixture and bacteria
Phage K and DRA88	Polystyrene	*S. aureus*	Complete elimination of the biomass after 48 h of treatment
Phages ISP, Romulus, and Remus	Polystyrene	*S. aureus*	Biofilm reduction of 37.8, 34.4, and 60.4% after24 h treatment when using phages ISP, Romulus, and Remus, respectively
Phages phiIPLA-RODI and phiIPLA-C1	Polystyrene	*S. aureus*	Reduction by 2 log units/well was achieved after8 h of treatment
Phage SAP-26	Polystyrene	*S. aureus*	Reduction of bacteria about 28% after phage treatment, while a synergistic effect with rifampicin allowed for a reduction of about 65%
Phage CP8 and CP30	Glass	*C. jejuni*	Reduction in the biofilm cell counts of 1–3 log units/cm^2
Phage KH1	Stainless steel	*E. coli* O157:H7	Reduction of 1.2 log units per coupon after 4 days of treatment at 4 °C
BEC8 (phage mixture)	Stainless steel, ceramic tile, and high density polyethylene	*E. coli* O157:H7	Reduction of the biofilm cell counts to undetectable levels after 1 h of treatment at 37, 23, and 12 °C
Phage mixture	Spinach harvester blade	*E. coli* O157:H7	Reduction of biofilm cell counts by 4.5 log units per blade after 2 h of treatment
Phage T4	Polystyrene	*E. coli* O157:H7	Complete elimination of the biomass after phage treatment combined with cefotaxime

(continued)

Table 10.3 (continued)

Phage or phage protein	Scope of application	Bacteria	Efficacy of the treatment
Endolysin from phage phi11	Polystyrene	*S. aureus*	Complete elimination of the biomass after 2 h of treatment at 37 °C
Endolysin SAL-2	Polystyrene	*S. aureus*	Reduction of the biomass after 2 h of treatment at 37 °C
EndolysinLysH5	Polystyrene	*S. aureus*	Reduction of biofilm cell counts by 1–3 log units after 3 h of treatment
Domain CHAPK derived from endolysin LysK	Polystyrene	*S. aureus*	Complete elimination of the biomass after 4 h of incubation Complete inhibition of biofilm formation was achieved
Chimeric lysinClyH	Polystyrene	*S. aureus*	Reduction of the biomass in more than 60% after 30 min of treatment
Endolysin Lys68	Polystyrene	*S. typhimurium*	Reduction of biofilm cell counts by a 1 log unit after2 h of treatment in the presence of outer membrane permeabilizers
Exopolysaccharide depolymerase Dpo7	Polystyrene	*S. aureus*	Degradation of 30% of the polysaccharidic matrix of the biofilm

Reprinted from Gutiérrez D, Rodríguez-Rubio L, Martínez B, Rodríguez A, García P (2016) Bacteriophages as weapons against bacterial biofilms in the food industry. Front Microbiol 7:825

associated to tail-spike proteins which are components of the tail of many bacteriophages.

Some phages also produce lytic enzymes called virion-associated peptidoglycan hydrolases (VAPGHs) that play a critical role in the initial phase of the infection cycle. As cartooned in Fig. 10.3b, the formation of small holes by these enzymes in the bacterial cell wall facilitates entry of the phage genetic material into the cytoplasm. Double-stranded DNA phages encode lytic peptidoglycan hydrolases, named endolysins, which access the periplasmic space through holes formed by holins to disrupt the cell wall and lyse the host bacteria at the last step of the lytic infection cycle (Fig. 10.3c).

The *Lactococcus lactis* biofilm structure was disrupted by phage c2 through water channels and cell clusters, while T4 phages infected surface-associated *E. coli*. Three phages LiMN4L, LiMN4p, and LiMN17 effectively disrupted *L. monocytogenes* biofilms on stainless steel coupons. Treatment with the single phages reduced the adhered bacterial cells up to 3 log units, whereas treatment with the phage cocktail reduced cell counts to undetectable levels after 75 min (Ganegama-Arachchi et al. 2013).

Staphylococcal phage K was reported to prevent *S. aureus* biofilm formation upon incubation for 48 h and showed significant time-dependent removal of bacteria by the phage cocktail (10^9 pfu/mL) with the highest being at 72 h at 37 °C (Kelly et al. 2012). A similar result was obtained using phage K combined with another staphylococcal phage, DRA88 (MOI10), to treat established biofilms produced by three *S. aureus* isolates, which were significantly reduced after 4 h and completely removed after 48 h at 37 °C (Alves et al. 2014). Other staphylococcal phages such as ISP, Romulus, and Remus applied individually at 10^9 phages per polystyrene peg effectively degraded the *S. aureus* PS47 biofilm after 24 h up to 37.8%, 34.4%, and 60.4%, respectively (Vandersteegen et al. 2013). Similarly, various other bacteriophages isolated for effective biofilm removal can be seen in Table 10.3 (Gutiérrez et al. 2016).

10.3.3 Nanotechnology-Driven Solutions

Nanoparticles with antimicrobial properties are being used widely for designing biofilm-resistant surfaces, microbicidal bandages, biofilm inhibitory surface coatings, and drug delivery systems. In this section, various metal and metal oxide nanoparticle-based antibiofilm biomaterials are discussed.

10.3.3.1 Gold Nanoparticles

Gold nanoparticles (AuNPs) have attractive physicochemical and optoelectronic properties which have been widely exploited for drug delivery, targeted therapies, biomedical imaging, and biosensors (Ghosh et al. 2018, 2019; Kitture and Ghosh 2019; Kale et al. 2017; Ghosh 2019). Although various physical and chemical methods are available for the synthesis of AuNPs, recently, biological methods have attracted wide attention due to their more environmentally benign nature (Ghosh 2018; Jamdade et al. 2019; Rokade et al. 2017, 2018; Bhagwat et al. 2018). Various medicinal plants like *Dioscorea bulbifera*, *Gnidia glauca*, *Gloriosa superba*, *Platanus orientalis*, *Barleria prionitis*, *Litchi chinensis*, *Adiantum philippense*, and *Dioscorea oppositifolia* were reported to synthesize biocompatible AuNPs with various shapes and sizes (Ghosh et al. 2011, 2012b, 2016a, b, c, d; Shende et al. 2017, 2018; Sant et al. 2013).

Gold nanoparticles (AuNPs) synthesized by the *Woodfordia fruticosa* (Kurz) flower aqueous extract (WfAe) was reported to curb bacterial infections caused by inhibiting and disrupting the biofilm formed at wound sites (Raghuwanshi et al. 2017). *W. fruticosa* possesses a rich phytochemistry that includes quercetin-3-*O*-oxylopyranoside, myricetin-3-*O*-*O*-galloyl-D-galactopyranoside, myricetin-3-*O*-arbinopyranoside, ellagic acid, and woodfordin A, B, and C along with five oligomers woodfordin E, F, G, H, and I, which may play an important role in the reduction of metal ions to their corresponding nanoparticles and thereafter stabilization. Initially, *W. fruticosa* flowers were shade dried for 15 days and pulverized into powder, which was then suspended in distilled water and heated at 50 °C for 20 min. This extract was filtered and added to $HAuCl_4$ followed by mixing at 25 °C with

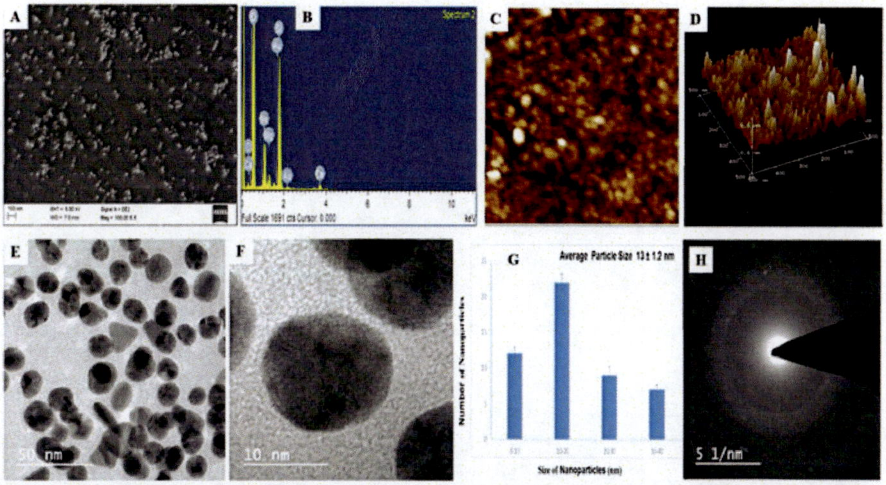

Fig. 10.4 The morphology of AuNPs synthesized using the *W. fruticosa* (Kurz) flower aqueous extract. (**a**) FESEM images. (**b**) EDX spectra. (**c**) AFM 2D image. (**d**) AFM 3D images and micrographs of HR-TEM analysis at (**e**) 50 nm resolution and (**f**) 10 nm resolution. (**g**) Histogram of particle size distribution. (**h**) SAED pattern of the synthesized AuNPs. (Reprinted with permission from Raghuwanshi N, Kumari P, Srivastava AK, Vashisth P, Yadav TC, Prasad R, Pruthi V (2017) Synergistic effects of *Woodfordia fruticosa* gold nanoparticles in preventing microbial adhesion and accelerating wound healing in Wistar albino rats in vivo. Mater Sci Eng C 80: 252–262. Copyright © 2017 Elsevier B.V)

continuous stirring (120 rpm). Figure 10.4 shows that the resulting AuNPs were spherical in nature with a size range of 10–20 nm. Pathogenic biofilms were effectively reduced by phytogenic AuNPs. At a concentration of 256 μg/mL AuNPs, biofilm reduction up to 96.7% and 92.2% was observed in *Candida albicans* and *Candida neoformans*, respectively, as seen in Fig. 10.5. Preformed biofilms of both *C. albicans* and *C. neoformans* were also equally susceptible to AuNPs. Further, bioreduced AuNPs were also checked for in vivo wound healing activity on Wistar albino rats where a total of 30 healthy Wistar albino rats (200 ± 30 g) were divided equally ($n = 6$) into 5 groups as follows: Group I, Control considered as untreated; Group II, Carbopol® 934 ointment base treated; Group III, Standard, povidone-iodine 5% (w/w) ointment treated; Group IV, 2% plant extract (WfAe)-Carbopol® 934; and Group V, 1% AuNPs (WfAuNPs)-Carbopol® 934 treated. On the 12th day, the percentage wound closure was found to be 67.20 ± 0.26%, 69.18 ± 0.27%, 96.01 ± 0.21%, 81.79 ± 0.22%, and 93.80 ± 0.15% for Groups I, II, III, IV, and V, respectively, as evident from Fig. 10.6 (Raghuwanshi et al. 2017).

Hu et al. (2017) reported surface-adaptive AuNPs with effective adherence and enhanced photothermal ablation against methicillin-resistant *Staphylococcus aureus* (MRSA) biofilms. The mixed charged zwitterionic gold nanostructures (abbreviated as AuNP-N-C) with 14 nm diameters were fabricated using mixed self-assembled monolayers (SAMs) consisting of strong electrolytic (10-mercaptodecyl) trimethylammonium bromide ($HS\text{-}C_{10}\text{-}N_4$) and weak electrolytic

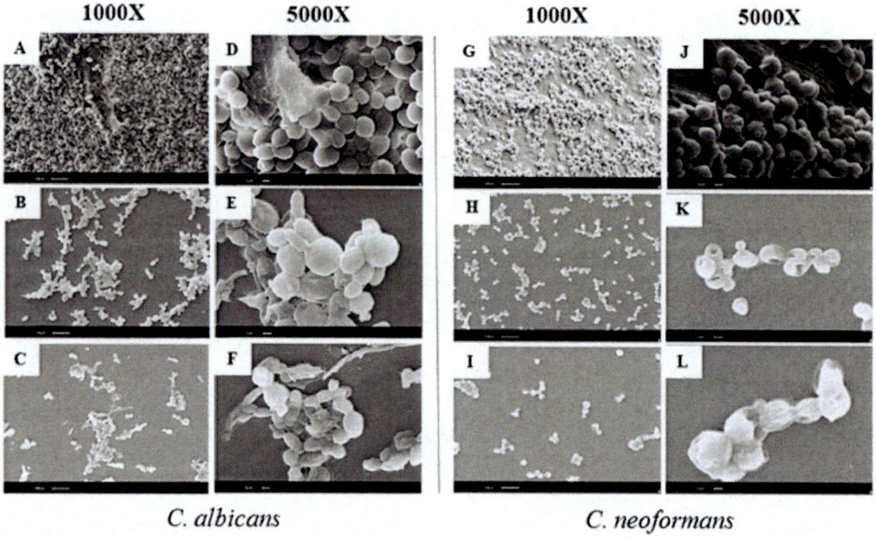

Fig. 10.5 SEM images of biofilm inhibitory potential of AuNPs synthesized using the *W. fruticosa* (Kurz) flower aqueous extract against *C. albicans* (**a–f**) and *C. neoformans* (**g–l**). (Reprinted with permission from Raghuwanshi N, Kumari P, Srivastava AK, Vashisth P, Yadav TC, Prasad R, Pruthi V (2017) Synergistic effects of *Woodfordia fruticosa* gold nanoparticles in preventing microbial adhesion and accelerating wound healing in Wistar albino rats in vivo. Mater Sci Eng C 80: 252–262. Copyright © 2017 Elsevier B.V)

11-mercaptoundecanoic acid (HS-C_{10}-COOH). Further, the 14 nm AuNPs were modified with the SAMs of HS-C_{10}-N_4 and strong electrolytic 10-mercaptodecanesulfonic acid (HS-C_{10}-SO_3H) which were abbreviated as AuNP-N-S. MRSA biofilms are associated with an acidic condition (pH~ 5.5) in which the negatively charged surfaces on AuNP-N-C are converted to the positively charged surfaces resulting in efficient electrostatic adherence to the bacterial surfaces having a negative charge. Upon addition of the AuNP-N-C into the acidic MRSA biofilm, a rapid and distinct color change from wine red to charcoal gray was noticed, indicating the pH-induced aggregation of AuNP-N-C as depicted in Fig. 10.7a. Moreover, the retention quantities of AuNP-N-C (maximum being 60%) were influenced by its concentration and duration of contact time with the MRSA biofilm as seen in Fig. 10.7b, c. However, in the control, no obvious adherence and aggregation of AuNP-N-S in the MRSA biofilm was observed regardless of its concentration or contact time intervals (Fig. 10.7b–d). Further, photothermal ablation of the MRSA biofilm could be achieved using aggregated AuNPs within the biofilm when irradiated with near infrared (NIR) light. However, the dispersed AuNPs barely damaged the surrounding healthy tissues. In order to test the in vivo bactericidal effect, the AuNP solutions were injected into the subcutaneous abscess created in a rabbit model via the local infection of MRSA and then was irradiated by NIR light. The AuNP-N-C were injected in MRSA biofilm at the subcutaneous abscess created in a rabbit model which exhibited a rapid temperature

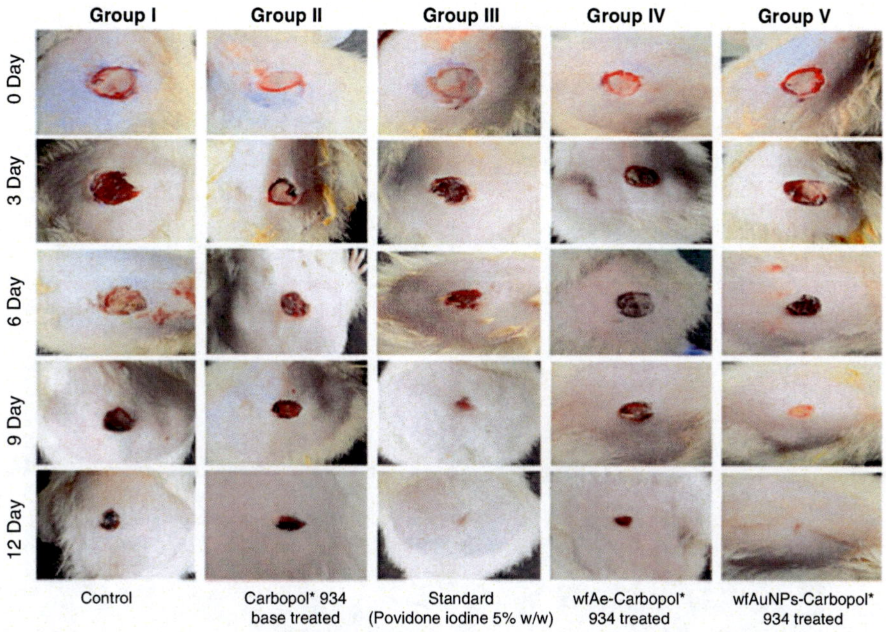

Fig. 10.6 Wound healing assessment of AuNPs synthesized using the *W. fruticosa* (Kurz) flower aqueous extract in terms of full thickness excision wounds in Wistar albino rats. Macroscopic appearance of wound closure (Group I to Group V) at 0, 3, 6, 9, and 12 days after treatment with different formulations. (Reprinted with permission from Raghuwanshi N, Kumari P, Srivastava AK, Vashisth P, Yadav TC, Prasad R, Pruthi V (2017) Synergistic effects of *Woodfordia fruticosa* gold nanoparticles in preventing microbial adhesion and accelerating wound healing in Wistar albino rats in vivo. Mater Sci Eng C 80: 252–262. Copyright © 2017 Elsevier B.V)

increase and remained at about 55 °C on irradiation by NIR light as seen in Fig. 10.8a. At this high temperature, significant bactericidal activity was observed. However, no rise in temperature and associated damages were observed in the tissue surrounding the MRSA biofilms. Severe inflammation and abscess were noticed on the untreated rabbits (Fig. 10.8b). No inflammation was observed on the scarfskin or dermis even after treatment with AuNP-N-C under NIR light for 7 days indicating effective killing of bacteria at the infection site that was further confirmed by counting colony-forming units as evident from Fig. 10.8c, d (Hu et al. 2017).

10.3.3.2 Silver Nanoparticles

Silver nanoparticles (AgNPs) are reported for their potent antimicrobial activity either individually or in combination with various antibiotics. AgNPs generate oxidative stress in the microbes resulting in high bactericidal activity. Biologically synthesized AgNPs employing *D. bulbifera* and *G. glauca* were found to be more very stable, monodispersed, and biocidal (Joshi et al. 2019; Shinde et al. 2018; Ghosh et al. 2012c). The phytochemicals like phenolics, flavonoids, citric acid, and ascorbic acid present in plant extracts can reduce the Ag^{+} ions to AgNPs while

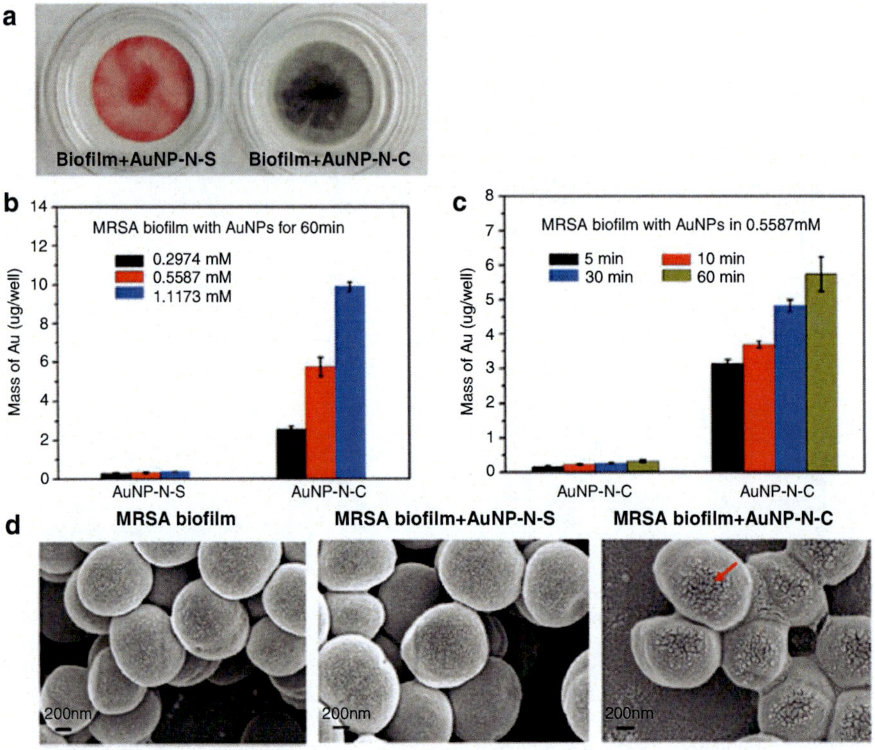

Fig. 10.7 (**a**) Digital images of AuNP-N-S (left, 0.5587 mM, 100 μL) and AuNP-N-C (right, 0.5587 mM, 100uL) when added to the MRSA biofilm at once. Aggregation quantities of AuNPs (**b**) with different concentrations for 60 min and (**c**) with the concentration of 0.5587 mM for different time intervals in MRSA biofilms evaluated by ICP-MS and (**d**) SEM images of MRSA biofilms treated with nothing, AuNP-N-S solution, and AuNP-N-C solution (0.5587 mM, 100 μL) for 60 min (scale bar in images is 200 nm). (Reprinted with permission from Hu D, Li H, Wang B, Ye Z, Lei W, Jia F, Jin Q, Ren K-F, Ji J (2017) Surface-adaptive gold nanoparticles with effective adherence and enhanced photothermal ablation of methicillin-resistant *Staphylococcus aureus* biofilm. ACS Nano 11: 9330–9339. Copyright © 2017, American Chemical Society)

starch, cellulose, proteins, and saponins can act as capping agents. Various chemical modifications were applied for the enhancement of biocompatibility of nanosilver.

Dai et al. (2017) reported the fabrication of cationic ligands functionalized on AgNPs with antibacterial and antibiofilm activity. AgNPs were capped with various cationic polymer ligands with different structural configurations (linear, 4-arm, and 8-arm), molecular weights, and side-chain lengths. The biocompatibility of the biomaterials was enhanced by using a macroinitiator, polyethylene glycol (PEG), that polymerized with 2-(dimethylamino)ethyl methacrylate (DMAEMA). Alkyl bromide further quaternized the tertiary amino groups of the polymer. Antimicrobial activity of the resulting series of AgNPs@polymers increased with the increasing arm number (from linear to 8-arm) of the star-shaped cationic polymer. Figure 10.9a shows that, although AgNPs displayed a characteristic absorption peak at 396 nm in

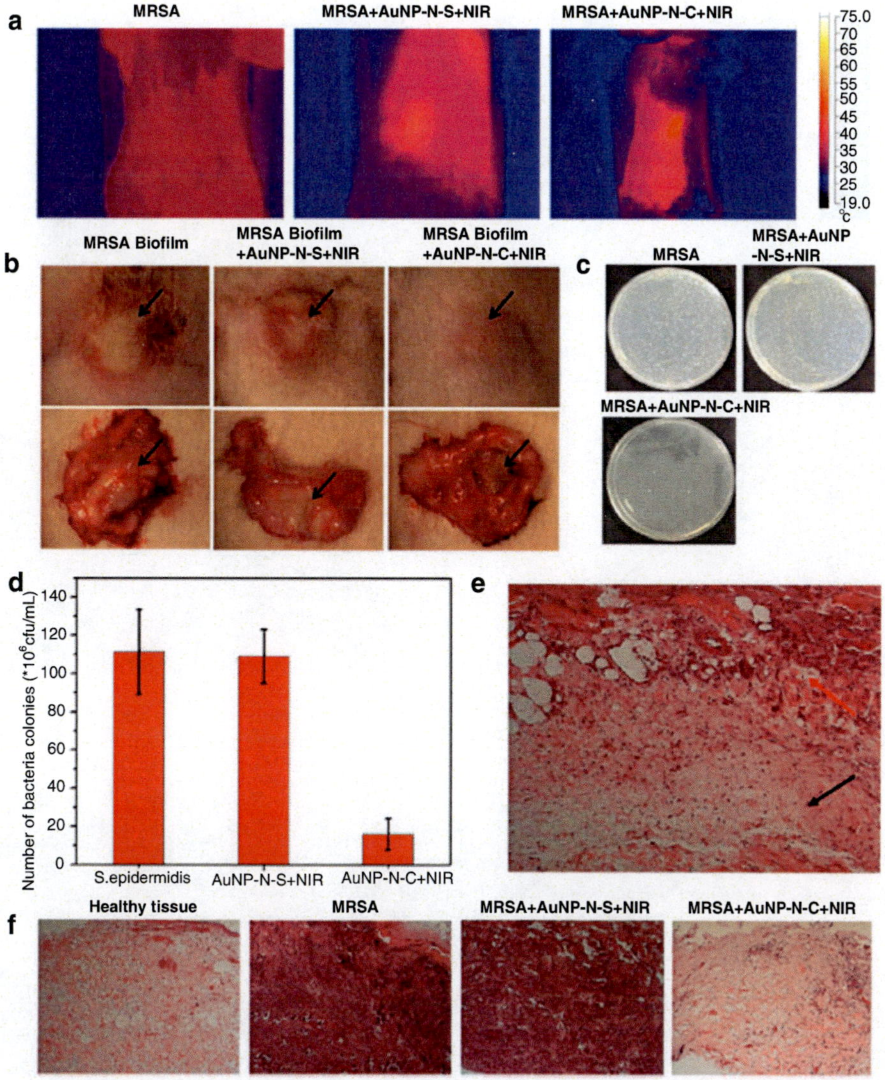

Fig. 10.8 (**a**) Thermographic images of rabbits treated with nothing, AuNP-N-S under NIR light, and AuNP-N-C under NIR light. (**b**) Digital photographs of MRSA infection sites on scarfskin and dermis with different treatments. (**c**) Photomicrographs of bacterial colony-forming units, obtained from infected tissue of rabbits treated under various experimental conditions. (**d**) Related quantitative results of a standard plate counting assay ($n = 6$). (**e**) Histological photomicrographs of MRSA biofilms (red arrow) and its surrounding tissues (black arrow) that were treated with AuNP-N-C under NIR light on day 7 after treatment. All sections were stained with H&E. (**f**) Histological photomicrographs of the skin tissue sections of infected rabbits with different treatments (on day 7). (Reprinted with permission from Hu D, Li H, Wang B, Ye Z, Lei W, Jia F, Jin Q, Ren K-F, Ji J (2017) Surface-adaptive gold nanoparticles with effective adherence and enhanced photothermal ablation of methicillin-resistant *Staphylococcus aureus* biofilm. ACS Nano 11: 9330–9339. Copyright © 2017, American Chemical Society)

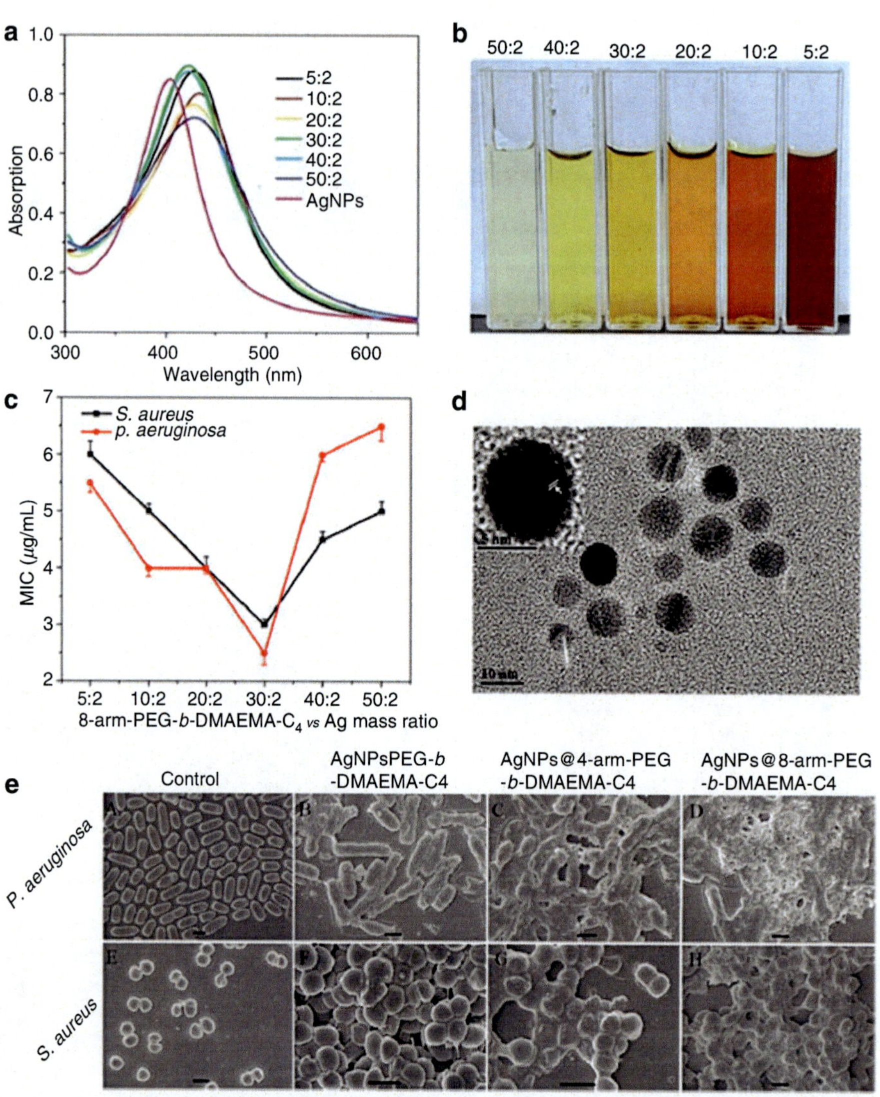

Fig. 10.9 Characterization and antibiofilm activity of AgNPs@8-arm-PEG-*b*-DMAEMA-C_4 with different mass ratios of 8-arm-PEG-*b*-DMAEMA-C_4 to AgNPs: (**a**) UV-visible spectroscopy spectra of the nanocomposites; (**b**) photographs of the nanocomposites dispersed in aqueous solution; (**c**) MIC value of the nanocomposites; (**d**) TEM image of the nanocomposites; (**e**) SEM images of bacteria before and after being treated with AgNPs@polymers: control (**a**, **e**); AgNPs@PEG-*b*-DAMEAM-C_4 (**b**, **f**); AgNPs@4-arm-PEG-*b*-DAMEAM-C_4 (**c**, **g**); and AgNPs@8-arm-PEG-*b*-DAMEAM-C_4 (**d**, **h**). (The scale bar is 1 μm.) (Reprinted with permission from Dai X, Chen X, Zhao J, Zhao Yu, Guo Q, Zhang T, Chu C, Zhang X, Li C (2017) Structure-activity relationship of membrane-targeting cationic ligands on a silver nanoparticle surface in an antibiotic-resistant antibacterial and antibiofilm activity assay. ACS Appl Mater Interfaces 9: 13837–13848. Copyright © 2017 American Chemical Society)

its UV-visible spectra, a red shift was exhibited by the nanocomposites as indicated by their absorption peak at approximately 410–430 nm. This shift was attributed to the variability in the deposition of the 8-arm-PEG-*b*-DMAEMA-C_4 onto the AgNP surface which was also reflected in the color change from yellow to mahogany with the increase in the AgNP concentration as depicted in Fig. 10.9b. All of the nanocomposites with 8-arm-PEG-*b*-DMAEMA-C_4 as the supporting material were excellently dispersed in an aqueous environment. Figure 10.9c shows that the MIC assay confirmed that the antibacterial activity of AgNPs@8-arm-PEG-*b*-DMAEMA-C_4 (30:2) against *S. aureus* and *P. aeruginosa* was stronger than that of the nanocomposites with other mass ratios. Figure 10.9d indicates that the AgNPs@8-arm-PEG-*b*-DMAEMA-C_4 (30:2) were spherical with an average diameter of 8 nm. The AgNPs@8-arm-PEG-*b*-DMAEMA-C_4 (64μg/mL) could eradicate 80% of the established antibiotic-resistant bacterial biofilms, which displayed a stronger antibiofilm activity than AgNPs and 8-arm-PEG-*b*-DMAEMA-C_4. As shown in Fig. 10.9e, untreated bacteria exhibited a smooth surface while irreversible disruption of the membrane structures was observed in the bacterial cells treated with AgNPs@8-arm-PEG-*b*-DAMEAM-C_4 that eventually led to bacterial death.

10.3.3.3 Gold-Silver Bimetallic Nanoparticles

Biologically synthesized gold core silver shell nanoparticles ($Au_{core}Ag_{shell}$NPs) were reported to show potent antibiofilm activity against bacterial pathogens. The *Dioscorea bulbifera* tuber extract (5 mL) was reacted with 95 mL of aqueous solution containing 1 mM $HAuCl_4$ and 0.7 mM of $AgNO_3$ followed by incubation at 50 °C for 5 h. The resulting $Au_{core}Ag_{shell}$NPs had a 9 nm inner gold core that was covered by a silver shell giving a total particle diameter up to 15 nm. This nanomaterial exhibited the highest biofilm inhibition up to 83.68 ± 0.09% against *A. baumannii*. Further, bacterial biofilms of *P. aeruginosa*, *E. coli*, and *S. aureus* were inhibited up to 18.93 ± 1.94%, 22.33 ± 0.56%, and 30.70 ± 1.33%, respectively (Ghosh et al. 2015a).

Similarly, the aqueous root extract of *P. zeylanica* (PZRE) rich in plumbagin, flavonoids, citric acid, sucrose, glucose, fructose, and starch was reported to synthesize AgNPs, AuNPs, and AgAuNPs showing absorbance maxima at 440 nm, 570 nm, and 540 nm, respectively. Optimization studies revealed that the maximum synthesis of AgNPs was achieved at 50 °C with 5 mM $AgNO_3$ within 4.5 h. Likewise, the AuNPs were optimally synthesized with 0.7 mM $HAuCl_4$ within 5 h at 50 °C. The synthesis of bimetallic AgAuNPs was completed within 90 min when 0.7 mM $AgNO_3$ and $HAuCl_4$ were reacted with PZRE. Bioreduced AgNPs were monodispersed with an average size of 60 nm while AuNPs were between 20 and 30 nm. AgAuNPs were mostly blunt-ended hexagons with a size of 90 nm. Phytogenic nanoparticles showed superior antimicrobial and antibiofilm activity against *E. coli*, *A. baumannii*, *S. aureus*, and a mixed culture of *A. baumannii* and *S. aureus*. AgAuNPs showed biofilm inhibition up to 99%, 98%, and 94% against *S. aureus*, *E. coli*, and *A. baumannii* (Salunke et al. 2014).

10.3.3.4 Nanocomposites and Magnetic Nanomaterials

There are now a variety of nanomaterials used to control bacterial biofilm-related infections. These are typically produced in a guided bottom-up approach where molecular compounds (monomers or polymers) are guided through the synthesis procedures to interact, incorporate, or encapsulate antimicrobials into composites that provide drug dosing and delivery. We typically see such materials as coatings on medical devices to prevent biofilms or for the direct antimicrobial/drug delivery itself. Many of these approaches overlap between antimicrobial and antitumor treatment goals as there are many analogies and processes similar between them (Lambert et al. 2011). Examples of such materials include lipid, detergents, or other amphiphilic molecules that would generate micelles or liposomes that would encapsulate a drug or active nanoparticle (Yu et al. 2016; Angelova et al. 2017). Hydrogels have also been explored, and mesoporous silica materials have become a common lattice to work with (Bernardos et al. 2019). Various polymers and cross-linking chemistries for encapsulation for both active and passive antibiofilm coatings have been explored, from nanocellulose backbones to silica polymers to various plastics (Balaure and Grumezescu 2020). Chemistries are in place to associate antibiotics, quaternary cations, and cationic peptides as well a metal ions or metal nanoparticles. Even carbon nanomaterials alone or conjugated to antibiotics have been explored (Maas 2016).

This materials topic is now quite extensive and beyond a complete description here. However, below we choose to describe a new novel approach of iron oxide nanoparticles to illustrate the direction of the field.

Silver-SPION Nanocomposite

Among various types of nanomaterials, iron oxide nanoparticles (IONPs) are well known for their promising therapeutic efficacy and diagnostic applications. Thus, rationally engineered IONPs with tailored physicochemical and surface properties have been developed for magnetic resonance imaging (MRI), drug delivery, magnetic hyperthermia, and in vitro diagnostics. Similarly, these magnetic nanomaterials can be surface functionalized by several bioactive molecules (like diosgenin and curcumin), drugs, antibodies, and targeting ligands for enhanced theranostic applications (Ghosh et al. 2015b; Kitture et al. 2012). Similarly, metal doping is considered as a significant strategy for synergistic enhancement for biomedical applications of individual metals. Thus, nanocomposites with silver, copper, platinum, palladium, zinc oxide, molybdenum, etc. have shown higher therapeutic properties compared to individual metal nanoparticles (Robkhob et al. 2020; Karmakar et al. 2020; Adersh et al. 2015; Ghosh et al. 2015c, d; Kitture et al. 2015).

Durmus and Webster (2013) reported that silver in combination of other metal oxides like superparamagnetic iron oxide nanoparticles (SPIONs) can also eradicate antibiotic-resistant biofilms of methicillin-resistant *S. aureus* (MRSA). Initially the SPIONs were synthesized using a variation of the high temperature reflux method in triethylene glycol (TREG). After mixing 2 mmol of iron (III) acetylacetone in 25 mL of TREG, the solution was heated gradually to 278 °C with constant stirring for 30 min. The synthesized SPIONs were then cooled slowly to room temperature and conjugated with 4 mmol of dimercaptosuccinic acid (DMSA), which was used as a capping agent and precipitated in 200 mL of 100% ethanol saturated with ammonium sulfate ($(NH_4)_2SO_4$) and then were isolated using a strong magnetic field. Conjugation with silver was brought about by diluting 10 mg/mL SPIONs solution in ethanol (1:10) with 1 mmol of $AgNO_3$ followed by overnight incubation with shaking (90 rpm). SPIONs had an average size of 10.07 ± 1.54 nm, while the silver-conjugated SPIONs showed a larger mean hydrodynamic diameter of 176.5 ± 27 nm. It is important to note that a 30% decrease in biofilm mass was achieved when treating the MRSA biofilms with 1 mg/mL silver-conjugated SPIONs. A magnetic field further enhanced biofilm eradication properties as a 47% reduction in MRSA biofilms was observed upon treatment with 1 mg/mL of silver-conjugated SPION in the presence of an external field. This increase in the antibacterial activity of silver-conjugated SPIONs in the presence of an external magnetic field might be attributed to improve penetration of the nanoparticles within the biofilm matrix. This novel nanomaterial can be strategically used for targeting the site of infection for disruption and eradication of the bacterial biofilm.

Carboxylate Functionalized SPIONs

In another study, Leuba et al. (2013) reported carboxylate functionalized SPIONs for the reduction of biofilm formation by *S. aureus*. The SPIONs were pre-synthesized using a high temperature reflux method in triethylene glycol (TREG) as discussed in the earlier section. Carboxylate functionalized SPIONs were synthesized by reacting succinic anhydride and the amine functionalized SPIONs in a tetrahydrofuran (THF) solution followed by rinsing with THF the next day to remove any unreacted succinic anhydride. The resulting carboxylate functionalized SPIONs were between 10 and 20 nm in diameter that greatly affected the growth of *S. aureus* after the initial biofilm formation over the 24 h time period. After the 24 h period, carboxylate functionalized SPION had a 33.5% decrease in bacterial growth compared to the untreated *S. aureus* growth, while the unfunctionalized SPION resulted in only an 18.7% decrease.

SPION-Encapsulated Polymersome

Geilich et al. (2017) reported the biofilm-eradicating potential of iron oxide-encapsulating polymersomes (IOPs) which was synthesized employing a two-phase self-assembly process as depicted in Fig. 10.10a. Initially, 100 mL of oleic acid-coated iron oxide nanocrystals (5 ± 2.5 nm) were resuspended in 1 mL of tetrahydrofuran under nitrogen gas. Next, 10 mg of the methoxy poly(ethylene glycol)-poly(D)-(L)-lactic acid copolymer was subsequently dissolved in the SPION suspension and dispersed via ultrasonication which was then injected through a syringe atomizer into 10 mL of a rapidly stirring solution of methicillin sodium salt solution. The resulting IOPs showed a narrow size distribution of 83 ± 6 nm as seen in Fig. 10.10b. Neither SPIONs alone nor drug (methicillin)-free IOPs succeeded in eliminating more than 60% of the live cell content while IOPs containing $\geq$ 40 μg/mL SPION and 20 μg/mL of methicillin exhibited complete

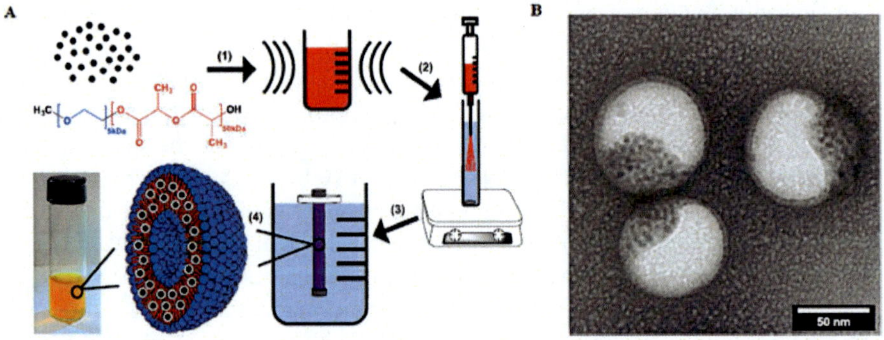

Fig. 10.10 Synthesis of iron oxide-encapsulating polymersomes. (**a**) 5 nm monodisperse hydrophobic SPIONs were combined with a mPEG-PDLLA copolymer in an organic solvent and ultrasonicated to create a uniform suspension (*1*). This organic phase is injected through an atomizer into an actively stirring aqueous phase containing PBS and methicillin (*2*). Finally, the mixture was dialyzed against pure PBS to remove the organic solvent and unencapsulated drug (*3*). The resulting polymersome solution was highly stable and translucent orange (*4*); (**b**) Representative transmission electron micrograph of hydrophobic SPIONs embedded inside the polymersome bilayer. (Reprinted with permission from Geilich BM, Gelfat I, Sridhar S, van de Ven AL, Webster TJ (2017) Superparamagnetic iron oxide-encapsulating polymersome nanocarriers for biofilm eradication. Biomaterials 119: 78–85. Copyright © 2016 Elsevier Ltd.)

eradication of MRSA biofilms. Deformity in the extracellular polymeric matrix architecture led to enhanced biofilm eradication.

10.4 Conclusions and Future Perspectives

Biofilm-associated diseases due to multidrug-resistant bacterial pathogens have severely challenged the available therapeutics for human healthcare. Hence, novel antimicrobial strategies are constantly being developed to address biofilm-associated nosocomial infections. As indicated in this chapter, novel phytochemicals from medicinal plants may be potential quorum-sensing inhibitors, metal chelators, and biofilm efflux pump inhibitors that can be used as complementary and alternative therapy. At the same time, physical-, chemical-, and bacteriophage-mediated methods for biofilm inhibition and disruption are found to be beneficial over the use of high doses of antibiotics. However, reduced penetration of therapeutic agents within the highly compact and dense biofilm matrix limits the efficacy of available as well as newly developed antibiofilm therapeutic agents. Nanoscale metal and metal oxide particles can prove to be beneficial as they have greater penetrability and higher damaging capacity. Silver-based nanostructures generate more reactive oxygen species (ROS) resulting in oxidative stress damaging nucleic acid, proteins, and even cell boundaries. Interestingly, magnetic particles can be targeted using a magnetic field for the delivery of drugs to the site of infection for controlling bacterial biofilm. Further, anti-adhesive and polymer grafted biocidal nanoparticles with ROS- and nitric oxide (NO)-releasing properties can be used for surface coating on medical devices. Coating of nanoparticles on medical devices susceptible to biofilm formation will not only improve their mechanical (e.g., hardness, stress, and Young's modulus) and tribological properties (e.g., wear resistance, adhesion, and friction) but also decrease the risk of nosocomial infections. Meanwhile, improved specificity, safety, and efficiency alone or in combination with other antimicrobial agents are also required to be developed for effective inhibition and eradication of bacterial biofilms.

Acknowledgments Dr. Sougata Ghosh acknowledges the Department of Science and Technology (DST), Ministry of Science and Technology, Government of India, and Jawaharlal Nehru Centre for an Advanced Scientific Research, India, for funding under the Post-doctoral Overseas Fellowship in Nano Science and Technology (Ref. JNC/AO/A.0610.1(4) 2019-2260 dated August 19, 2019).

References

Achermann Y, Vogt M, Leunig M, Wüst J, Trampuz A (2010) Improved diagnosis of periprosthetic joint infection by multiplex PCR of sonication fluid from removed implants. J Clin Microbiol 48 (4):1208–1214

Adersh A, Kulkarni AR, Ghosh S, More P, Chopade BA, Gandhi MN (2015) Surface defect rich ZnO quantum dots as antioxidant inhibiting α-amylase and α-glucosidase: a potential anti-diabetic nanomedicine. **J Mater Chem B** 3:4597–4606

Alkawash MA, Soothill JS, Schiller NL (2006) Alginate lyase enhances antibiotic killing of mucoid *Pseudomonas aeruginosa* in biofilms. APMIS 114(2):131–138

Alves DR, Gaudion A, Bean JE, Perez Esteban P, Arnot TC, Harper DR, Kot W, Hansen LH, Enright MC, Jenkins ATA (2014) Combined use of bacteriophage K and a novel bacteriophage to reduce *Staphylococcus aureus* biofilm formation. Appl Environ Microbiol 80:6694–6703

Angelova A, Garamus VM, Angelov B, Tian Z, Li Y, Zou A (2017) Advances in structural design of lipid-based nanoparticle carriers for delivery of macromolecular drugs, phytochemicals and anti-tumor agents. Adv Colloid Interf Sci 249:331–345

Arciola CR, Campoccia D, Montanaro L (2018) Implant infections: adhesion, biofilm formation and immune evasion. Nat Rev Microbiol 16:397–409

Balaure PC, Grumezescu AM (2020) Recent advances in surface nanoengineering for biofilm prevention and control. Part I: molecular basis of biofilm recalcitrance. Passive anti-biofouling nanocoatings. Nanomaterials (Basel) 10(6):1230

Bernardos A, Piacenza E, Sancenon F, Mehrdad H, Maleki A, Turner RJ, Martinez-Manez R (2019) Mesoporous silica-based materials with bactericidal properties. Small 15:1900669

Berne C, Ellison C, Ducret A, Brun YV (2018) Bacterial adhesion at the single-cell level. Nat Rev Microbiol 16:616–627

Bhagwat TR, Joshi KA, Parihar VS, Asok A, Bellare J, Ghosh S (2018) Biogenic copper nanoparticles from medicinal plants as novel antidiabetic nanomedicine. World J Pharm Res 7 (4):183–196

Bjerkan G, Witsø E, Bergh K (2009) Sonication is superior to scraping for retrieval of bacteria in biofilm on titanium and steel surfaces in vitro. Acta Orthop 80(2):245–250

Bodet C, Grenier D, Chandad F, Ofek I, Steinberg D, Weiss EI (2008) Potential oral health benefits of cranberry. Crit Rev Food Sci Nutr 48:672–680

Booth SC, Workentine ML, Wen J, Shaykhutdinov R, Vogel HJ, Ceri H, Turner RJ, Weljie AM (2011) Differences in metabolism between the biofilm and planktonic response to metal stress. J Proteome Res 10:3190–3199

Brackman G, Defoirdt T, Miyamoto C, Bossier P, Van Calenbergh S, Nelis H, Coenye T (2008) Cinnamaldehyde and cinnamaldehyde derivatives reduce virulence in *Vibrio* spp. by decreasing

the DNA-binding activity of the quorum sensing response regulator LuxR. BMC Microbiol 8:149
Corrigan RM, Grundling A (2013) Cyclic di-AMP: another second messenger enters the fray. Nat Rev Microbiol 11:413–524
Dai X, Chen X, Zhao J, Zhao Y, Guo Q, Zhang T, Chu C, Zhang X, Li C (2017) Structure−activity relationship of membrane-targeting cationic ligands on a silver nanoparticle surface in an antibiotic-resistant antibacterial and antibiofilm activity assay. ACS Appl Mater Interfaces 9:13837-−13848
Delcaru C, Alexandru I, Podgoreanu P, Grosu M, Stavropoulos E, Chifiriuc MC, Lazar V (2016) Microbial biofilms in urinary tract infections and prostatitis: etiology, pathogenicity and combating strategies. Pathogens 5:65
Donlan RM (2009) Preventing biofilms of clinically relevant organisms using bacteriophage. Trends Microbiol 17(2):66–72
Dotsch A, Eckweiler D, Schniederjans M, Zimmermann A, Jensen V, Scharfe M, Geffers R, Haussler S (2012) The *Pseudomonas aeruginosa* transcriptome in Planktonic cultures and static biofilms using RNA sequencing. PLoS One 7:e31092
Durmus NG, Webster TJ (2013) Eradicating antibiotic-resistant biofilms with silver conjugated superparamagnetic iron oxide nanoparticles. Adv Healthc Mater 2:165–171
Eruardt M (2016) Strategies to block bacterial pathogenesis by interference with motility and chemotaxis. Curr Top Microbiol Immunol 398:185–205
Feraco D, Blaha M, Khan S, Green JM, Plotkin BJ (2016) Host environmental signals and effects on biofilm formation. Microb Pathog 99:253–263
Fey PD, Olson ME (2010) Current concepts in biofilm formation of *Staphylococcus epidermidis*. Future Microbiol 5:917–933
Flemming H-C, Wingender J (2010) The biofilm matrix. Nat Rev Microbiol 8:623–633
Flemming H-C, Wingender J, Szewzyk U, Steinberg P, Rice SA, Kjelleberg S (2016) Biofilms: an emergent from of bacterial life. Nat Rev Microbiol 14:563–575
Ganegama-Arachchi GJ, Cridge AG, Dias-Wanigasekera BM, Cruz CD, McIntyre L, Liu R, Flint SH, Mutukumira AN (2013) Effectiveness of phages in the decontamination of *Listeria monocytogenes* adhered to clean stainless steel, stainless steel coated with fish protein, and as a biofilm. J Ind Microbiol Biotechnol 40:1105–1116
Geilich BM, Gelfat I, Sridhar S, van de Ven AL, Webster TJ (2017) Superparamagnetic iron oxide-encapsulating polymersome nanocarriers for biofilm eradication. Biomaterials 119:78–85
Ghosh S (2018) Copper and palladium nanostructures: a bacteriogenic approach. Appl Microbiol Biotechnol 101(18):7693–7701
Ghosh S (2019) Mesoporous silica based nano drug delivery system synthesis, characterization and applications. In: Mohapatra SS, Ranjan S, Dasgupta N, Mishra RK, Thomas S (eds) Nanocarriers for drug delivery. Elsevier, Amsterdam, pp 285–317
Ghosh S, Chopade BA (2017) *Dioscorea bulbifera*: a model system in nanobiotechnology and medicine. LAP LAMBERT Academic Publishing GmbH & Co. KG, Germany
Ghosh S, Patil S, Ahire M, Kitture R, Jabgunde A, Kale S, Pardesi K, Bellare J, Dhavale DD, Chopade BA (2011) Synthesis of gold nano-anisotrops using *Dioscorea bulbifera* tuber extract. J Nanomater 2011:354793. https://doi.org/10.1155/2011/354793
Ghosh S, Ahire M, Patil S, Jabgunde A, Bhat Dusane M, Joshi BN, Pardesi K, Jachak S, Dhavale DD, Chopade BA (2012a) Antidiabetic activity of *Gnidia glauca* and *Dioscorea bulbifera*: potent amylase and glucosidase inhibitors. Evid Based Complement Alternat Med 2012:929051. https://doi.org/10.1155/2012/929051
Ghosh S, Patil S, Ahire M, Kitture R, Gurav DD, Jabgunde AM, Kale S, Pardesi K, Shinde V, Bellare J, Dhavale DD, Chopade BA (2012b) *Gnidia glauca* flower extract mediated synthesis of gold nanoparticles and evaluation of its chemocatalytic potential. J Nanobiotechnol 10:17
Ghosh S, Patil S, Ahire M, Kitture R, Jabgunde A, Kale S, Pardesi K, Cameotra SS, Bellare J, Dhavale DD, Chopade BA (2012c) Synthesis of silver nanoparticles using *Dioscorea bulbifera*

tuber extract and evaluation of its synergistic potential in combination with antimicrobial agents. Int J Nanomedicine 7:483–496

Ghosh S, Derle A, Ahire M, More P, Jagtap S, Phadatare SD, Patil AB, Jabgunde AM, Sharma GK, Shinde VS, Pardesi K, Dhavale DD, Chopade BA (2013) Phytochemical analysis and free radical scavenging activity of medicinal plants *Gnidia glauca* and *Dioscorea bulbifera*. PLoS One 8(12):e82529

Ghosh S, Jagtap S, More P, Shete UJ, Maheshwari NO, Rao SJ, Kitture R, Kale S, Bellare J, Patil S, Pal JK, Chopade BA (2015a) *Dioscorea bulbifera* mediated synthesis of novel $Au_{core}Ag_{shell}$ nanoparticles with potent antibiofilm and antileishmanial activity. J Nanomater 2015:562938

Ghosh S, More P, Derle A, Kitture R, Kale T, Gorain M, Avasthi A, Markad P, Kundu GC, Kale S, Dhavale DD, Bellare J, Chopade BA (2015b) Diosgenin functionalized iron oxide nanoparticles as novel nanomaterial against breast cancer. J Nanosci Nanotechnol 15(12):9464–9472

Ghosh S, More P, Nitnavare R, Jagtap S, Chippalkatti R, Derle A, Kitture R, Asok A, Kale S, Singh S, Shaikh ML, Ramanamurthy B, Bellare J, Chopade BA (2015c) Antidiabetic and antioxidant properties of copper nanoparticles synthesized by medicinal plant *Dioscorea bulbifera*. J Nanomed Nanotechnol S6:7

Ghosh S, Nitnavare R, Dewle A, Tomar GB, Chippalkatti R, More P, Kitture R, Kale S, Bellare J, Chopade BA (2015d) Novel platinum-palladium bimetallic nanoparticles synthesized by *Dioscorea bulbifera*: anticancer and antioxidant activities. Int J Nanomedicine 10 (1):7477–7490

Ghosh S, Patil S, Chopade NB, Luikham S, Kitture R, Gurav DD, Patil AB, Phadatare SD, Sontakke V, Kale S, Shinde V, Bellare J, Chopade BA (2016a) *Gnidia glauca* leaf and stem extract mediated synthesis of gold nanocatalysts with free radical scavenging potential. J Nanomed Nanotechnol 7:358

Ghosh S, Harke AN, Chacko MJ, Gurav SP, Joshi KA, Dhepe A, Dewle A, Tomar GB, Kitture R, Parihar VS, Banerjee K, Kamble N, Bellare J, Chopade BA (2016b) *Gloriosa superba* mediated synthesis of silver and gold nanoparticles for anticancer applications. J Nanomed Nanotechnol 7:4

Ghosh S, Chacko MJ, Harke AN, Gurav SP, Joshi KA, Dhepe A, Kulkarni AS, Shinde VS, Parihar VS, Asok A, Banerjee K, Kamble N, Bellare J, Chopade BA (2016c) *Barleria prionitis* leaf mediated synthesis of silver and gold nanocatalysts. J Nanomed Nanotechnol 7:4

Ghosh S, Gurav SP, Harke AN, Chacko MJ, Joshi KA, Dhepe A, Charolkar C, Shinde VS, Kitture R, Parihar VS, Banerjee K, Kamble N, Bellare J, Chopade BA (2016d) *Dioscorea oppositifolia* mediated synthesis of gold and silver nanoparticles with catalytic activity. J Nanomed Nanotechnol 7:5

Ghosh S, Sanghavi S, Sancheti P (2018) Metallic biomaterial for bone support and replacement. In: Balakrishnan P, Sreekala MS, Thomas S (eds) Fundamental biomaterials: metals, vol 2. Woodhead Publishing., Elsevier, Amsterdam, pp 139–165

Ghosh S, Patil PD, Kitture RD (2019) Physically responsive nanostructures in breast cancer theranostics. In: Thorat ND, Bauer J (eds) External field and radiation stimulated breast cancer nanotheranostics. IOP Publishing Ltd, UK

Gutiérrez D, Rodríguez-Rubio L, Martínez B, Rodríguez A, García P (2016) Bacteriophages as weapons against bacterial biofilms in the food industry. Front Microbiol 7:825

Han Q, Song X, Zhang Z, Fu J, Wang X, Malakar PK, Liu H, Pan Y, Zhao Y (2017) Removal of food borne pathogen biofilms by acidic electrolyzed water. Front Microbiol 8:988

Harrison JJ, Ceri H, Turner RJ (2007) Multimetal resistance and tolerance in microbial biofilms. Nat Rev Microbiol 5:928–938

Harrison JJ, Stremick CA, Turner RJ, Allan ND, Olson ME, Ceri H (2010) Microtiter susceptibility testing of microbes growing on peg lids: a miniaturized biofilm model for high-throughput screening. Nat Protoc 5:1236–1254

Hauryliuk V, Atkinson GC, Murakami KS, Tenson T, Gerdes K (2015) Recent functional insights into the role of (p)ppGpp in bacterial physiology. Nat Rev Microbiol 13:298–309

He J, Wang S, Wu T, Cao Y, Xu X, Zhou X (2013) Effects of ginkgoneolic acid on the growth, acidogenicity, adherence, and biofilm of *Streptococcus mutans* in vitro. Folia Microbiol 58:147–153

Hu D, Li H, Wang B, Ye Z, Lei W, Jia F, Jin Q, Ren K-F, Ji J (2017) Surface-adaptive gold nanoparticles with effective adherence and enhanced photothermal ablation of methicillin-resistant *Staphylococcus aureus* biofilm. ACS Nano 11:9330-–9339

Jamdade DA, Rajpali D, Joshi KA, Kitture R, Kulkarni AS, Shinde VS, Bellare J, Babiya KR, Ghosh S (2019) *Gnidia glauca* and *Plumbago zeylanica* mediated synthesis of novel copper nanoparticles as promising antidiabetic agents. Adv Pharmacol Sci 2019:9080279

Jenal U, Reinders A, Lori C (2017) Cyclic di- GMP: second messenger extraordinaire. Nat Rev Microbiol 15:271–284

Jia K, Wang G, Liang L, Wang M, Wang H, Xu X (2017) Preliminary transcriptome analysis of mature biofilm and planktonic cells of *Salmonella enteritidis* exposure to acid stress. Front Microbiol 8:1861

Joshi KA, Ghosh S, Dhepe A (2019) Green synthesis of antimicrobial nanosilver using *in-vitro* cultured *Dioscorea bulbifera*. Asian J Org Med Chem 4(4):222–227

Jung H-S, Ehlers MM, Lombaard H, Redelinghuys MJ, Kock MM (2017) Etiology of bacterial vaginosis and polymicrobial biofilm formation. Crit Rev Microbiol 43:651–667

Kale SN, Kitture R, Ghosh S, Chopade BA, Yakhmi JV (2017) Nanomaterials as enhanced antimicrobial agents/activity enhancer for transdermal applications: a review. In: Grumezescu AM (ed) Antimicrobial architectonics from synthesis to applications. Elsevier, Amsterdam, pp 279–321

Karmakar S, Ghosh S, Kumbhakar P (2020) Enhanced sunlight driven photocatalytic and antibacterial activity of flower-like ZnO@MoS_2 nanocomposite. J Nanopart Res 22:11

Kelly D, McAuliffe O, Ross RP, Coffey A (2012) Prevention of *Staphylococcus aureus* biofilm formation and reduction in established biofilm density using a combination of phage K and modified derivatives. Lett Appl Microbiol 54:286–291

Kim SW, Chang IM, Oh KB (2002) Inhibition of the bacterial surface protein anchoring transpeptidase sortase by medicinal plants. Biosci Biotechnol Biochem 66:2751–2754

Kitture R, Ghosh S (2019) Hybrid nanostructures for *in-vivo* imaging. In: Bohara RA, Thorat N (eds) Hybrid nanostructures for cancer theranostics. Elsevier, Amsterdam, pp 173–208

Kitture R, Ghosh S, Kulkarni P, Liu XL, Maity D, Patil SI, Jun D, Dushing Y, Laware SL, Chopade BA, Kale SN (2012) Fe_3O_4-citrate-curcumin: promising conjugates for superoxide scavenging, tumor suppression and cancer hyperthermia. J Appl Phys 111:064702–064707

Kitture R, Chordiya K, Gaware S, Ghosh S, More PA, Kulkarni P, Chopade BA, Kale SN (2015) ZnO nanoparticles-red sandalwood conjugate: a promising anti-diabetic agent. J Nanosci Nanotechnol 15:4046–4051

Lambert G, Estévez-Salmeron L, Oh S, Liao D, Emerson BM, Tlsty TD, Austin RH (2011) An analogy between the evolution of drug resistance in bacterial communities and malignant tissues. Nat Rev Cancer 11(5):375–382

Lanter BB, Sauer K, Davies DG (2014) Bacteria present in carotid arterial plaques are found as biofilm deposits which may contribute to enhanced risk of plaque rupture. *MBio* 5:e01206–e01214

Lebeaux D, Ghigo J-M, Beloin C (2014) Biofilm-related infections: bridging the gap between clinical management and fundamental aspects of recalcitrance toward antibiotics. Microbial Mol Biol Rev 78:510–543

Lee JH, Regmi SC, Kim JA, Cho MH, Yun H, Lee CS, Lee J (2011) Apple flavonoid phloretin inhibits *Escherichia coli* O157:H7 biofilm formation and ameliorates colon inflammation in rats. Infect Immun 79:4819–4827

Lemire JJ, Kalan L, Gugala N, Bradu A, Turner RJ (2017) Silver oxynitrate—an efficacious compound for the prevention and eradication of dual—species biofilms. Biofouling 33:460–469

Leuba KD, Durmus NG, Taylor EN, Webster TJ (2013) Short communication: carboxylate functionalized superparamagnetic iron oxide nanoparticles (SPION) for the reduction of *S. aureus* growth post biofilm formation. Int J Nanomedicine 8:731–736

Lewis K (2010) Persister cells. Annu Rev Microbiol 64:357–372

Lu L, Hu W, Tian Z, Yuan D, Yi G, Zhou Y, Cheng Q, Zhu J, Li M (2019) Developing natural products as potential anti-biofilm agents. Chin Med 14:11

Maas M (2016) Carbon nanomaterials as antibacterial colloids. Materials (Basel) 9(8):617

Malik A, Mohammad Z, Ahmad J (2013) The diabetic foot infections: biofilms and antimicrobial resistance. Diab Metab Syndr 7(2):101–107

Meireles A, Borges A, Giaouris E, Simões M (2016) The current knowledge on the application of anti-biofilm enzymes in the food industry. Food Res Int 86:140–146

Monds RD, O'Toole GA (2008) The developmental model of microbial biofilms: ten years of a paradigm up for review. Trends Microbiol 17:73–87

Mulani MS, Kamble EE, Kumkar SN, Tawre MS, Pardesi KR (2019) Emerging strategies to combat ESKAPE pathogens in the era of antimicrobial resistance: a review. Front Microbiol 10:539

Nadell CD, Xavier JB, Foster KR (2009) The sociobiology of biofilms. FEMS Microbiol Rev 33:206–240

Ono K, Oka R, Toyofuku M, Sakaguchi A, Hamada M, Yoshida S, Nomura N (2014) cAMP signaling affects irreversible attachment during biofilm formation by *Pseudomonas aeruginosa* PAO1. Microbes Environ 29:104–106

Orgaz B, Neufeld RJ, SanJose C (2007) Single-step biofilm removal with delayed release encapsulated Pronase mixed with soluble enzymes. Enzym Microb Technol 40(5):1045–1051

Peng X, Zhang Y, Bai G, Zhou X, Wu H (2016) Cyclic di-AMP mediates biofilm formation. Mol Microbiol 99:945–959

Persson T, Hansen TH, Rasmussen TB, Skinderso ME, Givskov M, Nielsen J (2005) Rational design and synthesis of new quorum-sensing inhibitors derived from acylated homoserine lactones and natural products from garlic. Org Biomol Chem 3:253–262

Petrova OE, Sauer K (2012) Sticky situations: key components that control bacterial surface attachment. J Bacteriol 194:2413–2425

Raghuwanshi N, Kumari P, Srivastava AK, Vashisth P, Yadav TC, Prasad R, Pruthi V (2017) Synergistic effects of *Woodfordia fruticosa* gold nanoparticles in preventing microbial adhesion and accelerating wound healing in Wistar albino rats in vivo. Mater Sci Eng C 80:252–262

Robkhob P, Ghosh S, Bellare J, Jamdade D, Tang IM, Thongmee S (2020) Effect of silver doping on antidiabetic and antioxidant potential of ZnO nanorods. J Trace Elem Med Biol 126448:58

Rohacek M, Weisser M, Kobza R, Schoenenberger AW, Pfyffer GE, Frei R, Erne P, Trampuz A (2010) Bacterial colonization and infection of electrophysiological cardiac devices detected with sonication and swab culture. Circulation 121(15):1691–1697

Rokade SS, Joshi KA, Mahajan K, Tomar G, Dubal DS, Parihar VS, Kitture R, Bellare J, Ghosh S (2017) Novel anticancer platinum and palladium nanoparticles from *Barleria prionitis*. Glob J Nanomed 2(5):555600

Rokade S, Joshi K, Mahajan K, Patil S, Tomar G, Dubal D, Parihar VS, Kitture R, Bellare JR, Ghosh S (2018) *Gloriosa superba* mediated synthesis of platinum and palladium nanoparticles for induction of apoptosis in breast cancer. Bioinorg Chem Appl 2018:4924186

Salunke GR, Ghosh S, Santosh RJ, Khade S, Vashisth P, Kale T, Chopade S, Pruthi V, Kundu G, Bellare JR, Chopade BA (2014) Rapid efficient synthesis and characterization of AgNPs, AuNPs and AgAuNPs from a medicinal plant, *Plumbago zeylanica* and their application in biofilm control. Int J Nanomedicine 9:2635–2653

Sant DG, Gujarathi TR, Harne SR, Ghosh S, Kitture R, Kale S, Chopade BA, Pardesi KR (2013) *Adiantum philippense* L. frond assisted rapid green synthesis of gold and silver nanoparticles. J Nanoparticles 2013:1–9

Schwering M, Song J, Louie M, Turner RJ, Ceri H (2013) Multi-species biofilms defined from drinking water microorganisms provide striking protection against chlorine disinfection. Biofouling 29:917–928

Shende S, Joshi KA, Kulkarni AS, Shinde VS, Parihar VS, Kitture R, Banerjee K, Kamble N, Bellare J, Ghosh S (2017) *Litchi chinensis* peel: a novel source for synthesis of gold and silver nanocatalysts. Glob J Nanomed 3(1):555603

Shende S, Joshi KA, Kulkarni AS, Charolkar C, Shinde VS, Parihar VS, Kitture R, Banerjee K, Kamble N, Bellare J, Ghosh S (2018) *Platanus orientalis* leaf mediated rapid synthesis of catalytic gold and silver nanoparticles. J Nanomed Nanotechnol 9:2

Shinde SS, Joshi KA, Patil S, Singh S, Kitture R, Bellare J, Ghosh S (2018) Green synthesis of silver nanoparticles using *Gnidia glauca* and computational evaluation of synergistic potential with antimicrobial drugs. World J Pharm Res 7(4):156–171

Subhadra B, Kim DH, Woo K, Surendran S, Choi CH (2018) Control of biofilm formation in healthcare: recent advances exploiting quorum-sensing interference strategies and multidrug efflux pump inhibitors. Materials 11(9):1676

Sun JL, Zhang SK, Chen JY, Han BZ (2012) Efficacy of acidic and basic electrolyzed water in eradicating *Staphylococcus aureus* biofilm. Can J Microbiol 58(8):448–454

Vandersteegen K, Kropinski AM, Nash JH, Noben JP, Hermans K, Lavigne R (2013) Romulus and Remus, two phage isolates representing a distinct clade within the Twortlike virus genus, display suitable properties for phage therapy applications. J Virol 87:3237–3247

Veenstra GJ, Cremers FF, van Dijk H, Fleer A (1996) Ultrastructural organization and regulation of a biomaterial adhesin of Staphylococcus epidermidis. J Bacteriol 178:537–541

Veerachamy S, Yarlagadda T, Manivasagam G, Yarlagadda PK (2014) Bacterial adherence and biofilm formation on medical implants: a review. Proc Inst Mech Eng H 228:1083–1099

Vestby LK, Gronseth T, Simm R, Nesse LL (2020) Bacterial biofilm and its role in the pathogenesis of disease. Antibiotics 9:59

Vikram A, Jesudhasan PR, Pillai SD, Patil BS (2012) Isolimonic acid interferes with *Escherichia coli* O157:H7 biofilm and TTSS in QseBC and QseA dependent fashion. BMC Microbiol 12:261

Walsh LJ (2020) Novel approaches to detect and treat biofilms within the root canals of teeth: a review. Antibiotics 9:129

Wilking JN, Zaburdaev V, De Volder M, Losick R, Brenner MP, Weitz DA (2013) Liquid transport facilitated by channels in *Bacillus subtilis* biofilms. Proc Natl Acad Sci U S A 110:848–852

Wolska KI, Grudniak AM, Rudnicka Z, Markowska K (2016) Genetic control of bacterial biofilms. J Appl Genet 57:225–238

Workentine ML, Harrison JJ, Weljie AM, Tran VA, Stenroos PU, Tremaroli V, Vogel HJ, Ceri H, Turner RJ (2010) Phenotypic and metabolic profiling of colony morphology variants evolved from *Pseudomonas fluorescens* biofilms. Environ Microbiol 12:1565–1577

Workentine ML, Wang S, Ceri H, Turner RJ (2013) Spatial distributions of *Pseudomonas fluorescens* variants in mixed-culture biofilms. BMC Microbiol 13:175

Young D, Morton R, Bartley J (2010) Therapeutic ultrasound as treatment for chronic rhinosinusitis: preliminary observations. J Laryngol Otol 124(5):495–499

Yu X, Trase I, Ren M, Duval K, Guo X, Chen Z (2016) Design of nanoparticle-based carriers for targeted drug delivery. J Nanomater 2016:1087250

Zhao G, Usui ML, Lippman SI, James GA, Stewart PS, Flecxkman P, Olerud JE (2013) Biofilms and inflammation in chronic wounds. Adv Wound Care (New Rochelle) 2:389–399

Zimmerli W, Sendi P (2017) Orthopaedic biofilm infections. APMIS 125(4):353–364

Biofilm: The Unknown Armor in the Arsenal of Bacteria: A Case Study

11

Anindita Ballav and Aihik Saha

Abstract

Hospital-acquired infections or nosocomial infections are an increasing cause of concern in critically ill patients. The major nosocomial infections include urinary tract infection (UTI), surgical site infection (SSI), ventilator-associated pneumonia (VAP), and catheter-related bloodstream infection (CRBSI). CRBSI has gained importance lately as central venous catheterization is becoming a routine procedure in critically ill patients. It is the preferred method of infusion for certain medicines like norepinephrine, potassium chloride infusion, parenteral nutrition, hemodialysis, plasmapheresis, and hemodynamic monitoring as well. But at the same time it carries a much higher risk of infection compared to other vascular access, leading to an increased number of hospital days and in turn expenditure. With this case study, we are discussing unusual pathogenesis of bacteria which help them to establish infection on medical device.

11.1 Introduction

The twenty-first century marked a rapid decline in deaths due to communicable diseases. With an arsenal of antibiotics and vaccines, the mortality due to communicable diseases came down. It saw a shift in the epidemiology from communicable

A. Ballav
Department of Microbiology, College of Medicine and Sagore Dutta Hospital, Kamarhati, Kolkata, West Bengal, India

A. Saha (✉)
Department of Biochemistry, Nil Ratan Sircar Medical College and Hospital, Kolkata, West Bengal, India

R. R. Ray et al. (eds.), *Biofilm-Mediated Diseases: Causes and Controls*, https://doi.org/10.1007/978-981-16-0745-5_11

diseases to noncommunicable one as the average life expectancy of individuals increased globally. Yet today doctors across the world fret about a febrile patient on any given day.

The advancements in medical science have also helped salvage patients which were previously considered moribund. At the same time, this has led to an increased duration of hospital stay and use of ventilators, catheters, stents, prosthesis, and other implantable devices. While these devices, on one hand, have revolutionized the way we treat patients today, on the downside they serve as a portal for entry of infections in the human body. Hospital-acquired infections or nosocomial infections are an increasing concern among critically ill patients.

WHO defines healthcare-associated infection (HAI) or nosocomial infections as the type of infections patients, who are not initially inflicted by those infections at time of admission to the hospital, acquire from hospitals or other healthcare facilities during the time of incubation at the hospital or at the time of discharge (Ducel et al. 2002). An estimated 7–12% of hospitalized patients suffer from nosocomial infections, with the highest prevalence among ICU patients and patients in acute surgical and orthopedic wards (Ducel et al. 2002). The scenario is worse in the developing nations due to a disproportionate burden on the healthcare system. Hospital-acquired infections divert the already scarce resources to HAI which might had been preventable.

The major nosocomial infections include UTI, SSI, VAP, and CRBSI, respectively. CRBSI has gained importance lately as central venous catheterization is becoming a routine procedure in critically ill patients. It is the preferred method of infusion for certain medicines like norepinephrine, potassium chloride infusion, parenteral nutrition, hemodialysis, plasmapheresis, and hemodynamic monitoring as well. But at the same time it carries a much higher risk of infection compared to other vascular access, leading to an increased number of hospital days and in turn expenditure.

CRBSI can be described as bacterial-associated infection originating from an intravenous catheter. Among the intravenous catheter, CVCs have 64 times more capacity to develop CRBSI than peripheral venous catheter. The colonization of bacterial cells occurs upon the surfaces such as tip of catheter, cutaneous tract within skin flora and lumen via contamination, hematogenous seeding of the catheter from another infected region, and catheter lumen contamination with infusate (Gahlot et al. 2014).

Over the years, higher animals have adapted themselves to live in groups or clusters. This communalism confers them with an evolutionary survival vantage over predators or any adverse situation. The same is seen in unicellular bacteria, in which, even though was known, its importance in the medical field has been unraveled over the past few decades with the rise in HAI. Once thought to exist only in sessile forms, the discovery of biofilms has helped us to gain insight into the pathogenesis of resistant infections.

11.2 Case History

A 64-year-old male presented to the ER of our hospital with a H/o fall about 3 h back followed by drowsiness and several bouts of vomiting. The drowsiness was progressively increasing as stated by the patient relatives. There was no history of loss of consciousness, convulsions, fever, diarrhea, and bladder or bowel incontinence. The patient is a known hypertensive for more than 10 years on regular medications (amlodipine and telmisartan) and a recently diagnosed diabetic on OHA (metformin). No history of any known cardiac, renal, or pulmonary disease was noted. The patient was predominantly a nonvegetarian with an addiction of smoking of more than 40 years.

11.3 On Examination

On arrival, the patient had a poor GCS of E1V2M4 and left-sided weakness. Pupils were b/l 2+ reacting to light. There was an extensor plantar response on the left side. Owing to the poor GCS and labored breathing, blood gas analysis was done which revealed respiratory acidosis. The patient was immediately intubated and put on ventilatory support (IPPV). A central venous catheter was secured in the right internal jugular vein. A normal ECG and a negative TropT ruled out any acute cardiac event. The patient had a CBG of 248 mg/dL. An urgent CT scan was ordered and revealed a right-sided ICH with IVH with a mild midline shift. In view of the midline shift and the ventriculomegaly due to the IVH, an emergency craniotomy and laminectomy were undertaken by the team of neurosurgeons.

After the surgery, the patient was started on empirical antibiotics (injection ceftriaxone 1 g IV BD) owing to the post-op status. The patient was being gradually weaned off the ventilatory support, owing to a satisfactory neurological improvement. But on the third post-op day, the patient had a drop in the GCS and continuous high fever. Blood investigations revealed increased total white blood cell count (neutrophilia), increased C-reactive protein, and procalcitonin. The chest radiograph did not reveal any major infiltration.

11.4 Lab Investigations

Test name	Report	Normal value
Total leucocyte count	21,400	6000–10,000/cumm
C-reactive protein	154.0	<5.0 mg/L
Procalcitonin	3.4	0.15 mg/mL

Routine examination of urine revealed a normal picture, but the Foley catheter was changed. Blood culture and urine cultures were sent. The scalp wound was inspected for localized infection. Three sets of blood culture were sent, two from the CVC catheter and one from a peripheral venipuncture. All cultures were performed using a conventional culture technique. Patient was empirically started on amikacin along with ceftriaxone. In urine culture, no pathogenic organism was found, but blood culture showed existence of pathogenic organism.

11.5 Diagnosis

From the above signs and symptoms, patient was suspected to have CRBSI (catheter-related blood stream infection) as there were no other local signs of infection. For the diagnosis of CRBSI, require a microbiological confirmation along with the clinical şigns and symptoms. For this, the CVC catheter was removed, and the catheter tip (5 cm segment) was sent for microbiological culture. The distal tip of the catheter was processed using Maki's roll plate semiquantitative culture technique. In this technique, the distal tip of the catheter was rolled back and forth on a blood agar plate, and the plate was incubated at 37 °C. After overnight incubation, colonies grown on that blood agar plate were counted, and further biochemical tests were performed for identification of that organism.

11.6 Identification of Microorganism

On blood agar plate, colonies were medium sized, rounded with entire margin, convex, gray white, opaque, and nonhemolytic. Gram stain was performed from those colonies, and it was found to be Gram-positive cocci. So for further confirmation, catalase test, slide test, and tube coagulase test were performed. It was found to be other *Staphylococcus* spp. according to CLSI guideline (previously named as coagulase-negative staphylococci). For further confirmation, urease test, phosphatase test, ornithine decarboxylase test, novobiocin susceptibility test, and polymyxin B susceptibility test were performed. The organism was confirmed to be *Staphylococcus epidermidis*, and it was the same organism which was isolated previously from the blood culture of that patient. Along with this, semiquantitative culture showed that there was >15 CFU on that blood agar plate.

From the above finding, patient was confirmed to have CRBSI.

11.7 Antimicrobial Susceptibility Testing

Antimicrobial susceptibility testing was performed using modified Kirby Bauer disk diffusion technique. According to CLSI (clinical laboratory and standard institute) guideline, antibiotics were tested. The antibiotics were as follows: penicillin (10 units), cefoxitin (30 μg), linezolid (30 μg), clindamycin (2 μg), erythromycin (15 μg), trimethoprim-sulfamethoxazole (1.25/23.75 μg), and ciprofloxacin (5 μg). Along with this, MIC (minimum inhibitory concentration) was determined for vancomycin.

For this susceptibility test, four to five colonies of the same morphology were selected and inoculated into broth. Inoculated broth was incubated at 37 °C for 3 h. After that, the turbidity of that broth culture was matched with 0.5 McFarland turbidity standard which gives a bacterial suspension of 1.5×10^8 CFU/mL. After achieving the required suspension, inoculum of *Staphylococcus epidermidis* was swabbed into Mueller-Hinton agar plate. The plate was incubated at 37 °C aerobically for 24 h. After which, the zone of inhibition around each disk was measured. *Staphylococcus aureus* ATCC 25923 was used as reference strain for quality control (Tille 2013).

It was found that organism was resistant to penicillin, cefoxitin, clindamycin, erythromycin, trimethoprim-sulfamethoxazole, ciprofloxacin, and vancomycin.

The patient was now started on linezolid.

11.7.1 Qualitative Biofilm Production Assay by Congo Red Method

After isolating the organism from central venous catheter and performing the antimicrobial susceptibility testing, other tests were performed to identify the biofilm producing nature of the organism. For this, a simple qualitative method was used which was described by Freeman et al. (1989). According to this method, an inoculum of *Staphylococcus epidermidis* was made and inoculated into a Congo red agar media (consists of BHIA with 5% sucrose and Congo red indicator). The plate was then incubated at 37 °C for 24 h. After 24 h of incubation, a black colony with dry crystalline consistency has appeared on that Congo red agar plate. That finding showed a strong evidence for its ability to form biofilm.

11.7.2 Quantitative Biofilm Detection Assay by Tube Method

This quantitative method for biofilm detection was described by Christensen et al. (1982). According to this method, 10 mL of tryptic soy broth with 1% glucose was inoculated with *Staphylococcus epidermidis* and incubated at 37 °C for 24 h. After 24 h, the tube was washed with phosphate-buffered saline and dried. Dried tube was stained with crystal violet (0.1%) for 7 min and then washed with distilled water for 5 min. Visible film lining the wall and bottom of the inner surface of the tube was indicative of biofilm formation.

The patient improved on the 4th day of the initiation of antibiotics. The fever subsided, and subsequently the patient was weaned off ventilator and extubated. The antibiotic was continued for 14 days.

11.8 Discussion

As already discussed, CVCs have now become a necessity in almost all critically ill patients. They are the most important cause of sepsis and bacteremia leading to both increased morbidity and mortality. It is often referred to as the quality indicator of infection control. Precautions like proper aseptic techniques during insertion, choosing the correct type of catheter, correct site of insertion, keeping the localized site clean and dry, and reducing the duration of the indwelling catheter are small steps which make a huge difference. Adequately trained interventionists and nursing staff also play a crucial role.

Parameswaran et al. (2011) in their study found out that 64% of the pathogens causing CRBSI were Gram-positive and 36% were Gram-negative. The common organisms included *Klebsiella*, *Staphylococcus*, *Acinetobacter*, and *Pseudomonas*. Recently coagulase-negative *staphylococci* have gained a lot of importance. Previously they were considered as normal flora of human being; over the last few decades, their role in pathogenicity has been clearly established. Though several different species of coagulase-negative *staphylococci* have been identified, few of them had significant role in human life. Fifty to seventy percent of nosocomial blood stream infections have been caused by them. Among different species, *Staphylococcus epidermidis* is the most frequently isolated organism from various clinical samples. From various studies, it has been documented that *Staphylococcus epidermidis* has a significant role in various infections such as urinary tract infections, surgical site infections, infections of various prosthetic devices, cerebrospinal fluid (CSF) shunt infections, peritoneal dialysis-related infections, and ophthalmic infections. When several studies have been conducted to find out the virulence factor of *Staphylococcus epidermidis*, it also has been found that this bacteria produce cell surface and extracellular macromolecules that initiate and subsequently enhance bacterial adhesion to foreign bodies, eventually responsible for biofilm formation.

In the human body, when plastic prosthetic devices are implanted, they become rapidly coated with serum proteins, and *Staphylococcus epidermidis* has the ability to bind with the fibronectin, fibrinogen, collagen, and vitronectin as well as to the plastic catheter material itself. For biofilm formation, there are five distinct stages, which are as follows: (1) reversible attachment, (2) irreversible attachment, (3) maturation I, (4) maturation II, (5) dispersion. For the initial stage, various adhesion molecules have been found on the cell surface of *Staphylococcus epidermidis*. These includes Aae and AltE (autolysin-adhesin), Ebp (elastin-binding protein), Embp (extracellular matrix-binding protein), Fbe (fibrinogen-binding protein), and GehD (glycerol ester hydrolase). These adhesion molecules bind with the serum proteins. For the succeeding stage, the microorganisms synthesize a polysaccharide intracellular adhesion (PIA) that crosslinks cells together within the developing biofilm. PIA is a linear polysaccharide which is composed of ß-1,6-linked 2-deoxy-2-amino-D-glucopyranosyl residues. In the maturation stage, microbial cells communicate with one another through autoinducer signals. These autoinducers facilitate quorum sensing. In the Gram-positive bacteria, autoinducer peptides are responsible for quorum sensing. In the final stage, biofilm cells may be dispersed: either by (a) shedding of daughter cells, (b) depletion of nutrient levels or quorum sensing, or (c) detachment of biofilm aggregates by physical forces. In the case of *Staphylococcus epidermidis*, phenol-soluble modulins disperse the outer biofilm layer which interact with neutrophils, causing degranulation and cell lysis. Now by forming the biofilm, these organisms have become the continuous source of infection, and they also developed increased antibiotic resistance which is a great concern in recent time. The mechanisms, by which the microorganisms developed antibiotic resistance within the biofilm, are as follows: (a) restricted penetration, (b) stress response, (c) altered microenvironment, and (d) persisters. In case of restricted penetration, substances in the EPS act as a diffusion barrier. The negatively charged EPS restricts permeation of positively charged molecules of the antibiotic. Due to altered microenvironment within the biofilm, there occurs depletion of nutrient and oxygen which creates an anaerobic environment. Within this anaerobic environment, microorganisms grow slowly. Antimicrobials have the ability to kill rapidly dividing cells. So for slowly growing organisms within the biofilm, they become ineffective. Not only these, but the presence of subpopulation of persister cells within the biofilm is also responsible for the development of antibiotic resistance.

Till now, for management of CRBSI, there are no standard guidelines. Depending on various research, several recommendations have been published. These are use of oral or intravenous (IV) antibiotics or an antibiotic lock technique (ALT) and removal and reinsertion of the CVC. As the biofilm has a complex mechanism, when considering staphylococcal biofilm treatment options, the ideal therapy should have several characteristics:

- Use of antibiotics with bactericidal activity against slow-growing stationary phase.
- Use of antibiotics for long duration.
- Use of antimicrobials with broad efficacy against all *staphylococci* and polymicrobial biofilm infection.

So now, to prevent this, we have to adapt some antibiofilm measures. Antibiofilm strategies should be based on:

(a) Prevention of microbial attachment and colonization.
(b) Modulating the development of the biofilm by interfering with the signal molecules.
(c) Disintegrating the matrix of the biofilm.

By coating the device surface with hydrophilic polymer, we can inhibit the microbial adhesion to the surface. Not only this, but to avoid the growth of already adhered and/or colonizing microorganisms, material bulk impregnation with one or two antimicrobial agents can be a promising approach. Now there is another approach which can be a promising one, which is quorum sensing inhibition. As we all know, quorum sensing is an important mechanism adopted by the mischievous microorganisms to establish a biofilm. So, quorum sensing inhibitor could be a promising therapeutic tool. Along with this, we can also rely on the use of combined drugs with different antimicrobial spectra for microbial killing within an established biofilm. Another approach is disaggregation of the biofilm matrix by use of substances like bacteriophage.

11.9 Conclusion

The importance of biofilms in medical devices and its role in infection have been well established. In the medical arena, it is estimated that about 65% of all bacterial infections are associated with bacterial biofilms. It leads to a major increase in the expenses for the treatment of the patient. On one hand, we shall take every step possible to prevent its occurrence; on the flipside, we should be always very vigilant to detect it at the earliest and treat it appropriately. But most importantly we need to develop ways to inhibit the biofilms.

References

Christensen GD, Simpson WA, Bisno AL, Beachey EH (1982) Adherence of slime producing strains of *Staphylococcus epidermidis* to smooth surfaces. Infect Immun 37:318–326

Ducel G, Fabry J, Nicolle L (2002) Introduction. In: Prevention of hospital acquired infection: a practical guide, 2nd edn. World Health Organisation WHO/CDS/CSR/EPH/2002, Geneva

Freeman DJ, Falkiner FR, Keane CT (1989) New method for detecting slime production by coagulase negative Staphylococci. J Clin Pathol 42:872–874

Gahlot R, Nigam C, Kumar V, Yadav G, Anupurba S (2014) Catheter related blood stream infection. Int J Crit Illn Inj Sci 4(2):162–167

Parameswaran R, Sherchan JB, Varma DM, Mukhopadhyay C, Vidyasagar S (2011) Intravascular catheter-related infections in an Indian tertiary care hospital. J Infect Dev 5:452–458

Tille PM (2013) Laboratory methods and strategies for antimicrobial susceptibility testing. In: Bailey & Scott's diagnostic microbiology, 13th edn. AACC, Washington, DC, pp 168–190